The series Lecture Notes in Computer Science (LNCS), including its subseries Lecture Notes in Artificial Intelligence (LNAI) and Lecture Notes in Bioinformatics (LNBI), has established itself as a medium for the publication of new developments in computer science and information technology research, teaching, and education.

LNCS enjoys close cooperation with the computer science R & D community, the series counts many renowned academics among its volume editors and paper authors, and collaborates with prestigious societies. Its mission is to serve this international community by providing an invaluable service, mainly focused on the publication of conference and workshop proceedings and postproceedings. LNCS commenced publication in 1973.

Clemens Grelck · Goran Mauša ·
João Paulo Fernandes · Sandi Ljubić
Editors

Promoting Sustainability as a Fundamental Driver in Software Development Training and Education

First International Summer School, SusTrainable 2022
Rijeka, Croatia, July 4–8, 2022
Revised Selected Papers

 Springer

Editors
Clemens Grelck
Friedrich Schiller University Jena
Jena, Germany

João Paulo Fernandes
New York University Abu Dhabi
Abu Dhabi, United Arab Emirates

Goran Mauša
University of Rijeka
Rijeka, Croatia

Sandi Ljubić
University of Rijeka
Rijeka, Croatia

ISSN 0302-9743   ISSN 1611-3349 (electronic)
Lecture Notes in Computer Science
ISBN 978-3-032-22277-0   ISBN 978-3-032-22278-7 (eBook)
https://doi.org/10.1007/978-3-032-22278-7

This Springer imprint is published by the registered company Springer Nature Switzerland AG
The registered company address is: Gewerbestrasse 11, 6330 Cham, Switzerland

If disposing of this product, please recycle the paper.

Preface

This volume presents the revised and extended lecture notes of the first SusTrainable Summer School on *Promoting Sustainability as a Fundamental Driver in Software Development Training and Education,* organized by the University of Rijeka, Faculty of Engineering, during the week from July 4 to July 8, 2022 [1]. The summer school was organized in the context of the Erasmus+ Strategic Partnership for Higher Education (project number 2020-1-PT01-KA203-078646 [2]), funded by the European Union and coordinated by the University of Coimbra, Portugal. It attracted 21 academics and 70 students, mostly doctoral candidates and early-career researchers, representing ten European institutions across the Netherlands, Portugal, Romania, Bulgaria, Slovakia, Hungary, and Croatia. We compiled an intensive program of 11 lectures accompanied by supervised practical lab sessions spread over five full days of the week. In addition to the academic program, a series of evening social activities was organized to promote informal discussions, foster collaboration, and support internationalization, including a night swim and dinner at a beach bar, a food truck festival near the city castle, a sunset boat trip on the Kvarner Gulf, and a guided city tour. The SusTrainable summer school loosely stood in the long tradition of the Central European Functional Programming Schools (CEFP), regularly held since 2005.

Sustainability has become a central concern in the development of modern societies and has received increasing attention at the global level. In this context, the information and communication technology (ICT) sector contributes a significant and steadily growing share to worldwide resource and energy consumption, a trend that is expected to continue in the coming years. Although energy consumption ultimately takes place at the hardware level, software determines the behavior and efficiency of ICT systems and thus has a substantial impact on their overall sustainability. Consequently, software engineering decisions play a crucial role in controlling and reducing the energy footprint of ICT infrastructures. The SusTrainable Strategic Partnership, therefore, promoted the integration of sustainability considerations across all aspects of software engineering practice, positioning sustainability as a fundamental design concern.

SusTrainable aimed to actively promote education of the next generation of software engineers to consider sustainability in all aspects of the software engineering process — our acronym *SusTrainable* means Training for Sustainability. We aimed to provide future generations of software engineers with essential skills to develop software that is not only functionally correct and easy to maintain but that at the same time uses the underlying hardware in the most energy-efficient way. Our objective was to train the future avant-garde of software engineers for the sustainable software and ICT that the knowledge-based, environmentally concerned societies of the twenty-first century demand. The majority of summer school participants were close to their transition from learning to doing and have meanwhile joined the European software engineering

---

[1] https://sustrainable.uniri.hr/.

[2] https://sustrainable.github.io/.

workforce. They will carry the ideas, concepts, and methods they learned at the summer schools into industrial software engineering practice across Europe and potentially world-wide. Furthermore, they will act as facilitators and multipliers alike for a sustainable future software-driven Europe. Conversely, we fostered the ubiquitous enthusiasm of the younger generation for sustainable development with a constructive, meaningful, and high-impact route forward.

During the second SusTrainable summer school, held in Coimbra, Portugal, during the summer of 2023, the SusTrainable consortium concluded that the scientific quality and pedagogical value of the presented material warranted formal publication and wide dissemination. Rather than pursuing an immediate proceedings volume, the consortium deliberately opted to explore appropriate publication venues that would ensure long-term visibility and alignment with the academic standards of the contributions. This process led to the identification of Springer Nature as a distinguished publishing partner. The resulting time gap between the summer school and the present volume reflects this intentional and quality-focused approach, as well as the substantial effort devoted to revising, extending, and rigorously reviewing the material for publication.

In the following, we thoroughly revised and extended the material and commenced a rigorous and constructive single-blind review process organized with the help of the EasyChair conference management software system. We formed a program committee consisting of the principal investigators of the 10 universities participating in the Strategic Partnership. In addition, we invited independent experts in relevant research areas, not affiliated with the SusTrainable Strategic Partnership, to serve as external reviewers. Each contribution was reviewed by at least one member of the program committee and one external reviewer. We used the relevant EasyChair features to manage conflicts of interest and to ensure the confidentiality of reviews, including submissions co-authored by the program committee chairs themselves.

In the first round, nine manuscripts were accepted for inclusion in this volume while two manuscripts embarked on a rigorous shepherding process with several iterations between the authors and the reviewers to reach the required scientific quality. Any communication between authors and shepherds was relayed via the program committee chairs to further safeguard confidentiality and objectivity in the process.

In the following, we briefly summarize each chapter and point out its relation to the state of the art.

Mart Lubbers and Pieter Koopman:

Green Computing for the Internet of Things

The number of devices in the Internet of Things (IoT) is increasing rapidly. While individual devices are typically small and not power-hungry, their sheer number results in a significant contribution to world-wide energy consumption. Furthermore, programming IoT systems is challenging due to high semantic friction. Although microcontrollers have low-power sleep modes, deciding when and for how long to enter these modes remains non-trivial. Task-Oriented Programming addresses both issues simultaneously. It offers a high-level declarative programming language for programming entire IoT systems from a single source. Moreover, due to the high level of abstraction and execution semantics, the scheduler can decide at runtime when and for how long the microcontroller should enter low-power sleep.

Štefan Korečko:

Verified Software in Web-Based Virtual Environments

From the perspective of environmental and economic sustainability, the use of online virtual prototypes and environments is highly beneficial when developing control software for autonomous devices. In addition to saving the resources needed to build a physical prototype, such approaches may also reduce travel cost, as both validation and evaluation take place in a virtual space where participants connect remotely. This chapter demonstrates the process of constructing such virtual prototypes and environments. The control software of a fictional cleaning robot is developed using formal methods and deployed into a virtual representation of the robot. The robot operates in a simple virtual environment, performing a cleaning operation, and multiple users can join the environment and interact with the robot.

Anikó Kopacz, Zoltán Tasnádi, and Lehel Csató:

Sustainable Programming in the Julia Language

The two-language dilemma is prevalent in developing systems with more involved mathematical concepts, such as finite element modeling, fluid dynamics in weather prediction, or digital twin systems. In these cases, mathematical programming, often done in Mathematica or Matlab, is used for prototyping. As these languages are not resource-efficient, one must transfer the prototype implementation to production code written in a resource-efficient language after the modeling part is completed. Thus, the development cycle requires two programming languages, a modeling language and an implementation language with different teams of programmer on either side. This procedure is not only costly and inefficient, but also raises both formal and practical concerns regarding the equivalence between the prototype and the production solution. In this chapter, the feasibility of the Julia programming language is evaluated as a potential approach to addressing the two-language dilemma. Julia provides features that make it an excellent candidate for both the mathematical modeling stage and production code generation. The features that enable Julia to achieve efficiency comparable to that of C or Java are highlighted. In addition, the main characteristics of the Julia language and the associated development cycle are presented through practical sessions, in which various problems are solved following mathematical programming concepts.

Tihana Galinac Grbac, Emili Puh, and Neven Grbac:

Algorithms for Sustainable System Topologies

This chapter is an introductory invitation to the world of emerging applications of algebraic topology in computer science. Geometry, topology, and algebraic topology in particular, have an emerging range of applications that are not limited to topological data analysis, but extend to the fields of sustainable and explainable artificial intelligence, topological approaches to the reliability of quantum computers, and topological aspects of robotic motion, to mention just a few. The chapter commences by outlining potential applications of algebraic topology to the study of software structure and reliability, followed by a gentle introduction to the most basic notions of algebraic topology associated with software graphs. The chapter concludes with a running example and accompanying lab exercises.

Marco Couto, Rui Rua, and João Saraiva:

Energy Debt: Foundations, Techniques and Tools

This chapter adapts the metaphor of technical debt to the context of green software. Energy debt is defined as the estimated additional energy cost incurred when executing a software system due to energy inefficiencies in the software source code, relative to a hypothetical energy-optimal and functionally equivalent implementation. To quantify energy debt, a catalog of energy-related code smells is compiled based on the current state-of-the-art literature on green software, together with the energy savings reported in quantitative studies. Finally, the E-Debitum tool is presented, which provides an energy debt estimation plugin for the SonarQube code quality and security tool.

Marko Njirjak, Erik Otović, and Goran Mauša:

Applying Soft Computing for Sustainability: The Effects of Feature Reduction on Machine Learning Model Energy Consumption

Energy efficiency in machine learning has become an increasingly important concern as modern models continue to grow in complexity and scale. While artificial intelligence has achieved remarkable success across domains such as computer vision, natural language processing, and data generation, its increasing computational demands pose sustainability challenges. Recent studies have investigated model optimization and hardware efficiency; however, feature selection remains a relatively underexplored approach to reducing energy consumption without compromising performance. This chapter builds upon previous work in therapeutic peptide prediction, demonstrating that reducing the number of input features can result in significant energy savings while maintaining predictive accuracy. By systematically analyzing energy consumption across different machine learning tasks, this study contributes to ongoing research on sustainable AI and provides practical insights for optimizing model training processes.

Luís Paquete:

Multicriteria Decision Methods for Sustainable Software Development

Software development involves the consideration of multiple criteria at different stages of the process, including sustainability-related aspects. Multi-criteria decision-aiding methods support decision-making in such contexts, particularly when conflicting objectives must be addressed. In software engineering, this is often the case when evaluating alternative projects or technologies, where cost or profit must be balanced against less easily quantifiable criteria such as team well-being and project sustainability. This chapter focuses on the application of multi-criteria decision-aiding methods to software development, with an emphasis on outranking approaches. It provides a brief overview of the main methods, discusses their strengths and limitations, and presents a didactic case study on ranking programming languages using outranking techniques.

Richárd Szalay and Zoltán Porkoláb:

Using Strong Types in C++ for Long-Term Code Management

This chapter discusses the importance of strong typing in long-term software maintenance. In many large-scale C++ projects, the type system is not used to its full extent, and trade-offs are often made. This typically manifests in developers overusing built-in and widely used library types, such as int or string, instead of types that express the domain of the represented values with a finer granularity. The consequences of such practices range from severe hindrance to code comprehension to subtle security vulnerabilities that may remain undetected for extended periods. Education is a crucial element in counteracting the gradual deterioration of software quality. By emphasizing the importance of stronger and safer type systems, more sustainable software development practices can be fostered. This chapter describes how education in strong typing was introduced in an advanced, master-level C++ university course. While certain principles of strong typing are actionable at early stages of education, hands-on experience in incrementally improving an existing, hard-to-understand code base has been shown to support deeper understanding of said code base. The chapter further discusses the step-by-step migration of an example software project toward the use of strong types.

Mihail Petrov and Vladimir Valkanov:

Teaching by Example – Temporal-Logic Automated Solutions for Sustainable Educational Progress

This chapter explores the dynamic nature of teaching and learning, which rarely follows a standardized approach. The effectiveness of a curriculum and the initial assessment of students by instructors do not necessarily determine final learning outcomes. Student performance may vary due to multiple factors, including gender, age, social background, prior experience, and preferred learning methods such as listening or reading. By analyzing these variables, this chapter aims to identify key factors that enhance student performance, leveraging automated solutions based on temporal logic to support sustainable educational progress.

Ana Oprescu, Kyrian Maat, Lukas Koedijk, Sander van Oostveen, and Stephan Kok:

Energy-Driven Software Engineering: A Tutorial

The EU Climate Law targets climate neutrality by 2050. At the same time, the knowledge economy is deemed essential to the prosperity of the European Union. Consequently, the energy footprint of software services remains an important topic for both research and education. This chapter aims to inspire students and professional programmers alike to engage in energy-aware software engineering in practice and education. Relevant background theory is presented, together with suggestions for further reading material. Previous research findings on Java, JavaScript, Python, PHP, Ruby, C, C++, and C# are summarized. A selection of programming languages and programs from the Computer Language Benchmarks Game (formerly known as the Great Computer Language Shootout) is examined to illustrate how the experimental setup can be used to assess whether certain programming languages are inherently more energy-efficient than others. The chapter further explores the extent to which implementation-level design choices impact the energy consumption of computer programs.

Jianhao Li and Viktória Zsók:

UActor: Group-Based Actor Model Middleware with UDP Protocol

This chapter presents a novel distributed communication middleware, named UActor. UActor leverages connectionless transport protocols with the actor model. Connection-oriented protocols are compared to telephone systems, whereas connectionless protocols are compared to postal services, arguing that their UDP-based, brokerless design aligns naturally with the decentralized principles of the actor model. The middleware unifies a range of communication patterns into a single, user-friendly framework, with the aims of improving development efficiency and supporting sustainability. Moreover, the energy consumption estimation built into UActor fosters energy-efficient design practices for future software systems.

To conclude this preface, we thank the local organizers from the Faculty of Engineering of the University of Rijeka as well as their teams for an excellent venue and the perfect organization of the summer school week. We particularly highlight the fact that the first SusTrainable summer school was originally planned to take place in Cluj-Napoca, Romania. However, due to the uncertainties of the unfolding war of Russia against Ukraine, we were forced to explore alternative locations further away. Thankfully, the team from Rijeka — professors Goran Mauša, Sandi Ljubić, and Daniela Kalafatovic, with PhD students Erik Otović, Marko Njirjak, and Arian Skoki — volunteered to organize the summer school on short notice.

We further thank all members of the program committee and our diligent external reviewers for their extensive work that greatly contributed to the scientific quality of this LNCS volume. We thank our enthusiastic lecturers for all their hard work and all further summer school participants, academic or student, for their engaged participation in all activities of the summer school. Last but not least, we thank the Erasmus+ program and the Erasmus National Agency of Portugal for their financial and organizational support of the SusTrainable Strategic Partnership for Higher Education as well as Springer Nature for the inclusion of this volume in their Lecture Notes in Computer Science series.

December 2025

Clemens Grelck
Goran Mauša
João Paulo Fernandes
Sandi Ljubić

Organization

Program Committee Chairs

Clemens Grelck	Friedrich Schiller University Jena, Germany
Goran Mauša	University of Rijeka, Croatia
João Paulo Fernandes	New York University Abu Dhabi, United Arab Emirates
Sandi Ljubić	University of Rijeka, Croatia

Program Committee

Lehel Csató	Babeş-Bolyai University Cluj-Napoca, Romania
Tihana Galinac Grbac	Juraj Dobrila University of Pula, Croatia
Pieter Koopman	Radboud University Nijmegen, The Netherlands
Štefan Korečko	Technical University of Košice, Slovakia
Mart Lubbers	Radboud University Nijmegen, The Netherlands
Ana Oprescu	University of Amsterdam, The Netherlands
Luís Paquete	University of Coimbra, Portugal
Zoltán Porkoláb	Eötvös Loránd University, Hungary
João Saraiva	University of Minho, Portugal
Elena Somova	University of Plovdiv "Paisii Hilendarski", Bulgaria
Csaba Szabó	Technical University of Košice, Slovakia
Viktória Zsók	Eötvös Loránd University, Hungary

External Reviewers

Wolfram Amme	Friedrich Schiller University Jena, Germany
Christopher Brown	University of St. Andrews, UK
António Casimiro	University of Lisbon, Portugal
Luís Cruz	Technical University of Delft, The Netherlands
Fredrik Enoksson	KTH Royal Institute of Technology, Sweden
Dalvan Griebler	Pontifical Catholic University of Rio Grande do Sul, Brazil
Georg Jäger	Freiberg University of Mining and Technology, Germany

Maja Hanne Kirkeby	Roskilde University, Denmark
Petar Rajković	University of Niš, Serbia
Gordana Rakić	University of Novi Sad, Serbia
Vadim Zaytsev	University of Twente, The Netherlands

Contents

Green Computing for the Internet
of Things

Mart Lubbers[(✉)] [ID] and Pieter Koopman [ID]

Institute for Computing and Information Sciences, Radboud University, Nijmegen,
The Netherlands
`mart@cs.ru.nl`, `pieter@cs.ru.nl`

Abstract. The Internet of Things (IoT), contains a huge number of
nodes and this amount is rapidly increasing. This makes it worthwhile
to invest in energy saving for the edge nodes even when they are not battery powered. Using cheap and energy efficient ocessor based hardware
for the edge nodes yields a significant contribution to green computing.
However, this hardware has major restriction on the processing power
and memory available. These restrictions impose substantial constraints
on the software executable on such nodes.

Task-oriented programming (TOP) offers a convenient way to program entire IoT applications from a single source at a high level of
abstraction. A tailor-made domain-specific dialect of TOP combines the
advantages of this paradigm with the constraints of ocessors.

In this paper we show that we can reduce the energy consumption of
the edge nodes significantly in TOP. The system switches automatically
to a low-power sleep mode when the current state of all tasks in the
system allow a nap. In addition, we introduce a high-level way to deal
with interrupts. This can replace the power hungry polling of sensors by
much greener waiting for an event in a low-power mode of the system.

Keywords: internet of things · green computing · task-oriented
programming · functional programming · Clean · mTask

This work received financial support through the Erasmus+ Strategic Partnership for
Higher Education *SusTrainable—Promoting Sustainability as a Fundamental Driver in
Software Development Training and Education* (project number 2020-1-PT01-KA203-
078646), funded by the European Union and coordinated by the University of Coimbra,
Portugal.
The information and views set out in this publication are those of the authors and do
not necessarily reflect the official opinion of the European Union. Neither the European Union institutions and bodies nor any person acting on their behalf may be held
responsible for the use which may be made of the information contained therein.

1   Introduction

The Internet of Things (IoT), is the collection on interconnected smart devices that senses and controls objects in the world. It consists of a wide variety of nodes including smart thermostats, lightbulbs, doorbells, human presence sensors and electricity meters. The number of IoT devices is huge. It is estimated at 14.4 billion at the end of 2022 and is expected to grow to 27 billion connected devices at the end of 2027.[1] As a reference the world population is nearly 8 billion persons in 2022 and is expected to grow with a few percent over this period.[2]

The number of nodes in the IoT is not only large, most of these nodes are also running 24/7. By the nature of the IoT there is a significant amount of communication between the devices. This makes it challenging to run the IoT in a sustainable way. Fortunately there are some opportunities to limit the energy consumption of the IoT. First, many tasks of the IoT nodes are rather simple. This implies that we can use a small and energy-efficient microcontroller based node instead of a solution that is based on a single-board computer, like a Raspberry Pi. Second, we can use edge computing to limit the amount of communication. For instance, a sensor reports only significant changes in its readings to a server instead of submitting each value. Third, many microcontrollers have facilities to reduce the power consumption. For instance they can switch off the WiFi-radio when there is temporarily no need for communication. When there is no need for any activity during some time the microcontroller can even switch to a low-power sleep mode. Finally, some sensors have the ability to wake up the microcontroller whenever required from a sleep state using hardware interrupts. This prevents regular waking up of the microcontroller to check whether something interesting is happening.

The IoT not only consumes a lot of energy, its applications help to save power by optimizing climate control systems in buildings, lighting based on occupancy, predictive maintenance to name just a few application areas. There is obviously a balance in the power it takes to operate the IoT and the energy savings and increase of comfort that is provided by the IoT. This makes it worthwhile to ensure that IoT systems themselves are sustainable.

To give an impression of the power consumption of IoT nodes we list some figures. A single-board ocessor based computer like the Raspberry Pi 4 consumes about 4W when it is idle and 6W when its four cores are active. This power consumption excludes the power used by peripherals and the WiFi communication. This is not bad compared to the 60 W consumed by the average 15.6 in laptop.[3] The popular ESP8266 microcontroller consumes a maximum of 0.5 W during active WiFi communication. When the WiFi is switched off, modem sleep, this is reduced to 0.05 W. When the microcontroller takes a nap, light sleep,

---

[1] https://iot-analytics.com/number-connected-iot-devicesaccessed 11th November, 2022.

[2] https://ourworldindata.org/world-population-growthaccessed 11th November, 2022.

[3] https://www.netbooknews.com/tips/how-many-watts-laptop-use/accessed   14th November, 2022.

it consumes only 0.001 W.[4] Moreover, a single-board computer is typical about ten times more expensive than a simple microcontroller board. There are microcontrollers that consume considerable less energy, but they are less common and usually more expensive.

Energy efficiency is obviously very important for battery operated IoT nodes. However, due to the huge number of devices it is important for the entire IoT. Operating 14.4 billion IoT nodes based on Raspberry Pi's for a year requires about 600TWh. As a reference, the annual productions of the single Dutch nuclear power plant is 4TWh.[5] This indicates that operating the IoT takes a serious amount of energy and reducing the power demand of individual nodes makes the system more sustainable.

These numbers make it very attractive to build IoT nodes with microcontrollers rather than with single-board computers such as the Raspberry Pi. However, there is a significant price to be paid. First of all the microcontrollers are much slower and have far less memory than the single-board computers. As a consequence these IoT nodes run typically without a full operating system (OS). Often there is just the user program. Sometime the user program is constructed with the help of a special purpose OS like FreeRTOS. This makes programming IoT systems and updates of these programs very complicated. Especially when several subtasks have to run on the IoT node. Without OS there is no support for multithreading. This implies that the user program has to interleave the subtasks. Modem sleep and light sleep have to be controlled by the user program containing those subtasks. The system can only go to a low-power state when this is possible in all subtasks.

Adding such a node to the IoT is also challenging. Due to the limitations of the microcontroller it is typically programmed in a stripped-down dialect of C++ or Python. For the communication it uses a protocol like MQTT on top of the normal internet protocols. The server that has to cooperate with the IoT node is typically written in high-level language like Java. Since the IoT node has a very limited amount of memory, the sensor reading is be stored in a database on the server. The user interface of the program either runs in a browser, using HTML and for instance JavaScript or WebAssembly, or in a smartphone app for Android and iOS. The program on the IoT node also has to take care about its own updates since there is no OS and physical access to the node for an update is usually too demanding. This implies that many programming languages, protocols and systems are involved in a single task of the IoT. These differences in data types and interaction are known as *semantic friction*. In general, this makes the development of IoT applications cumbersome and its maintenance even more challenging.

---

[4] https://www.espressif.com/sites/default/files/documentation/9b-esp8266-low_power_solutions__en.pdf accessed 11th November, 2022.

[5] https://www.rijksoverheid.nl/onderwerpen/duurzame-energie/opwekking-kernenergie accessed 15th November, 2022.

1.1 Tierless Program Stacks

These problems can be prevented by generating all required software for the entire system from a single source. Such tierless systems are known for web stacks, e.g. Links [4], Hop [15], and iTask [12]. These systems focus on webpages and the associated servers.

In this paper we introduce a tierless system focussed on task-oriented programming (TOP). As the name suggests this declarative formalism focusses on tasks. These pieces of work are executed by humans or machines, including the edge nodes of the IoT. Tasks are here either primitive tasks like reading a sensor, operating an actuator, and filling out a web form or subtasks composed by combinators to more complex tasks. For instance, we can compose tasks in parallel as an alternative for each other or subtasks that are both required without fixing their order. Subtasks can also be composed sequentially. In contrast to ordinary computations, tasks can inspect the current value of subtasks. They can decide to move on based on the current value of a subtask, even when that subtask is not finished yet.

The iTask system is the oldest implementation of TOP [12,14]. It creates a web server, the associated storage and tailor-made web pages for all users of the system. This web interface guides the users through their current tasks. To ensure good performance, a part of the task is executed in the browser rather than on the server. The entire system is built as a shallowly embedded domain-specific language (DSL), in the pure functional programming language Clean [1,13]. This implies that the iTask system is a set of functions and associated data types that are fully integrated with Clean. The tasks can use all data types and computations of Clean. Each iTask program is a single piece of software in Clean. This implies that the compiler checks all types and definitions as usual. The entire system with dynamic web pages, storage control, the web server and the required networking is generated from this single tierless source.

The TOP approach is also very suited to control the IoT edge devices. They run tasks reading sensors, controlling peripherals and communicating with the servers. However, the microcontrollers powering these devices are not powerful enough to run the iTask programs. Since these edge devices do not require to have an integrated web server nor support heavy computations, it is also not necessary to run a full iTask server there. The mTask system is created to solve this gap. It is an embedded DSL in Clean that is fully integrated with iTask. Just like tasks in iTask, tasks in mTask can be dynamically created in the hosting programming language. The task-oriented programs written in the mTask language are compiled at runtime to domain-specific bytecode that fits on the edge devices with very limited power. The device is only programmed once with a domain-specific tailor-made tiny OS. This bytecode is shipped to the edge device and executed by the OS. Since everything is in the same source we obtain the best of both worlds. The tierless approach ensures consistency of the entire software system, while the separate mTask DSL facilitates executions of distinguished parts of this system on cheap and energy-efficient hardware.

The tiny OS on the devices is tailor made to execute mTask programs. It can use this domain-specific knowledge for smart scheduling. The subtasks are only scheduled after a complete rewrite step in the mTask DSL. This implies that the subtasks are never interrupted during communication with the server or peripherals, nor while they are accessing shared memory. This makes it easy to interleave subtasks efficiently. Moreover, the OS knows what the tasks are doing and uses this knowledge to save energy. For instance, the system can switch to a light sleep mode when all subtasks are waiting in a delay. When the system detects that some tasks is checking a temperature sensor in a high frequency, these sensor reading can be delayed since it is unlikely that the temperature changes rapidly. This delay can enable more execution time for more useful tasks or switching to sleep mode when there are no urgent tasks to be done. The automatic delay of subtasks can be fine tuned within the mTask program whenever that is desired. In a traditional tiered system it is much harder to achieve similar energy savings. The programmer needs detailed knowledge of all subtask that are present dynamically and schedule the sleep explicitly. This is difficult and error prone, especially when new versions of the software are required.

The iTask/mTask system enables sustainability in various different interpretations. First, the tierless approach ensures programs that are much smaller and easier to maintain than traditional tiered programs [10]. Second, the mTask addition to this TOP system enable us to use energy efficient and cheap ocessor based hardware. Finally, the optimized scheduling of mTask tasks saves energy by switching to sleep mode automatically.

1.2   Structure of the Paper

Section 3 explains the mTask DSL and its implementation in the amount of detail required for this paper. For a more thorough overview of this language the reader is referred to [5, 6, 8, 9]. In Sect. 4 we explain how the interpreter can save energy by inspecting the current tasks. When these subtasks can be delayed, the entire system goes to a sleep mode to save energy. We can save even more energy when we replace polling of sensors by interrupts. The system can be sleeping until the interrupt wakes it up. In Sect. 6 we draw conclusions.

Appendix A explains how the software used in this paper can be installed. This software is required to compile and run the exercises provided in this paper. Appendix B shows a way to measure the actual poser consumption of a ocessors with an appropriate sensor and another microcontroller. This can be used to verify that techniques discussed here actually decrease the amount of energy used. The solutions to the exercises are provided in Appendix C.

2   Introducing Edge Computing by an Example

Traditionally, the first program that one writes when trying a new language is the so-called *Hello World!* program. This program has the single task of printing the text *Hello World!* to the screen and exiting again. It helps the programmer to become familiarised with the syntax of the language and to verify that the toolchain and runtime environment are working. Microcontrollers usually do not come with screens in the traditional sense. Nevertheless, almost always there is a built-in 1 pixel screen with a 1bit color depth, namely the on-board LED. The *Hello World!* equivalent for microcontrollers blinks this LED.

2.1   Traditional Edge Computing

Using Arduino's C++ dialect to create the blink program results in the code seen in Listing 1.1. Arduino programs are implemented as cyclic executives and hence, each program defines a `setup` and a `loop` function. The `setup` function is executed only once on boot, the `loop` function is continuously called afterwards and contains the event loop. In the blink example, the `setup` function only contains code for setting the General Purpose Input/Output (GPIO), pin to the correct mode. The `loop` function alternates the state of the pin representing the LED between `HIGH` and `LOW`, turning the LED off and on respectively. In between, it waits 500ms so that the blinking is actually visible for the human eye.

```
void setup() {
    pinMode(D2, OUTPUT);
}

void loop() {
    digitalWrite(D2, HIGH);
    delay(500);
    digitalWrite(D2, LOW);
    delay(500);
}
```

Listing 1.1. Blinking an LED in Arduino C++.

To execute such a program on a ocessor this C++ code is first compiled to machine code for that processor. Next the compiled code is stored in the persistent flash memory of the ocessor with help of a small bootloader available on the microcontroller. After a soft restart the `setup` function is executed once and the `loop` function is repeated as long as the processor runs. The flash memory has a specified life of 100 000 write cycles. So, you may need to be careful about how often you write to it. Many programs can be uploaded and executed in such a memory unit. As soon as you start updating this type of memory during program execution, it might wear out rather quickly.

2.2   Task-Oriented Edge Computing

The mTask/iTask variant of the *Hello World!* program looks very much like the Arduino variant and is as useful for testing the toolchain. Therefore, this first exercise is just to verify that your mTask installation is working and that you are able to compile and execute Clean/mTask programs.

```
1   module blink
2   import StdEnv, iTasks                          // Imports
3   import mTask.Interpret
4   import mTask.Interpret.Device.TCP
5
6   Start w = doTasks main w                       // Start the task engine
7
8   main :: Task Bool                              // Main task
9   main =            enterDeviceInfo
10      >>? \spec->withDevice spec (\dev->liftmTask blink dev)
11  where
12      enterDeviceInfo :: Task TCPSettings        // Ask which device to use
13      enterDeviceInfo
14        = enterInformation [] <<@ Label "Device information"
15
16  blink :: Main (MTask v Bool) | mtask v         // mTask task
17  blink = declarePin D4 PMOutput \ledPin->       // Builtin LED of D1 Mini
18      fun \blinkfun=(\state->                     // Recursive function to blink
19              delay (ms 500)                      // Wait for 500ms
20          >>|. writeD ledPin state                // Switch LED to state
21          >>|. blinkfun (Not state))              // Call with inverse state
22      In {main=blinkfun true}                     // Main program
```

Listing 1.2. Blinking an LED in mTask and iTask.

All Clean programs start with a module declaration, in this case the module name is `blink`. To work with the mTask library, some imports are required (see Line 2). As mTask is embedded in iTask, iTask's `doTasks` is called on Line 6 as the `Start` rule of the program. This function starts the iTask engine and execute the `main` task that is given as the second argument. The `main` task is defined at Line 8 and first asks the user to enter the device information (see Line 12). With this information, mTask's `withDevice` function is called that allows the program to interact safely with a connected microcontroller. Using `liftmTask`, an mTask task is lifted to iTask, i.e., it is compiled to bytecode, sent to the microcontroller and the result is observable in iTask. The `blink` task is an mTask task defined at Line 16 with the type `Main (MTask v Bool) | mtask v` that can be read as: *An mTask task of the type* Bool *parametric in the view v as long as this v implements the classes defined by the* `mtask` *class collection.* This has to do with the tagless-final embedding technique that is used to create the mTask language [2]. Do not worry when you do not understand all technical details.

The start expression of every mTask program is wrapped in a `main` record to assure that functions and sensors are only defined at the top level. First on Line 17 the GPIO pin D4, connected to the builtin LED is set to output mode. As mTask is hosted in a functional programming language, instead of using a loop, a recursive function is used. Line 18 shows this function (`blinkfun`), that has one boolean argument, the state. This function first waits for 500ms (Line 19), then writes the state to the pin to either turn on or turn off the LED (Line 20) and finally it calls itself recursively with the inverse of the state (Line 21). When calling this function at Line 22 with some initial state, it results in a blinking LED when executing.

Installation of all required software and suitable ocessors to execute the examples and exercises of this chapter is outlined in Sect. A. To measure the power consumption of the ocessor executing our programs one can use the machinery discussed in Appendix B.

Exercise 1 Hello world! (`blink.icl`)

Compile and run the `blink` module by running (see the readme for instructions).

```
nitrile build -only=blink./blink
```

When you have connected the device properly, it should connect to either one of the networks and show its address. Enter this address in the `enterDeviceInfo` task together with the default mTask port number 8123. If you then press *Continue* the light on top of the microcomputer should turn on and off according to the given frequency (500ms).

Exercise 2 Tailor-made blinking (`blinkparam.icl`)

Change the `blink` function so that it gets a parameter, an argument, i.e. the time between state changes as follows:

```
blink :: Int -> Main (MTask v Bool) | mtask v
blink wait = declarePin D4 PMOutput \d4->
```

This does require you to provide this extra argument as well at the call site. By using the `-&&-` combinator, the `enterDeviceInfo` can be combined with another task that asks the user for a time in ms. The result of this line will then be a tuple that can be pattern matched and passed on to the `blink` function as follows:

```
>>? \(spec, wait)->withDevice spec (\dev->liftmTask (blink wait) dev)
```

Finally, adapt the `delay` task in the mTask task so that it uses `wait` instead of a fixed number of ms.
Hint: `wait` *is of type* `Int` *and not of type* `v` `Int` *so you have to lift it to the mTask domain first by* `lit`.

---

3   The mTask DSL

Every mTask definition consists of a `main` record defining the initial task. In Listing 1.2 this is on Line 22. The task definition is constructed from basic tasks like `writeD` and `delay`. Declarations like `declarePin` and `fun` add user-defined elements to the task. Constructors like $\gg|.$ are used to compose tasks sequentially to compound tasks. This particular combinator denotes sequential composition. There are also combinators to combine subtasks in parallel.

Tasks produce a value after each rewrite step. Such a result can be `NoValue`, for instance when the task is waiting for input. Values can be either stable or unstable. As a rule of thumb; a result is unstable when it can be different in a next evaluation, like reading a sensor.

Tasks can also communicate via a shared data source (SDS). As the name suggests this is a container holding values that can be inspected and updated by various tasks. Every SDS is typed, it always hold a value of the assigned type. The semantics of TOP prevent race conditions for updating SDSs.

A full overview of the language is not needed here. Language elements are introduced as they are required.

3.1   Implementation of the mTask system

The main interpretation, or view/backend of the mTask language, is the bytecode compiler. With it, and a handful of integration functions and tasks, mTask tasks can be executed on microcontrollers and integrated in iTask as if they were regular iTask tasks. Furthermore, with a special language construct, SDSs can be shared between mTask and iTask programs as well.

When using the bytecode compiler interpretation in conjunction with the iTask integration, mTask is a heterogeneous DSL [16]. Some components—for example the RTS on the microcontroller—is unaware of the other components in the system, and it is executed on a completely different architecture. The mTask language is an enriched simply-typed λ-calculus with support for some basic types, arithmetic operations, function definitions, and a task language.

As mTask programs are constructed and compiled to bytecode at runtime, it is possible to tailor make the program according to the needs of the current state. For example, it may be possible to ask a time between state changes from the user and inline that in the blink program.

The generated bytecode is shipped dynamically to the corresponding interpreter running on a ocessor. Since we can ship multiple tasks to the interpreter, it is actually a light-weight OS tailor-made for mTask programs. The mTask interpreter is stored in flash memory of the ocessor. It is available after a power interruption. To prevent wear of the ocessor memory, the tasks to be executed and their current state are stored in volatile RAM. The interpreter removes the state of the tasks and all associated code once the task is no longer needed from the iTask host program. The user needs only to install the appropriate software, dynamic compilation, shipping and removing tasks is done automatically.

3.2   The mTask embedding

To leverage the type checker of the host language, types in the mTask language are expressed as types in the host language, to make the language type safe.

However, not all types in the host language are suitable for microcontrollers that may only have 2KiB of RAM so class constraints are therefore added to the DSL functions. The most used class constraint is the `type` class collection containing functions for serialization, printing, iTask constraints, etc. Many of these functions can be derived using generic programming. An even stronger restriction on types is defined for types that have a stack representation. This `basicType` class has instances for many Clean basic types such as `Int`, `Real` and `Bool`. The class constraints for values in mTask are omnipresent in all functions and therefore often omitted throughout the chapters for brevity and clarity (Table 1).

Table 1. Translation from Clean/mTask data types to C++ data types.

Clean/mTask	C++ type	№bits
Bool	**bool**	16
Char	**char**	16
Int	int16_t	16
Real	**float**	32
Long	int32_t	32
T = A | B | C	**enum**	16

Listing 1.3 contains the definitions for the auxiliary types and type constraints (such as `type` and `basicType`) that are used to construct mTask expressions. The mTask language interface consists of a core collection of type classes bundled in the type class `class mtask`. Every interpretation implements the type classes in the `mtask` class There are also mTask extensions that not every interpretation or backend implements such as peripherals and iTask integration. Those are specified separately as they are not contained in the `mtask` class.

```
class type t | iTask,··· ,fromByteCode, toByteCode t
class basicType t | type t where···

class mtask v | expr,··· , int, real, long v
```

Listing 1.3. Classes and class collections for the mTask language.

Sensors, SDSs, functions, etc. may only be defined at the top level. The `Main` type is used to distinguish the top level from the main expression. Some top level definitions, such as functions, are defined using HOAS [3,11]. To make their syntax friendlier, the `In` type—an infix tuple—is used to combine these top level definitions as can be seen in `someTask` (Listing 1.4).

```
:: Main a = { main :: a }
:: In a b = (In) infix 0 a b

someTask :: MTask v Int | mtask v & liftsds v & sensor1 v &···
```

```
someTask =
    sensor1 config1 \sns1->
    sensor2 config2 \sns2->
       sds       \s1  = initialValue
    In liftsds \s2  = someiTaskSDS
    In fun       \fun1= ( ··· )
    In fun       \fun2= ( ··· )
    In { main = mainexpr }
```

Listing 1.4. Example task and auxiliary types in the mTask language.

3.3 Communication with iTask

As shown in Listing 1.2 the function `liftmTask` lifts an mTask task to an iTask
task. The result of this task determines the value of the iTask and is automatically communicated. In Exercise 2 you learned how to communicate from iTask
to mTask via parameters.

Communication via parameters works fine, but for dynamic communication
tasks in mTask share information through SDSs with iTask. The same SDS is
available on the edge device as well as on the server. In mTask we can read
the value of an SDS by `getSds` and write a new value by `setSds`. In iTask we
have `get` and `set` as well as web-based editors to update a SDS interactively.
The machinery ensures that changes of the SDS in one world are automatically
mirrored in the other world. This allows for very dynamic behaviour, such as
setting the blinking frequency during the run time of the mTask task.

Exercise 3 Dynamic blinking behaviour (`blinkshare.icl`)

To create an SDS that is available globally, the `sharedStore` function is used:

```
delayShareI :: SimpleSDSLens Int
delayShareI = sharedStore "delay" 500
```

This SDS can be lifted to an mTask SDS using the `liftsds` construct as follows:

```
blink :: Main (MTask v Bool) | mtask, liftsds v
blink = declarePin D4 PMOutput \d4->
    liftsds \delayShareM=delayShareI
    In fun \blinkfun=(\x->
    ...
```

Adapt the `blinkfun` function so that it first reads the value of `delayShareM` and uses
it as the time to wait.

Hint: *Use `getSds` to read the value of a SDS. It yields an unstable value. So, you
have to use a sequential combinator that steps on an unstable value. The operator $\gg\sim$.
does exactly this.*

---

4   Green Computing with mTask

MTask offers abstractions for edge layer-specific details such as the heterogeneity of architectures, platforms and frameworks; peripheral access; and multitasking but also for energy consumption and scheduling. In mTask, tasks are implemented as a rewrite system, where the work is automatically segmented in small atomic bits and stored as a task tree. During each cycle, a single rewrite step is performed on all task trees, during rewriting, tasks do a bit of their work and progress steadily, allowing interleaved and seemingly parallel operation. After a loop, the RTS knows which task is waiting on which triggers and is thus able to determine the next execution time for each task. Utilising this information, the RTS can determine when it is possible and safe to sleep and choose the optimal sleep mode according to the sleeping time. For example, the RTS never attempts to sleep during an I^2C communication because I/O is always contained *within* a rewrite step.

An mTask program is dynamically transformed to bytecode. This bytecode and the initial mTask expression are shipped to an mTask IoT node. The mTask rewrite engine rewrites the current expression just a single rewrite step at a time at the task level. Other computations are evaluated eagerly. When subtasks are composed in parallel, all subtasks are rewritten unless the result of the first rewrite step makes the result of the other tasks superfluous. This task design ensures that all time critical communication with peripherals is within a single rewrite step. Even functions that produce an infinite number of sequences, and thus rewrite steps, as in the `blink` example, are perfectly fine. The mTask system does tail-call elimination to facilitate this.

This rewriting of tasks is very convenient; the system can inspect the current state of all mTask expressions after a rewrite step. The system can detect if sleeping of the current task expression is useful and how long this is possible. When all current subtasks in the system allow a sufficiently long break, the ocessor can dynamically be switched to a low-energy sleeping mode. The system wakes up after the determined interval and continues rewriting of the tasks. The `delay` primitive is an obvious spot where sleeping is possible. In the next section we argue that there are many other spots where energy saving by inserting a short delay is possible without changing the behaviour of the system.

4.1   Temperature Monitor

Edge nodes in the IoT are very suited to read sensors. A typical example is a temperature sensor. Such a device communicates with the sensor via a dedicated protocol over the GPIO pins of the microcontroller. In this example the temperature is measured with the digital humidity and temperature (DHT) sensor that communicates via the I^2C protocol. We do not need to know any details about the sensor or protocol. The `DHT` primitive of mTask defines a tailor-made `dht` object. The task `temperature dht` reads the current value from this object.

Exercise 4 Initial temperature monitor (`tempmon.icl`)

The appointed device contains a SHT30x DHT sensor that communicates with the ocessor using I^2C. While task values can be observed directly using the appropriate combinators (`>&>`, `>&*`), writing the value to an SDS is an easier option.

Create a globally available SDS to store the temperature.

Hint: *In mTask, temperature is measured in °C stored in a* **Real**.

To view this SDS during operation, create `viewTemperature` *task and run that in parallel with the* `liftmTask` *task:*

```
viewTemperature :: Task Real
viewTemperature = viewSharedInformation [] tempShareI
    <<@ Label "Current temperature (C)"
```

Finally, implement the temperature monitoring task.

```
tempmon :: Main (MTask v Real) | mtask, dht, liftsds v
tempmon = DHT (DHT_SHT (i2c 0x45)) \dht->
   liftsds \tempShareM=tempShareI
   In fun \tempfun=(\()->temperature dht
       >>~. \t->setSds tempShareM t
       >>|. tempfun ()
   ) In {main=tempfun ()}
```

As you can see, this version of the temperature monitoring application generates a lot of data traffic between the edge node and the server. Every update of the local SDS results in an automatic message to the server to keep both versions of the SDS in sync.

Exercise 5 Temperature monitor, second iteration (`tempmon2.icl`)

The temperature is not bound to change every 1ms so it's not required to measure it that often.

Adapt the program so that there is a 5 second delay in between the measurements. This is done by adding a **delay** somewhere.

4.2   Rewrite Intervals

Some mTask examples contain one or more explicit **delay** primitives, offering a natural place for the node executing it to pause. However, there are many mTask programs that just specify a repeated set of primitives. A typical example is the

```
thermostat :: Main (MTask v Bool) | mtask v
thermostat = DHT I2Caddr \dht->
   {main = rpeat (temperature dht >>~. \temp.
                  writeD builtInLED (goal <. temp))}
```

Listing 1.5. A basic thermostat task.

program that reads the temperature for a sensor and sets the system LED if the reading is below some given `goal`.

This program repeatedly reads the DHT sensor and sets the on-board LED based on the comparison with the `goal` as fast as possible on the mTask node. This is a perfect solution as long as we ignore the power consumption. The mTask machinery ensures that if there are other tasks running on the node that they are making progress. However, this solution is far from perfect when we take power consumption into account. In most applications, it is very unlikely that the temperature changes significantly within one minute, let alone within some ms. Hence, it is sufficient to repeat the measurement with an appropriate interval.

There are various ways to improve this program. The simplest solution is to add an explicit delay to the body of the repeat loop. A slightly more sophisticated option is to add a repetition period to the **rpeat** combinator. The combinator implementing this is called **rpeatEvery**. Both solutions rely on an explicit action of the programmer.

Fortunately, mTask also contains machinery to do this automatically. The key of this solution is to associate dynamically an evaluation interval with each task. The interval $\langle low, high \rangle$ indicates that the evaluation can be safely delayed by any number of ms within that range. Such an interval is just a hint for the RTS. It is not a guarantee that the evaluation takes place in the given interval. Other parts of the task expression can force an earlier evaluation of this part of the task. When the system is very busy with other work, the task might even be executed after the upper bound of the interval. The system calculates the rewrite rates from the current task expression. This has the advantage that the programmer does not have to deal with them and that they are available in each and every mTask program.

4.3 Basic Tasks

We start by assigning default rewrite rates to basic tasks. These rewrite rates reflect the expected change rates of sensors and other inputs. Basic tasks to one-shot set a value of a sensor or actuator usually have a rate of $\langle 0, 0 \rangle$, this is never delayed, e.g. writing to a GPIO pin. Basic tasks that continuously read a value or otherwise interact with a peripheral have default rewrite rates that fit standard usage of the sensor. Table 2 shows the default values for the basic tasks. Reading SDSs and fast sensors such as sound or light aim for a rewrite every 100 ms, medium slow sensors such as gesture sensors every 1000 ms and slow sensors such as temperature or air quality every 2000 ms.

Table 2. Default rewrite rates of basic tasks.

task	default interval
reading an SDS	⟨0 m sec, 2000 m sec⟩
slow sensor	⟨0 m sec, 2000 m sec⟩
medium sensor	⟨0 m sec, 1000 m sec⟩
fast sensor	⟨0 m sec, 100 m sec⟩

Exercise 6 Temperature monitor, third iteration (`tempmon3.icl`)

Adapt the program so that only changes bigger than half a °C are reported. Using the step combinator (`>>*.`) you can observe the value of a task and step when the predicate holds.

If the predicate does not match, the left hand side of the step combinator is scheduled for execution some other time, depending on the characteristics of the task. In case of the temperature sensor, the next execution is within two seconds.

Hint: *Use functions to do the heavy lifting. For example, to take the absolute value over real numbers you can define:*

```
In fun \abs=(\x->If (x >. lit 0.0) x (lit 0.0 -. x))
```

Furthermore, remember that multiparameter functions are always written with tuple notation. So a function that determines if two real numbers differ more than 0.5° C is defined as:

```
In fun \differsenough=(\(old, new)->abs (old -. new) >. lit 0.5)
```

4.4   Tweaking Rewrite Rates

A tailor-made ADT (see Listing 1.6) determines the timing intervals for which the value is determined at runtime but the constructor is known at compile time. During compilation, the constructor of the ADT is checked and code is generated accordingly. If it is **Default**, no extra code is generated. In the other cases, code is generated to wrap the task tree node in a *tune rate* node. In the case that there is a lower bound, i.e. the task must not be executed before this lower bound, an extra *rate limit* task tree node is generated that performs a no-op rewrite if the lower bound has not passed but caches the task value.

```
:: TimingInterval v
   = Default
   | BeforeMs (v Int)                 // yields ⟨0, x⟩
   | BeforeS  (v Int)                 // yields ⟨0, x × 1000⟩
   | ExactMs  (v Int)                 // yields ⟨x, x⟩
   | ExactS   (v Int)                 // yields ⟨x × 1000, x × 1000⟩
   | RangeMs  (v Int) (v Int)  // yields ⟨x, y⟩
   | RangeS   (v Int) (v Int)  // yields ⟨x × 1000, y × 1000⟩
```

Listing 1.6. The ADT for timing intervals in mTask.

5   Interrupts

Most microcontrollers have built-in support for processor interrupts. These interrupts are hard-wired signals that can interrupt the normal flow of the program to execute a small piece of code, the interrupt service routine (ISR). While the ISRs look like regular functions, they do come with some limitations. For example, they must be very short, in order not to miss future interrupts; can only do very limited I/O; cannot reliably check the clock; and they operate in their own stack, and thus communication must happen via global variables. After the execution of the ISR, the normal program flow is resumed. Interrupts are heavily used internally in the RTS of the microcontrollers to perform timing critical operations such as WiFi, I^2C, or SPI communication; completed ADC conversions, software timers; exception handling; etc.

Interrupts offer two substantial benefits: fewer missed events and better energy usage. Sometimes an external event such as a button press only occurs for a very small duration, making it possible to miss it due to it happening right between two polls. Using interrupts is not a fool-proof way of never missing an event. Events may still be missed if they occur during the execution of an ISR or while the microcontroller is still in the process of waking up from a triggered interrupt. There are also some sensors, such as the CCS811 air quality sensor, with support for triggering interrupts when a value exceeds a critical limit.

Table 3 shows the different types of interrupts.

Table 3. Overview of GPIO interrupt types.

type	triggers
change	input changes
falling	input becomes low
rising	input becomes high
low	input is low
high	input is high

5.1   Arduino Platform

Listing 1.7 shows an exemplatory program utilising interrupts written in
Arduino's C++ dialect. The example shows a debounced light switch for the
built-in LED connected to GPIO pin 13. When the user presses the button
connected to GPIO pin 11, the state of the LED changes. As buttons sometimes induce noise shortly after pressing, events within 30ms after pressing are
ignored. In between the button presses, the device goes into deep sleep using the
`LowPower` library.

Line 1 to Line 3 defines the pin and debounce constants. Line 5 defines the
current state of the LED, it is declared **volatile** to exempt it from compiler
optimisations because it is accessed in the interrupt handler. Line 6 flags whether
the program is in debounce state, i.e. events should be ignored for a short period
of time.

In the **setup** function (Line 8 to Line 13), the pinmode of the LED and
interrupt pins are set. Furthermore, the microcontroller is instructed to wake up
from sleep mode when a *rising* interrupt occurs on the interrupt pin and to call
the ISR at Line 22 to Line 26. This ISR checks if the program is in cooldown
state. If this is not the case, the state of the LED is toggled. In any case, the
program goes into cooldown state afterwards.

In the **loop** function, the microcontroller goes to low-power sleep immediately
and indefinitely. Only when an interrupt triggers, the program continues, writes
the state to the LED, waits for the debounce time, and finally disables the
`cooldown` state.

```
1   #define LEDPIN 13
2   #define INTERRUPTPIN 11
3   #define DEBOUNCE 30
4
5   volatile int state = LOW;
6   volatile bool cooldown = true;
7
8   void setup() {
9       pinMode(LEDPIN, OUTPUT);
10      pinMode(INTERRUPTPIN, INPUT);
11      LowPower.attachInterruptWakeup(
12          INTERRUPTPIN, buttonPressed, RISING);
13  }
14
15  void loop() {
16      LowPower.sleep();
17      digitalWrite(LEDPIN, state);
18      delay(DEBOUNCE);
19      cooldown = false;
20  }
21
22  void buttonPressed() {
23      if (!cooldown)
24          state = !state;
```

```
25      cooldown = true;
26  }
```

Listing 1.7. Light switch using interrupts in Arduino.

5.2   MTask language

Listing 1.8 shows the interrupt interface in mTask. The `interrupt` class contains a single function that, given an interrupt mode and a GPIO pin, produces a task that represents this interrupt. Lowercase variants of the various interrupt modes such as `change :== lit Change` are available as convenience macros.

```
class interrupt v where
    interrupt :: (v InterruptMode) (v p) -> MTask v Bool | pin p

:: InterruptMode = Change | Rising | Falling | Low | High
```

Listing 1.8. The interrupt interface in mTask.

When the mTask device executes this task, it installs an ISR and sets the rewrite rate of the task to infinity, $\langle \infty, \infty \rangle$. The interrupt handler is set up in such a way that the rewrite rate is changed to $\langle 0, 0 \rangle$ once the interrupt triggers. As a consequence, the task is executed on the next execution cycle.

The `pirSwitch` function in Listing 1.9 creates, given an interval in ms, a task that reacts to motion detection by a PIR sensor (connected to GPIO pin 0) by lighting the LED connected to GPIO pin 13 for the given interval. The system turns on the LED again when there is still motion detected after this interval. By changing the interrupt mode in this program text from `High` to `Rising` the system lights the LED only one interval when it detects motion no matter how long this signal is present at the PIR pin.

```
pirSwitch :: Int -> Main (MTask v Bool) | mtask v
pirSwitch =
    declarePin D13 PMOutput \ledpin->
    declarePin D0 PMInput \pirpin->
    {main = rpeat (     interrupt high pirpin
                  >>|. writeD ledpin false
                  >>|. delay (lit interval)
                  >>|. writeD ledpin true) }
```

Listing 1.9. Example of a toggle light switch using interrupts.

5.3   Motion Detection

If a person enters the room, the temperature is bound to change faster. Using the passive infrared (PIR) sensor, motion can be detected. If the PIR detects motion, the GPIO pin it connects to is high for a couple of seconds. Polling this pin continuosly to check whether there is motion consumes a lot of energy. Luckily, using an *high* interrupt handler we get notified when the pin is high, i.e. when there is motion.

Exercise 8 Temperature monitor, motion detection (`tempmon4.icl`)

Adapt the temperature monitor so that the temperature is only measured every minute. In addition, if motion is detected, record the motion and take an extra temperature measurement.

1. Adapt the temperature function so that it measures only once every minute (this is done using `temperature`).
2. Add the declaration of the PIR sensor to the top level of the mTask task using the PIR functions:

```
tempmon = PIR D3 \pir->
```

3. Add an extra SDS to record the number of movements. E.g.

```
movementShareI :: SimpleSDSLens Int
```

4. Extend the main task using a parallel combinator so that it not only calls `tempfun` but also `motionfun`.
5. Extend `motionfun` so that it also measures the temperature when detecting motion.

---

6   Conclusion

This paper shows how we enable green computing for the IoT. A first important step is to replace the single-board computers, like a Raspberry Pi, driving the edge nodes by an energy-efficient microcontroller, like the ESP8266. These microcontrollers consume an order of magnitude less energy and are an order of magnitude less expensive than the full-fledged single-board computers. For battery powered nodes the advantages are obvious. Given the enormous and growing amount of devices, it is also worthwhile to use microcontrollers for nodes that have a permanent power supply.

Microcontrollers have very limited amounts of memory and processing power. This implies that they typically have no OS and cannot be programmed like normal computers. Typically there is only a single program executing at the microcontroller. When several independent subtasks must be executed, this program must explicitly schedule these tasks. The mTask system discussed here enables the execution of high-level TOP programs on these tiny devices. This system is fully integrated within the iTask DSL that enables multi-user web-based TOP. This implies that the host language ensures type safety and can generate the boilerplate code for storing values and communication between the edge nodes and the server. The task to be executed on an edge node is dynamically compiled to bytecode, shipped to the edge node, and interpreted there by a tailor-made feather-light TOP OS.

To ensure progress of all subtasks shipped to such an edge node, it implements small step rewriting of these tasks. After each and every rewrite step the state is stored to enable the rewriting of all other subtasks on the edge node. This implies that the mTask system is able to inspect the state of all subtasks. When this inspection shows that all subtasks can be delayed sufficiently, the entire system is switched to a low-power sleep state. To increase the possibilities to pause processing mTask implements heuristics. For instance a temperature sensor is by default used only once a second since we assume that its value does not change significantly during this interval. The user can tweak this reading delay at a high level of abstraction to adjust the balance between responsiveness and energy uses for a specific application.

Hardware supported interrupts enable the system to wait in a low-energy state until an event occurs. This is way more power friendly than the usual polling for the event. Moreover, it greatly reduces the chance of missing the event. The mTask system also offers these interrupts at a convenient high abstraction level that is much easier to use than the traditional interrupt service routines.

These high-level energy savings achieve a reduction of the power consumption of another order of magnitude with no or very limited effort from the programmer. Hence, we consider these extensions of the mTask system a very valuable contribution to green computing of the IoT.

A   Installing the Software

To execute the iTask programs listed in this chapter one needs the functional programming language Clean and the iTask library. These systems use the `nitrile` package manager and their installation instructions can be found at https:// clean-lang.org/. The Clean/`nitrile` ecosystem are under constant change. For 64-bit linux machines, an out-of-the-box working setup can be found here including desktop application to simulate an mTask microcontroller [7].

To prepare a microcontroller for the use with mTask, the domain-specific OS must be installed. A list of appropriate devices and the required software is available at https://gitlab.com/mtask/client, the examples from these lecture notes are tested on client version v0.1.1.

B   Measuring Power Consumption

For the power measurements we recommend a INA260 or INA226 current sensor connected to a separate Wemos D1 mini controller. A program for the Wemos D1 mini measuring the energy consumed can be found in Listing 1.10. This program sends the actual energy consumption to the serial output and supports both current sensors. Two different versions of the Wemos D1 mini microcontroller can be used, Fig. 1 shows wiring instructions.

The Arduino IDE's serial plotter is used to graph of the energy consumption. Connect only the leftmost microcontroller, the one on the breadboard, with an USB cable to your PC! This microcontroller should run a power monitor program. The second ocessor is executing mTask programs. Follow the instructions from Appendix A to install the mTask system on this ocessor. To connect it to the SHT30x DHT temperature shield used in the exercises it might be convenient to stick the rightmost ocessor with the temperature sensor on a Wemos triple base. This assumes that you use the Wemos temperature shield for the sensor. The code should work for any SHT30x DHT sensor with I^2C communication.

```cpp
#include <Wire.h>

//Comment if you use the INA 226 breakout board
#define INA260

#ifdef INA260
#include <Adafruit_INA260.h>
Adafruit_INA260 INA = Adafruit_INA260();
#define READPOWER readPower
#else

#include <INA226_WE.h>
#define I2C_ADDRESS 0x40
INA226_WE INA = INA226_WE(I2C_ADDRESS);
#define READPOWER getBusPower
#endif

void setup() {
    Serial.begin(115200);
    Wire.begin();
#ifdef INA260
    if (!INA.begin()) {
        Serial.println("Couldn't find INA260 chip");
        while(1);
    }
    INA.setAveragingCount(INA260_COUNT_4);
#else
    INA.init();
    INA.setAverage(AVERAGE_4);
#endif
}

void loop() {
    Serial.println("Power(W)");
    for (int i = 0; i < 100; i += 1) {
        Serial.printf("%.3f\n", INA.READPOWER());
        delay(100);
    }
}
```

Listing 1.10. Arduino C++ sketch for the power monitor.

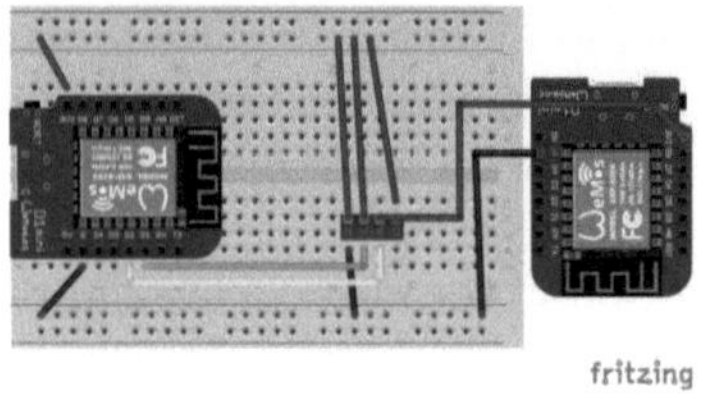 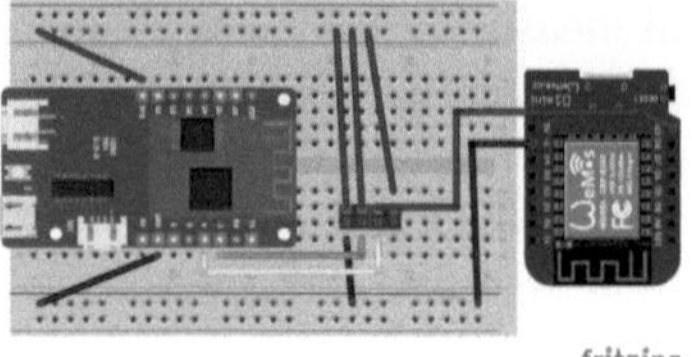

Fig. 1. Wiring instructions for the powermonitor using the Wemos D1 mini (left) or pro variant (right).

C   Solutions

This appendix contains a possible solution for the exercises in this paper. Other solutions might be fine, like usual in programming.

Solution 1 Hello world! (`blink.icl`)

The full program was given in Exercise 1.

Solution 2 Tailor-made blinking (`blinkparam.icl`)

```
import StdEnv, iTasks
import mTask.Interpret
import mTask.Interpret.Device.TCP

Start w = doTasks main w

main :: Task Bool
main =                    enterDeviceInfo -&&- enterDelayTime
    >>? \(spec, wait)->withDevice spec (\dev->liftmTask (blink wait) dev)
where
    enterDeviceInfo :: Task TCPSettings
    enterDeviceInfo = enterInformation [] <<@ Label "Device information"

    enterDelayTime :: Task Int
    enterDelayTime = enterInformation [] <<@ Label "Time between state
    change (ms)"

blink :: Int -> Main (MTask v Bool) | mtask v
blink wait = declarePin D4 PMOutput \d4->
    fun \blinkfun=(\x->
            delay (lit wait)
        >>|. writeD d4 x
        >>|. blinkfun (Not x))
    In {main=blinkfun true}
```

Solution 3 Dynamic blinking behaviour (`blinkshare.icl`)

```
import StdEnv, iTasks
import mTask.Interpret
import mTask.Interpret.Device.TCP

Start w = doTasks main w

delayShareI :: SimpleSDSLens Int
delayShareI = sharedStore "delay" 500

main :: Task Bool
main =          enterDeviceInfo
    >>? \spec->withDevice spec (\dev->
            liftmTask blink dev
        -|| updateSharedInformation [] delayShareI
    )
where
    enterDeviceInfo :: Task TCPSettings
    enterDeviceInfo = enterInformation [] <<@ Label "Device information"

blink :: Main (MTask v Bool) | mtask, liftsds v
blink = declarePin D4 PMOutput \d4->
    liftsds \delayShareM=delayShareI
    In fun \blinkfun=(\x->
            getSds delayShareM
        >>~. \wait->delay wait
        >>|. writeD d4 x
        >>|. blinkfun (Not x))
    In {main=blinkfun true}
```

Solution 4 Initial temperature monitor (`tempmon.icl`)

```
import StdEnv, iTasks
import mTask.Interpret
import mTask.Interpret.Device.TCP

Start w = doTasks main w

tempShareI :: SimpleSDSLens Real
tempShareI = sharedStore "temp" 0.0

main :: Task Real
main =          enterDeviceInfo
    >>? \spec->withDevice spec (\dev->
            liftmTask tempmon dev
        -|| viewTemperature
    )
where
    enterDeviceInfo :: Task TCPSettings
    enterDeviceInfo = enterInformation [] <<@ Label "Device information"

    viewTemperature :: Task Real
    viewTemperature = viewSharedInformation [] tempShareI
        <<@ Label "Current temperature (C)"

tempmon :: Main (MTask v Real) | mtask, dht, liftsds v
tempmon = DHT (DHT_SHT (i2c 0x45)) \dht->
    liftsds \tempShareM=tempShareI
    In fun \tempfun=(\()->temperature dht
        >>~. \t->setSds tempShareM t
        >>|. tempfun ()
    ) In {main=tempfun ()}
```

Solution 5 Temperature monitor, second iteration (`tempmon2.icl`)

```
import StdEnv, iTasks
import mTask.Interpret
import mTask.Interpret.Device.TCP

Start w = doTasks main w

tempShareI :: SimpleSDSLens Real
tempShareI = sharedStore "temp" 0.0

main :: Task Real
main =            enterDeviceInfo
    >>? \spec->withDevice spec (\dev->
            liftmTask tempmon dev
        -|| viewTemperature
    )
where
    enterDeviceInfo :: Task TCPSettings
    enterDeviceInfo = enterInformation [] <<@ Label "Device information"

    viewTemperature :: Task Real
    viewTemperature = viewSharedInformation [] tempShareI
        <<@ Label "Current temperature (C)"

tempmon :: Main (MTask v Real) | mtask, dht, liftsds v
tempmon = DHT (DHT_SHT (i2c 0x45)) \dht->
    liftsds \tempShareM=tempShareI
    In fun \temp=(\()->temperature dht
        >>~. \t->setSds tempShareM t
        >>|. delay (ms 5000)
        >>|. temp ()
    ) In {main=temp ()}
```

Solution 6 Temperature monitor, third iteration (`tempmon3.icl`)

```
import StdEnv, iTasks
import mTask.Interpret
import mTask.Interpret.Device.TCP

Start w = doTasks main w

tempShareI :: SimpleSDSLens Real
tempShareI = sharedStore "temp" 0.0

main :: Task Real
main =          enterDeviceInfo
    >>? \spec->withDevice spec (\dev->
            liftmTask tempmon dev
        -|| viewTemperature
    )
where
    enterDeviceInfo :: Task TCPSettings
    enterDeviceInfo = enterInformation [] <<@ Label "Device information"

    viewTemperature :: Task Real
    viewTemperature = viewSharedInformation [] tempShareI
        <<@ Label "Current temperature (C)"

tempmon :: Main (MTask v Real) | mtask, dht, liftsds v & fun (v Real, v
    Real) v
tempmon = DHT (DHT_SHT (i2c 0x45)) \dht->
    liftsds \tempShareM=tempShareI
    In fun \abs=(\x->If (x >. lit 0.0) x (lit 0.0 -. x))
    In fun \differsenough=(\(old, new)->abs (old -. new) >. lit 0.5)
    In fun \tempfun=(\oldtemp->temperature dht
        >>*. [IfValue (\newtemp->differsenough (oldtemp, newtemp))
                \newtemp->setSds tempShareM newtemp]
        >>=. \newtemp->tempfun newtemp
    ) In {main=tempfun (lit 0.0)}
```

Solution 7 Temperature monitor, Motion detection (`tempmon4.icl`)

```
import StdEnv, iTasks
import mTask.Interpret
import mTask.Interpret.Device.TCP

Start w = doTasks main w

movementShareI :: SimpleSDSLens Int
movementShareI = sharedStore "movement" 0

tempShareI :: SimpleSDSLens Real
tempShareI = sharedStore "temp" 0.0

main :: Task Int
main =          enterDeviceInfo
    >>? \spec->withDevice spec (\dev->
            liftmTask tempmon dev
        -|| viewMovement
    )
where
    enterDeviceInfo :: Task TCPSettings
    enterDeviceInfo = enterInformation [] <<@ Label "Device information"

    viewMovement :: Task Int
    viewMovement = (viewSharedInformation [] movementShareI <<@ Label "
    No. of detected movements")
        -|| (viewSharedInformation [] tempShareI <<@ Label "Temperature")

tempmon :: Main (MTask v Int) | mtask, PIR, dht, liftsds v
tempmon = PIR D3 \pir->
    DHT (DHT_SHT (i2c 0x45)) \dht->
    liftsds \movementShareM=movementShareI
    In liftsds \tempShareM=tempShareI
    In fun \motionfun=(\()->
            interrupt rising pir
        >>|. updSds movementShareM ((+.)(lit 1))
        >>|. temperature dht
        >>~. \newtemp->setSds tempShareM newtemp
        >>|. motionfun ())
    In fun \abs=(\x->If (x >. lit 0.0) x (lit (0.0) -. x))
    In fun \differsenough=(\(old, new)->abs (old -. new) >. lit 0.5)
    In fun \tempfun=(\oldtemp->temperature` (BeforeSec (lit 60)) dht
        >>*. [IfValue (\newtemp->differsenough (oldtemp, newtemp))
                \newtemp->setSds tempShareM newtemp]
        >>=. \newtemp->tempfun newtemp)
    In {main= motionfun () .||. tempfun (lit 0.0)}
```

References

1. Brus, T.H., van Eekelen, M.C.J.D., van Leer, M.O., Plasmeijer, M.J.: Clean A language for functional graph rewriting. In: Kahn, G. (ed.) Functional Programming Languages and Computer Architecture, pp. 364–384. Springer, Heidelberg (1987)
2. Carette, J., Kiselyov, O., Shan, C.C.: Finally tagless, partially evaluated: Tagless staged interpreters for simpler typed languages. J. Funct. Program. **19**(5), 509 543 (2009). https://doi.org/10.1017/S0956796809007205
3. Chlipala, A.: Parametric higher-order abstract syntax for mechanized semantics. In: Proceedings of the 13th ACM SIGPLAN International Conference on Functional Programming, ICFP 2008, Victoria, BC, Canada, pp. 143–156. ACM, New York, NY, USA (2008). https://doi.org/10.1145/1411204.1411226
4. Cooper, E., Lindley, S., Wadler, P., Yallop, J.: Links: web programming without tiers. In: de Boer, F.S., Bonsangue, M.M., Graf, S., de Roever, W.P. (eds.) Formal Methods for Components and Objects, pp. 266–296. Springer, Heidelberg (2007)
5. Koopman, P., Lubbers, M., Plasmeijer, R.: A task-based DSL for microcomputers. In: Proceedings of the Real World Domain Specific Languages Workshop 2018 on - RWDSL2018, pp. 1–11. ACM Press, Vienna, Austria (2018). https://doi.org/10.1145/3183895.3183902. http://dl.acm.org/citation.cfm?doid=3183895.3183902
6. Lubbers, M., Koopman, P., Plasmeijer, R.: Multitasking on microcontrollers using task oriented programming. In: 2019 42nd International Convention on Information and Communication Technology, Electronics and Microelectronics (MIPRO), pp. 1587 1592. Opatija, Croatia (2019). https://doi.org/10.23919/MIPRO.2019.8756711
7. Lubbers, M., Koopman, P.: Code for the lecture notes: "Green Computing for the Internet of Things" (2023). https://doi.org/10.5281/zenodo.7643316
8. Lubbers, M., Koopman, P., Plasmeijer, R.: Interpreting task oriented programs on tiny computers. In: Stutterheim, J., Chin, W.N. (eds.) Proceedings of the 31st Symposium on Implementation and Application of Functional Languages, Singapore, Singapore, p. 12. IFL 19, ACM, New York, NY, USA (2021). https://doi.org/10.1145/3412932.3412936
9. Lubbers, M., Koopman, P., Plasmeijer, R.: Writing internet of things applications with task oriented programming. In: Porkoláb, Z., Zsók, V. (eds.) Composability, Comprehensibility and Correctness of Working Software, pp. 3–52. Springer, Cham (2023). https://doi.org/10.1007/978-3-031-42833-3_1
10. Lubbers, M., Koopman, P., Ramsingh, A., Singer, J., Trinder, P.: Tiered versus Tierless IoT stacks: comparing smart campus software architectures. In: Proceedings of the 10th International Conference on the Internet of Things, Malm , Sweden. IoT 20, ACM, Malm (2020). https://doi.org/10.1145/3410992.3411002

11. Pfenning, F., Elliott, C.: Higher-order abstract syntax. In: Proceedings of the ACM SIGPLAN 1988 Conference on Programming Language Design and Implementation, Atlanta, Georgia, USA, pp. 199–208. PLDI 88, ACM, New York, NY, USA (1988). https://doi.org/10.1145/53990.54010
12. Plasmeijer, R., Achten, P., Koopman, P.: iTasks: executable specifications of interactive work flow systems for the web. ACM SIGPLAN Not. **42**(9), 141–152 (2007)
13. Plasmeijer, R., van Eekelen, M., van Groningen, J.: Clean Language Report version 3.1. Technical report, Institute for Computing and Information Sciences, Nijmegen (2021)
14. Plasmeijer, R., Lijnse, B., Michels, S., Achten, P., Koopman, P.: Task-oriented programming in a pure functional language. In: Proceedings of the 14th Symposium on Principles and Practice of Declarative Programming, Leuven, Belgium, pp. 195–206. PPDP 12, ACM, New York, NY, USA (2012). https://doi.org/10.1145/2370776.2370801
15. Serrano, M., Gallesio, E., Loitsch, F.: Hop: a language for programming the web 2.0. In: OOPSLA Companion, pp. 975–985. ACM, Portland, Oregon, USA (2006)
16. Tratt, L.: Domain specific language implementation via compile-time meta- programming. ACM Trans. Program. Lang. Syst. **30**(6) (2008). https://doi.org/10.1145/1391956.1391958

Verified Software in Web-Based Virtual Environments

Štefan Korečko[(✉)] [iD]

Department of Computers and Informatics, Faculty of Electrical Engineering and
Informatics, Technical University of Košice, Košice, Slovakia
`stefan.korecko@tuke.sk`

Abstract. In verified software development, we use formal methods to
verify the software correctness with respect to its formally specified properties. However, there are no formal means to verify the properties themselves against informal requirements. Instead, validation via animation
of formal specifications is used. To improve the validation, an animat-
able formal specification or an executable prototype derived from it can
be embedded into a 3D virtual environment, resembling the one where
the system will be used. Utilization of virtual environments instead of
real ones contributes to the environmental and economic sustainability
by saving resources. This paper presents a hypothetical case study that
demonstrates how an executable prototype can be developed using the B-
Method formal method and subsequently embedded into a collaborative
virtual environment for validation purposes. The prototype developed is
a control program of a fictional cleaning robot. Its critical property is
to avoid harming people and other living beings by its operation. The
virtual environment is created using A-Frame, a software framework for
web-based virtual reality.

Keywords: formal methods · verified software development ·
B-Method · virtual environment · virtual reality · sustainability

---

This work received financial support through the Erasmus+ Strategic Partnership for
Higher Education *SusTrainable—Promoting Sustainability as a Fundamental Driver in
Software Development Training and Education* (project number 2020-1-PT01-KA203-
078646), funded by the European Union and coordinated by the University of Coimbra,
Portugal.
The information and views set out in this publication are those of the author and do
not necessarily reflect the official opinion of the European Union. Neither the European Union institutions and bodies nor any person acting on their behalf may be held
responsible for the use which may be made of the information contained therein.

C. Grelck et al. (eds.), *Promoting Sustainability as a Fundamental Driver in Software Development
Training and Education*, LNCS 15670, pp. 29–54, 2026.
https://doi.org/10.1007/978-3-032-22278-7_2

1   Introduction

Formal methods for software development, such as B-Method [1,20], Event-B [2] or VDM [8], allow to verify the correctness of a formally specified system with respect to its properties. However, all the effort put into the verification process would be wasted if the properties themselves are incorrect. To validate the properties against informal requirements on the system, animation (execution) of formal specifications is used. The animation is supported by tools, called animators. The animations are usually performed via a command-line interface while some animators also allow simple visualizations.

In some cases, for example, where the software to be developed is a control system of an autonomous entity that performs its duties in a physical space, it is beneficial if the animation, and corresponding validation, occur in a 3D virtual environment that resembles the entity and the environment where it is about to operate. Provided that the virtual representation of the entity is connected to the formally developed software, the properties of the software can be validated in interaction with the virtual environment. The entity may also interact with real users (people), represented by avatars in the virtual environment. Thanks to the recent developments in virtual reality (VR) hardware, VR headsets in particular, a full immersion of a user to the virtual environment can be provided together with a realistic capture of his or her movements.

With respect to the established definitions [9,10], this approach contributes to the environmental and economic sustainability. This is because it promotes creating virtual representations and prototypes instead of real ones, thus saving valuable resources. It may also save potential traveling costs, as the validation and evaluation take place in a virtual space, where the participants can connect remotely. Such position of VR with respect to the sustainability is also recognized in [11], where the VR role in manufacturing is seen in product design evaluation and virtual prototyping. There are several implementations of similar approaches to the VR utilization, for example a welding process design [5] or a disassembly line development [19].

In this paper, we present a hypothetical case study that demonstrates utilization of the approach to verified software development with formal methods. The study revolves around a fictional cleaning robot control software development and a simple virtual environment, where a virtual representation of the robot will be placed to emulate the cleaning operation. The verified software development utilizes B-Method [1,20] and the virtual environment is constructed using A-Frame [21], a web-based VR software framework. The primary purpose of the study is educational, where it tries to present two interesting topics to the audience, namely verified software development with formal methods and web-based collaborative virtual environments, and how these can be combined with benefits for the sustainability. The study is suitable for courses of various duration and has already been used in such manner. The approach is not limited to robotics in any way. The autonomous robot case has been chosen as it makes easy to show the importance of software correctness and to formulate corresponding critical properties.

2   Validation in Virtual Environments for Verified Software Development

In essence, the process of verified software development (VSD) with validation in virtual environments (VE) proceeds as shown in Fig. 1. For VSD, we assume utilization of a formal method (FM) that allows:

- To write an abstract formal specification (FS) of the software system. This specification may not necessarily be executable. It should define the state (data) and behavior (operations) of the system.
- To formalize the properties of the system and verify the abstract FS against them. The verification should prove that the properties hold in every state of the specified system. Usually, it is carried out by model checking or theorem proving.
- To refine the abstract FS into an implementable one and verify the implementable FS against the formalized properties. This verification should assure that the implementable FS maintains the original properties together with additional ones, introduced during the refinement process.

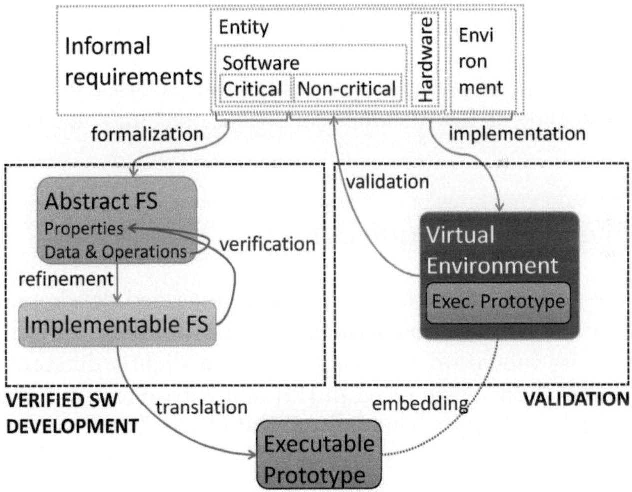

Fig. 1. Verified software development with validation in virtual environments process in general.

The whole process starts with informal requirements on the autonomous entity (e.g. a cleaning robot) to be developed and the environment where it is supposed to operate. Regarding the entity, on the basis of the requirements, we can identify:

- Safety-critical software. The part[1] of the control software, which failure may cause critical damage to the entity and the surrounding environment. In the rest of the paper we will call this part safety critical controller (SCC).
- Non-critical software. The rest of the control software.
- Hardware. Physical appearance and properties of the entity.

Compared to the standard development methods, VSD is more costly, so it is used for the SCC only. Here, a FM meeting the requirements specified above is used to write an abstract formal specification of the SCC and refine it into an implementable form. An inherent part of VSD is verification: the corresponding part of the informal requirements is formalized into properties of the abstract formal specification and its refinements. Subsequently, these specifications are proved against the properties. Finally, a code generator is used to translate the implementable specification into a source code in a general purpose programming language (GPPL), creating an executable prototype of the SCC.

In parallel with the SCC development, the remaining informal requirements are used to create a virtual environment (VE), where the executable prototype will be validated and evaluated. Tools such as 3D editors or game engines, together with 3D model repositories, should be used to minimize the effort needed for the VE creation. The executable prototype is then embedded into the virtual environment, where it can be validated and evaluated. The purpose of the validation is to check whether the informal requirements have been properly understood and formalized, that is whether the prototype has the right properties. An additional evaluation or testing may be used to fine-tune parameters of the prototype or test it under conditions that may not been considered originally.

3   Cleaning Robot Requirements

The case study demonstrates the process with a verified development of a control software of a fictional autonomous cleaning robot and creation of a virtual environment for its validation and evaluation. In a slightly different visual form, the case has been briefly introduced in [12] and subsequently implemented and utilized in [14] for usability evaluation of related web-based virtual (extended) reality components.

Let us name the cleaning robot *cBot* and assume that the informal requirements define its behavior as follows: The robot has a list of dirty locations to clean and it proceeds to clean them one by one. After cleaning all the dirty locations, it goes to a parking location and switches to a standby mode. The robot performs its job in areas with some "citizens" (i.e. people and other living beings), so it is critical to prevent it from hurting them. To detect the citizens, the robot uses a circular sensor array (Fig. 2) with eight sensors, each covering a

---

[1] In some fields, more than one level of criticality can be considered. For the sake of simplicity, here we consider just one level, i.e. everything the hypothetical project management decides to develop with formal methods goes to SCC.

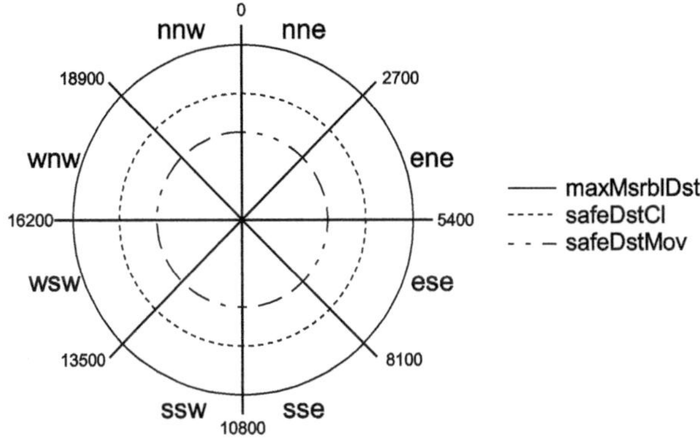

Fig. 2. Proximity sensors arrangement of the cleaning robot.

45° (2700 min) region, north-north-east (nne) to north-north-west (nnw). Each sensor returns a value equal to the distance to a nearest living being, detected in the corresponding region. If no living being is detected in the region, then the returned value is equal to the maximum distance, measurable by the array (maxMsrblDst in Fig. 2). The robot may hurt someone when moving or cleaning. The requirements reflect this by two safety critical properties:

SP.1 The cleaning cannot start or continue if a citizen gets as close or closer to the robot as the minimum safe cleaning distance (safeDstCl).

SP.2 The robot cannot move if a citizen gets as close or closer to its front as the minimum safe movement distance (safeDstMov).

Moreover, the requirements state that safeDstMov $\leq$ safeDstCl $\leq$ maxMsrblDst. The front of the robot is defined as three adjacent regions, with the region containing the angle where the robot is heading in the middle. For example, if the angle is 30° (1800 minutes) then the front consists of nnw, nne and ene. To keep the case simple, we assume that the robot operates in a perfectly flat environment and all citizens move on the ground and are high enough to be detected by the array. This means that their vertical position does not need to be considered.

4   B-Method

B-Method (B) [1], also called "classical B" or "B for software" [18], is a state-based model-oriented formal method for software development. It consists of a specification language, called *B language*, and a development process. The B language is used for a software components specification and combines the Zermelo-Fraenkel set theory for data definition and handling and the Guarded Command Language [7] for flow control. Development of a software system in B starts with writing an abstract formal specification of the system in B language. The specification is a collection of components, called *B-machines*. Each B-machine consists of several clauses, most of them are optional. The mandatory clause is MACHINE, containing the name M of the component and an optional list p of its formal parameters. If the parameters are present, then it is followed by the clause CONSTRAINTS with a predicate C, defining properties of p. Another clause the B components in this article use is DEFINITIONS with a list of macro definitions in the form $dname == dcont$, where $dname$ is the name of a macro and $dcont$ its content. When the component is processed, each occurrence of $dname$ is replaced by $dcont$. The clause CONCRETE_CONSTANTS lists constants k and CONCRETE_VARIABLES state variables v of the machine. Properties Bh of k are defined in the PROPERTIES clause and properties I of v in the INVARIANT clause. INITIALISATION clause holds an operation T that established the initial state of the machine by assigning values to its state variables. Predicates C, Bh and I must at least define types of the corresponding data elements. The clause OPERATIONS contains operations of the machine. An operation may modify the state variables and, in general, has the form

$$y\texttt{<-}\!\texttt{-}op(\mathsf{x}) = S,$$

where y is the list of its output parameters, x of its input parameters, op is the name of the operation and S its body. The x, y and op form the header of the operation. The initialisation T and bodies of operations contain commands, called *generalised substitutions* (GS). GS used in our case study are

- skip (empty command: do nothing),
- $x := e$ (assignment of a value of an expression e to a variable x),
- $x :: R$ (assignment of a value selected randomly from a set R to a variable x),
- $S_1 \| S_2$ (parallel composition: do GS S_1 and GS S_2 at once),
- PRE E THEN S_1 END (preconditioning: execute S_1 if E holds, otherwise do not terminate),
- IF E THEN S_1 ELSE S_2 END (conditional statement: if E holds, do S_1, otherwise do S_2. It can be extended by ELSIF keyword).
- CASE e OF EITHER v_1 THEN S_1 OR v_2 THEN S_2 ...ELSE S_d END (case statement: if $e = v_1$ do S_1, if $e = v_2$ do S_2, ..., otherwise do S_d)
- VAR v IN S_1 END (local variable: do S_1 with local variables listed in v) and
- S_1 ; S_2 (sequential composition: do S_1 then S_2).

After the abstract formal specification is created, an *internal consistency* of its components is formally proved. A B-machine is internally consistent if the predicates C, Bh and I guarantee an existence of meaningful data and the initialisation T establishes and all the operations preserve the invariant I. This ability to prove that properties specified in I hold in every state of the system is very important when defining safety-critical systems as we can include conditions that prevent failures in I and the proof ensures us that the system will never violate them.

The development process of B also supports a step-wise refinement of the abstract specification into a concrete, implementable one, which can be automatically translated to a GPPL. There are two additional components used

Table 1. Selected B language operators.

Operator(s)	Meaning
:	belongs to a set ($\in$)
..	integer interval (e.g. 0..7 is an interval from 0 to 7, including 0 and 7)
+ - /	integer addition, subtraction and division
*	product (integer or Cartesian)
mod	remainder after integer division
f(a)	value of the function f in a (e.g. if f contains a pair (0,3) then f(0) is 3)
& =>	logical operators conjunction and implication
!	universal quantifier ($\forall$)

during the process; *refinement* (RR) and *implementation* (II). Their structure is similar to B-machine (MM). A RR or II can refine only one MM or RR but one MM or RR can be refined by more RR or II. The component it refines is specified in the REFINES clause. In B, to refine means to modify data or operations while parameters and operation headers remain the same. The invariant of RR or II not only defines the properties of its variables but also their relation to the variables of the component it refines. This means that the consistency of each refinement step is proved by proving the consistency of the components created in the step. A specification in B usually consists of several components, bound together by so-called *composition mechanisms*. Two of them are used in our case study: INCLUDES (usable in MM and RR) and IMPORTS (usable in II). Both allow to call operations and read variables of an accessed component in the accessing one. B language contains numerous operators that can be used in expressions and predicates, some are listed in Table 1.

B-Method is well supported by software tools. The primary development environment is Atelier B[2], where B specifications can be written, type checked and their internal consistency proved. In addition, it provides translation of II to C, ADA, C++ and Ladder. Other available tools include the animator and model checker ProB[3] and the BKPI compiler [13] to Java, C#, JavaScript and TypeScript, developed at the home institution of the author. The description of B-Method presented here is far from being complete. An interested reader can find more information in [1, 20] and the manuals shipped with Atelier B.

4.1   Advantages and Alternatives

The primary reasons behind the choice of B-Method for this case study are that

- it was developed specifically for software development,

---

[2] http://www.atelierb.eu/en/.
[3] https://prob.hhu.de/.

- it provides a selection of contemporary, ready-to-use and freely available software tools supporting the whole development process, including formal proofs, and
- it has an impressive industrial use record [4,15,18], albeit predominantly in the railway domain.

When looking at the development of B-Method, one may ask why we chose it instead of what seems to be its successor âĂŞ the Event-B [2]. Event-B differs from B-Method in its focus on system modeling instead of software development. As it follows from the comparisons [3,18], the specifications in both Event-B and B-Method consist of machines, but those in Event-B contain events instead of operations. And these events cannot contain if-then-else statements, loops or sequences. In addition, output parameters are not allowed and it is impossible to call one event from another. The composition mechanisms are reduced to a read-only access to data elements. These restrictions make it hard to write software specifications in Event-B for a developer accustomed to common GPPL. On the other hand, they noticeably simplify the proofs and solve some inconsistencies found in B-Method. It should be also noted than the refinement process is less restricted in Event-B than in B-Method and there are code generators to several GPPL for Event-B, too. Event-B is supported by its own tool, called Rodin[4] and Atelier-B (with some syntax differences).

Regarding the alternatives outside of the "B landscape", several formal methods can be considered. Those utilized in the industry include VDM [8] and the Perfect Developer [6], both supported by software tools and offering code generation to selected GPPL. However, to our knowledge, these generators do not include JavaScript, which exclude them from being directly usable in this case study. The same is true for another promising alternative, which is the Coq Proof Assistant and its Gallina specification language. These were later [17] extended with the FreeSpec framework. While originally focusing on hardware, a more recent iteration of FreeSpec [16] enables formal verification and implementation of modular software systems within the Coq ecosystem. The code generation is supported, but primarily to functional languages.

5   Verified Development of Safety Critical Controller

Considering the requirements defined in Sect. 3, we can identify the safety critical controller (SCC) of *cBot* as the part making the decision about the next activity of the robot while maintaining the properties SP.1 and SP.2. We define its interface as an operation updateAndEvaluate, with 11 input and 4 output parameters (Fig. 3).

The control software of *cBot*, including SCC, will then operate by repeating the following procedure:

---

[4] The Rodin platform is available from https://www.event-b.org/.

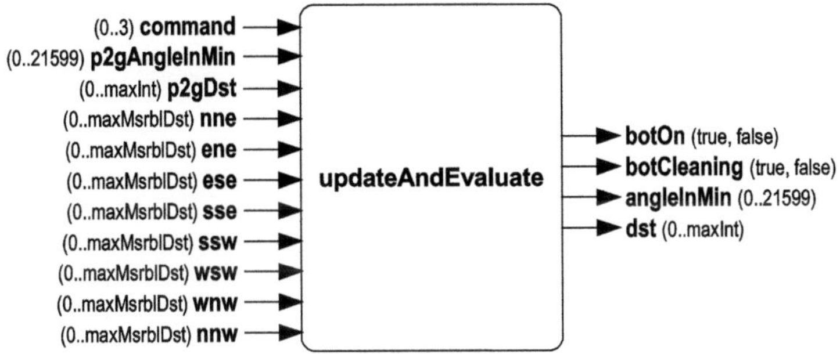

Fig. 3. SCC interface with value ranges of its parameters.

1. On the basis of the current activity of *cBot* and status of the locations to clean, determine the next activity of *cBot* by setting values of the variables command, p2gAngleInMin and p2gDst, where
   - command is the next activity of *cBot*. It may be to go to a desired destination and then clean (value 1), do not clean but stay on (2) or switch to standby (3). There is also a fourth option, do not go anywhere and switch to standby mode immediately (0).
   - p2gAngleInMin is the direction from the current position of *cBot* to the desired destination, given as a compass angle.
   - p2gDst is the distance from the current position of *cBot* to the desired destination.
2. Use the sensor array to scan the area for citizens and store the readings to variables nne, ene, ... nnw.
3. Use SCC to determine the next action by calling updateAndEvaluate with input parameters set to the values, gathered in steps 1 and 2.
4. Carry out the next activity of *cBot* according to the updateAndEvaluate. The output is defined by four parameters
   - botOn - *cBot* is turned on (true) or in a stand-by mode (false).
   - botCleaning - *cBot* is cleaning (true) or not (false).
   - angleInMin - compass angle to a destination of *cBot*.
   - dst - distance to the destination of *cBot*.

The maximum value for the compass angles is 21599 (359°59′) and the distances are in milimeters.

5.1   Safety-Critical Controller in B-Method

The specification of the SCC in B consists of three B-machines, each refined in one step to the corresponding implementation (Fig. 4).

The only component not accessing another one is ProximSensors (Fig. 5), which handles the sensor array readings. It has only one variable, sensorsData, for storing the sensors readings. The variable is an array of the size 8, indexed

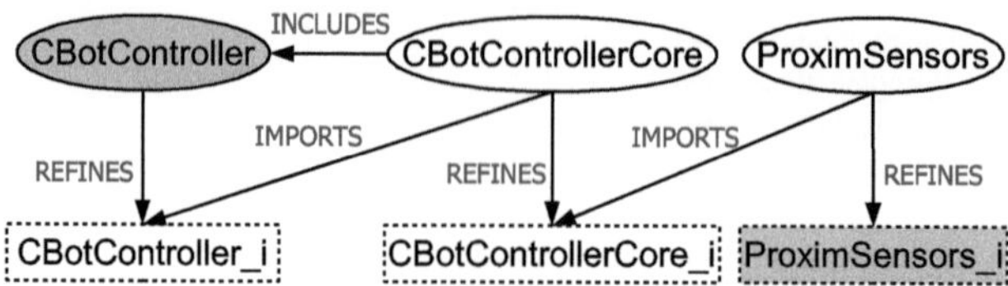

Fig. 4. B components hierarchy. Grey background indicates that the component code is not shown in this paper. (Color figure online)

```
MACHINE ProximSensors(maxMsrblDst)
CONSTRAINTS maxMsrblDst:NAT
CONCRETE_VARIABLES sensorsData
INVARIANT sensorsData: (0..7)-->(0..maxMsrblDst)
INITIALISATION sensorsData:= (0..7)*{0}

OPERATIONS
  readSensors(nne, ene, ese, sse, ssw, wsw, wnw, nnw)=
    PRE nne:0..maxMsrblDst & ... & nnw:0..maxMsrblDst
    THEN sensorsData:={0|->nne,1|->ene,2|->ese,3|->sse,4|->ssw,5|->wsw,6|->wnw,7|->nnw} END;

  bb<--allClear(dst)= PRE dst:NAT THEN
    IF !xx.((xx:(0..7)) => sensorsData(xx)>dst) THEN bb:=TRUE ELSE bb:=FALSE END END;

  bb<--frontClear(angle,dst)= PRE angle:0..21599 & dst:NAT  THEN
    IF (sensorsData(((angle / 2700)+7) mod 8)>dst & sensorsData(angle / 2700)>dst &
        sensorsData(((angle / 2700)+1) mod 8)>dst)
      THEN bb:=TRUE ELSE bb:=FALSE END END
END
```

Fig. 5. Specification of the machine ProximSensors in B language.

from 0, which members belong to the integer interval from 0 to maxMsrblDst. All of them are set to 0 during the initialisation. The machine has three operations. The first one, readSensors, stores new readings to sensorsData, allClear(dst) returns *true* if all values in sensorsData are greater than its input parameter (dst) and frontClear(angle,dst) returns *true* if the values in sensorsData for the region to which the parameter angle belongs and two adjacent regions are greater than its parameter dst. The differences between ProximSensors and its implementation ProximSensors_i are only those forced by the restrictions on GS use in B components.

Next two components define the core functionality of the SCC. The machine CBotControllerCore (Fig. 6) specifies its interface, i.e. parameters and operation headers. The first three parameters correspond to the values introduced in Fig. 2 and cleanRange represents the maximum distance from a location to clean, where it still makes sense to have the cleaning process turned on.

The core functionality itself is fully defined in CBotControllerCore_i (Fig. 7 and Fig. 8), the most complex component of the SCC. It also includes the safety properties SP.1 and SP.2. The positive integer constants EV_STEP and EV_START are here for a fine adjustment of an *evasive action* the robot takes

```
MACHINE CBotControllerCore(maxMsrblDst, safeDstCl, safeDstMov, cleanRange)
CONSTRAINTS maxMsrblDst:NAT & safeDstCl:NAT & safeDstMov:NAT & cleanRange:NAT &
            safeDstMov>0 & safeDstMov <= safeDstCl & safeDstCl <= maxMsrblDst

OPERATIONS
   botOn , botCleaning , angleInMin , dst <-- getInstr4Bot =
     BEGIN botOn ::BOOL || botCleaning ::BOOL || angleInMin::0..21599 || dst :: NAT END;

   readGoalPosAndSensors(angleInMin,dst,nne, ene, ese, sse, ssw, wsw, wnw, nnw) =
     PRE angleInMin:0..21599 & dst:NAT & nne:0..maxMsrblDst & ene:0..maxMsrblDst & ...  & nnw:0..maxMsrblDst
     THEN skip END;

   standBy = BEGIN skip END;

   goToPosAndClean = BEGIN skip END;

   goToPosAndNoClean = BEGIN skip END
END
```

Fig. 6. Specification of the machine CBotControllerCore in B language.

when someone is too close to the front of it when moving. Exact values of the
constants are given in the VALUES clause. The evasive action is quite simple:
if possible, the robot turns 90° to the left or right from its destination and
moves in this direction for a certain period of time. The period is defined by a
countdown from EV_START down to 0, which decreases by EV_STEP during
each call of the controller. The machine has 10 state variables. First two hold
the current operational status of the robot: whether it is standby or on (isOn)
and whether it is cleaning or not (isCleaning). Variables willBeOn and willClean
contain similar information about the next action of the robot. The rest of the
state variables form 3 pairs. Each consist of a compass angle and a distance and
defines certain destination relatively to the current position of the robot. The
pair goalAngle and goalDst stores the desired destination, i.e. where the robot
should go, pos2goAngle and pos2goDst a destination chosen by the controller
and evAngle and evDst a destination for the evasive action. The variable evDst
also serves as a countdown timer during the evasive action and, thus, indicates
whether the evasive action is active (if evDst> 0). The first two lines of the
INVARIANT clause define types of the variables and the rest (in italic) are safety
properties: 3rd line is SP.1 and 4th and 5th line are SP.2. The purpose of the
macro definitions is to simplify the operations of the component. What is given
in the definitions can be alternatively defined in separate components (such as
ProximSensors), but this will make the consistency proofs more complicated.

There are six operations in CBotControllerCore_i. The first one, getInstr4Bot,
is a "getter", which returns values of variables with instructions about the robot's
next action. It also updates the variables isOn and isCleaning. On the other hand,
readGoalPosAndSensors is a "setter"; it updates corresponding variables of this
component and of the imported machine ProximSensors by the desired destination and sensor array readings. The operation standBy sets the state variables
in order to turn the robot to the standby mode, goToPosAndClean evaluates the
situation when the goal is to go to the desired location and clean it and sets the

```
IMPLEMENTATION CBotControllerCore_i(maxMsrblDst, safeDstCl, safeDstMov, cleanRange)
REFINES CBotControllerCore
IMPORTS ProximSensors(maxMsrblDst)

CONCRETE_CONSTANTS EV_STEP, EV_START
PROPERTIES EV_STEP:NAT & EV_START:NAT & EV_STEP>0 & EV_START>0 &  EV_START mod EV_STEP=0
VALUES EV_STEP=1; EV_START=100

CONCRETE_VARIABLES isOn, isCleaning, willBeOn, willClean,
                   goalAngle, goalDst, pos2goAngle, pos2goDst, evAngle, evDst

INVARIANT isOn: BOOL & isCleaning: BOOL & willBeOn: BOOL & willClean: BOOL & goalAngle:0..21599 & goalDst:NAT &
          pos2goAngle:0..21599 & pos2goDst:NAT & evAngle:0..21599 & evDst:0..EV_START &
          ((willClean=TRUE) => (!xx.((xx:(0..7)) => sensorsData(xx)>safeDstCl))) &
          ((pos2goDst>0)=>(sensorsData(((pos2goAngle / 2700)+7) mod 8)>safeDstMov &
          sensorsData(pos2goAngle / 2700)>safeDstMov & sensorsData(((pos2goAngle / 2700)+1) mod 8)>safeDstMov))

DEFINITIONS
   acceptDesired == pos2goDst:=goalDst; pos2goAngle:=goalAngle;
   stop == pos2goDst:=0; pos2goAngle:=0; evDst:=0; evAngle:=0;
   goalTrnLeftAngle==(goalAngle+16200) mod 21600;
   goalTrnRightAngle==(goalAngle+5400) mod 21600;
   evadeLeft == evDst:=EV_START; evAngle:=goalTrnLeftAngle; pos2goDst:=evDst; pos2goAngle:=evAngle;
   evadeRight== evDst:=EV_START; evAngle:=goalTrnRightAngle; pos2goDst:=evDst; pos2goAngle:=evAngle;
   continueEvasion == pos2goDst:=evDst; pos2goAngle:=evAngle;
                      IF (evDst>=EV_STEP) THEN evDst:=evDst-EV_STEP END;
   goStopOrEvade== IF(goalDst=0) THEN acceptDesired ELSE IF(evDst=0) THEN
                       VAR frClr, leftClr, rightClr IN
                         frClr<--frontClear(goalAngle,safeDstMov);
                         leftClr<--frontClear(goalTrnLeftAngle,safeDstMov);
                         rightClr<--frontClear(goalTrnRightAngle,safeDstMov);
                         IF(frClr=TRUE) THEN acceptDesired; evDst:=0 ELSE
                           IF(leftClr=TRUE) THEN evadeLeft  ELSIF(rightClr=TRUE) THEN evadeRight ELSE stop END
                         END END
                     ELSE
                       VAR frEvClr IN   rEvClr<--frontClear(evAngle,safeDstMov);
                                     IF(frEvClr=TRUE) THEN continueEvasion ELSE stop END
                       END
                     END END

INITIALISATION isOn:=FALSE ; isCleaning:=FALSE; willBeOn:=FALSE; willClean:=FALSE; goalAngle:=0; goalDst:=0;
               pos2goAngle:=0; pos2goDst:=0; evDst:=0; evAngle:=0
```

Fig. 7. First part of the specification of the implementation CBotControllerCore_i in B language.

state variables accordingly. The **goToPosAndNoClean** does the same when the goal is to go to the desired location without cleaning it.

The machine **CBotController** defines the agreed interface of SCC (Fig. 3) in the same non-deterministic fashion as **CBotControllerCore**. How the operation is implemented by calling corresponding operations from **CBotControllerCore** is defined in **CBotController_i** (Fig. 9).

It should be noted that the B specification of the SCC shown here is a result of several iterations between Atelier-B, B-KPI and a more complex version of the VE presented in Sect. 6. For example, early versions of the SCC only stopped the robot when an obstacle had been detected. The evasive action (the definitions in Fig. 7) was added later.

6   Virtual Environment

As outlined in Sect. 2, the rest of the control software, the virtual representation of *cBot*, the surrounding environment and its citizens are captured in the virtual environment (VE), where the SCC is embedded. For this case study, we chose a web-based VE, that is a VE that runs directly in a web browser. There are two main reasons behind this decision. First, for a practical use of the proposed approach to VSD, a web-based VE means an easier access for potential evaluators as the only things they need is a web browser, a URL and a set of instructions. No additional software installation is required. Second, for the educational use, the corresponding software project can be hosted on a platform like Glitch[5], which provides both web hosting and an integrated development environment.

```
OPERATIONS
  botOn , botCleaning , angleInMin , dst <-- getInstr4Bot= BEGIN
    isOn:=willBeOn; isCleaning:= willClean;
    botOn := willBeOn; botCleaning := willClean;  angleInMin := pos2goAngle; dst := pos2goDst   END;

  readGoalPosAndSensors(angleInMin,dst,nne, ene, ese, sse, ssw, wsw, wnw, nnw)= BEGIN
    IF 0<=angleInMin & angleInMin<=21599 & 0<=nne & nne<= maxMsrblDst & ... & 0<=nnw & nnw<= maxMsrblDst
    THEN goalAngle:=angleInMin;  goalDst:=dst; readSensors(nne,ene,ese,sse,ssw,wsw,wnw,nnw)
    ELSE goalAngle:=0; goalDst:=0; readSensors(0,0,0,0,0,0,0,0) END;
    willBeOn:=FALSE; willClean:=FALSE; pos2goDst:=0; pos2goAngle:=0   END;

  standBy= BEGIN
    isOn:=FALSE ; isCleaning:=FALSE; willBeOn:=FALSE; willClean:=FALSE; goalAngle:=0; goalDst:=0;
    pos2goAngle:=0; pos2goDst:=0; readSensors(0,0,0,0,0,0,0,0); evDst:=0    END;

  goToPosAndClean= BEGIN willBeOn:=TRUE;
   VAR allClr IN
     allClr<--allClear(safeDstCl);
     IF(isCleaning=TRUE) THEN
       IF(allClr=FALSE) THEN stop ELSE goStopOrEvade END;
       IF(allClr=FALSE or goalDst>cleanRange) THEN willClean:=FALSE ELSE willClean:=TRUE END
     ELSE
       IF(goalDst<=cleanRange & allClr=TRUE) THEN willClean:=TRUE ELSE willClean:=FALSE END;
       goStopOrEvade
     END
   END END;

  goToPosAndNoClean= BEGIN willBeOn:=TRUE; IF(isCleaning=TRUE) THEN willClean:= FALSE END;
                  goStopOrEvade END
END
```

Fig. 8. Second part of CBotControllerCore_i in B language.

6.1   A-Frame and Alternatives

Our VE is implemented in *A-Frame* [21], a software framework for VR web applications, built on a JavaScript-based WebGL engine, called three.js [23]. A-Frame provides at least two notable advantages over three.js. First, virtual environments (scenes) can be easily constructed using a set of special HTML elements. Second,

---

[5] https://glitch.com/.

```
IMPLEMENTATION CBotController_i(maxMsrblDst, safeDstCl, safeDstMov, cleanRange)
REFINES CBotController
IMPORTS CBotControllerCore(maxMsrblDst, safeDstCl, safeDstMov, cleanRange)

OPERATIONS
  botOn, botCleaning, angleInMin, dst <-- updateAndEvaluate( command,p2gAngleInMin, p2gDst,
                                              nne, ene, ese, sse, ssw, wsw, wnw, nnw)=
    IF command:0..3 & p2gAngleInMin:0..21599 & p2gDst>=0 & nne:0..maxMsrblDst & ... nnw:0..maxMsrblDst THEN
      readGoalPosAndSensors(p2gAngleInMin,p2gDst,nne, ene, ese, sse, ssw, wsw, wnw, nnw);
      CASE command OF
        EITHER 0 THEN standBy
        OR 1 THEN goToPosAndClean
        OR 2 THEN goToPosAndNoClean
        OR 3 THEN IF p2gDst=0 THEN standby ELSE goToPosAndNoClean END
        ELSE standBy END
      END;
      botOn , botCleaning , angleInMin , dst <-- getInstr4Bot
    ELSE botOn:=FALSE; botCleaning:=FALSE; angleInMin:=0; dst:=0 END
END
```

Fig. 9. Specification of the implementation CBotController_i in B language.

A-Frame supports a systematic and component-based approach to VE building by implementing the Entity Component System (ECS) software architectural pattern. We also use *Networked-Aframe (NAF)* [22], an extension of A-Frame, which allows multiple users to share the same VE. Software built with respect to the ECS pattern consists of entities, composed of components that define their properties and functionality. Systems are processes that act on multiple entities at once. Which entities a system affects depends on components the entities possess. According to [21], entities in A-Frame represent objects from which VE are composed. Entities are defined using the a-entity HTML element and serve as containers for components. Components have a form of HTML attributes and define properties (position, rotation, geometry (shape), dimensions, etc.) and behavior of entities. An entity without components is meaningless, it does not even have a physical appearance in VE. Systems are attributes of a-scene, a top level HTML element of the corresponding VE (scene). The functionality of both components and systems is written in JavaScript.

A-Frame is not the only option for a web-based VE. One can use the corresponding lower-level JavaScript libraries, such as three.js for the VE and Socket.IO for synchronizing its shared instances. Some JavaScript frameworks also offer their own wrappers or renderers for three.js (e.g. React three fiber for React). If the VE does not need to be web-based, game engines (Unity, Unreal, ...) can be used instead. Of course, simulation software tools specific for the domain of the SCC under development should be considered, too.

6.2   Virtual Environment Structure and Design

The directory structure of our VE, including all the HTML and client-side JavaScript files, can be seen in Fig. 10. Server-side scripts are omitted, because they are exactly the same as in the NAF introductory example, available from

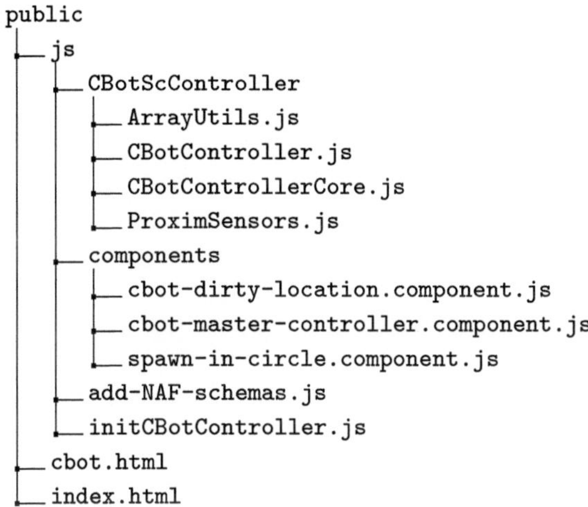

Fig. 10. Directory structure of the virtual environment implementation (client-side), including the SCC.

[22]. The composition and appearance of the VE is defined in the HTML files. There are two of them, `cbot.html` for the *cBot* and `index.html` for a citizen. Registration of shared entities is handled by the `add-NAF-schemas.js` script. The `CBotScController` directory holds the SCC, translated from the implementation components in B language by the B-KPI compiler. It also contains `ArrayUtils.js`, which is provided by the B-KPI and implements auxiliary functions for array handling, necessary for the translated SCC. The SCC is loaded by the `initCBotController.js` script. Finally, the `components` directory holds scripts of A-Frame components that implement the non-critical part of the controller and interactions between the *cBot* and the VE (i.e. citizens and positions to clean).

To keep the case study simple, the VE design is minimalistic. In Fig. 11, we can see the VE with one *cBot* and two citizens. This means that two instances of `index.html` and one instance of `cbot.html`, all hosted on the same server, are simultaneously open in web browsers[6]. As Fig. 11 is captured from the point of view of the first citizen, only the second one is visible. It is the green sphere-shaped head with eyes on the left. This appearance is taken from the NAF introductory example and individual citizens differ only in the color of the sphere. The desert-like environment with pyramids in the background is rendered using the environment[7] system by Supermedium. The *cBot* is on the right side and has a form of a cylinder-shaped head with eyes. We do not deal with the real appearance of the *cBot* here, so let's just assume that its shape and dimensions correspond to the cylinder that represents it in the VE. Its color is red as it is

---

[6] Not necessarily on the same device.

[7] https://github.com/supermedium/aframe-environment-component.

Fig. 11. Virtual environment with one citizen (left) and *cBot* (right, cleaning a dirty location) visible. The locations to clean are brown, albeit they appear almost red due to the illumination.

cleaning a dirty position. Other possible colors are orange (when moving) and grey (when on standby). There are four locations to clean in the VE, rendered as (reddish) brown rectangular plates on the ground. The cleaning process is indicated by increasing the transparency of the corresponding plate. The one under the *cBot* in Fig. 11 is almost clean. When the cleaning ends, the plate disappears completely. The grey circular plate in the middle is the parking location for the *cBot*.

6.3   Virtual Environment as A-Frame Scene

The code of `cbot.html`, which defines the VE (scene) for the *cBot*, can be seen in Listing 1. Some unimportant or repetitive parts are omitted and replaced by three dots (...). We also omit closing tags of those script elements that have no content. The description of the code can be found below Listing 1.

Listing 1. cbot.html (slightly reduced).

```
 1  <html>
 2    <head>
 3      <title>cBot — Verififed  controller  in  VE</title>   ...
 4
 5      <script  src="  ...  /1.3.0/ aframe . min . js ">
 6      <script  src=" ... aframe—environment—component . min . js ">
 7      <script  src="  ...  socket . io . slim . js ">
 8      <script  src="/ easyrtc / easyrtc . js ">
 9      <script  src="  ...  networked—aframe . min . js ">
10
11      <script  src=" js / add—NAF—schemas . js ">
12      <script  type="module"  src=" js / initCBotController . js ">
13      <script
14        src="/ js / components / cbot—dirty —location . component . js ">
15      <script  src=
```

```
16      "/js/components/cbot-master-controller.component.js">
17
18  </head>
19  <body>
20   <a-scene environment="preset: egypt"
21    networked-scene=" room: basic; debug: false;">
22    <a-assets>
23
24     <template id="template-avatar-citizen">
25      <a-entity class="avatarCitizen">
26       <a-sphere class="head" scale="0.45 0.5 0.4">
27       </a-sphere>
28       <a-entity class="face" position="0 0.05 0">
29        <a-sphere class="eye" color="#efefef"
30         position="0.16 0.1 -0.35" scale="0.12 0.12 0.12">
31         <a-sphere class="pupil" color="#000"
32          position="0 0 -1" scale="0.2 0.2 0.2">
33         </a-sphere>
34        </a-sphere>
35        <a-sphere class="eye" ...>
36         <a-sphere class="pupil" ...></a-sphere>
37        </a-sphere>
38       </a-entity>
39      </a-entity>
40     </template>
41
42     <template id="template-avatar-cbot">
43      <a-entity class="avatarCBot">
44       <a-cylinder class="cbot-head" height="1.5"
45        radius="0.45" color="#5e548c"></a-cylinder>
46       <a-entity class="face" position="0 0.05 0">
47        ...
48       </a-entity>
49      </a-entity>
50     </template>
51
52     <template id="dirty-location">
53      <a-entity scale="1 0.2 1" material="color: red"
54       geometry="primitive: box"></a-entity>
55     </template>
56
57     <template id="parking-location">
58      <a-entity scale="1 0.2 1" material="color: grey"
59       geometry="primitive: cylinder"></a-entity>
60     </template>
61    </a-assets>
62
63    <a-entity
64     id="player" geometry="primitive: box" camera
65     networked="template:#template-avatar-cbot;
```

```
66                        attachTemplateToLocal: false;"
67      cbot-master-controller >
68       <a-cylinder class="cbot-head" height="1.5"
69         radius="0.45" visible="false"></a-cylinder>
70      </a-entity>
71
72      <a-entity position="4 0 -4" cbot-dirty-location
73        networked="template: #dirty-location" ></a-entity>
74      <a-entity position="4 0 4" ... ></a-entity>
75      <a-entity position="-4 0 4" ... ></a-entity>
76      <a-entity position="-4 0 -4" ... ></a-entity>
77      <a-entity position="0 0 0"
78        networked="template: #parking-location"></a-entity>
79
80      </a-scene>
81    </body>
82  </html>
```

As in the case of an ordinary HTML file, `cbot.html` starts with the head element (lines 2–18 in Listing 1). Its most important task is to load necessary JavaScript files. The scripts loaded on lines (l.) 5–9 are third-party ones, necessary for A-Frame and NAF to function. The remaining ones (l. 11–16) are specific for this VE. They are shown and described later in this paper.

The body element (l. 19–81) contains a sole a-scene element (l. 20–80) with the whole VE for the *cBot*. The element a-scene belongs to the ones defined by A-Frame. The names of all such elements start with the prefix "a-". Our a-scene element includes two systems. The first one is environment (l. 20). It defines the desert-like background. The second one, the networked-scene (l. 21), is from NAF and manages shared entities.

The first element inside the a-scene is a-assets (l. 22–61), where pre-loaded resources, used in the scene, are defined. In our case, it consists solely of templates (the template element) of shared entity types. These templates are used by NAF to replicate shared entities in each scene loaded from the same server. In `cbot.html`, we have four templates. The first one (l. 24–40) defines the appearance of the citizen. It consists of five spheres, one for the head, two for the eyes and another two for the pupils of the eyes. We use the a-sphere element for the spheres. It is one of A-Frame elements for graphical primitives. These elements are in fact shortcuts for specific combinations of a-entity and components. Another such element is a-cylinder, used in the second template (l. 42–50) that determines how the *cBot* looks. The remaining two templates are for the plates indicating locations to clean (l. 52–55) and the parking location (l. 57–60) of the *cBot*.

Only after the a-assets we have the actual entities (l. 63–78) that populate the VE of the *cBot*. All of them are shared, so they replicate in every VE opened from the same server simultaneously with this one. The first one is *cBot* (l. 63–70), followed by four locations to clean (l. 72–76) and the parking location (l. 77–78). To be shared, they possess the networked component with the template parameter to indicate the corresponding template. We can see that the values of the template parameter match the id's of the template elements from l. 24–60.

The file index.html defines the VE with a citizen. We will not show its code here as it differs only in a few details from cBot.html. The most significant differences are that index.html

- does not load the *cBot*-specific scripts from l. 12–16 in Listing 1. Instead, it loads a script that randomly chooses the initial position of the citizen (spawn-in-circle.component.js).
- uses the citizen template (id = template-avatar-citizen) for the player entity (a-entity element with id = player). In addition, the player entity does not have the cbot-master-controller component. Instead, it has components wasd-controls and look-controls, allowing the user to move and rotate the entity with keyboard and mouse.
- the entities for dirty and parking positions (l. 72–78 in Listing 1) are excluded.

Listing 2. A part of add-NAF-schemas.js with registration of the citizen template.

```
 1 NAF.schemas.getComponentsOriginal =
 2    NAF.schemas.getComponents;
 3 NAF.schemas.getComponents = (template) => {
 4
 5 if(!NAF.schemas.hasTemplate("#template-avatar-citizen")){
 6   NAF.schemas.add({
 7     template: "#template-avatar-citizen",
 8     components: [ "position", "rotation",
 9      { selector: ".head", component: "material",
10        property: "color", }, ],
11   });
12 }
13
14 const components =
15    NAF.schemas.getComponentsOriginal(template);
16 return components;
17 };
```

For the entity sharing to work properly, we must also register so-called schemas, which specify what elements and components of these entities are shared. In our case, this is done in the add-NAF-schemas.js script. Listing 2 shows a part of the script with registration of the first template (id= template-avatar-citizen, l. 5–12 in Listing 2) only. The full version continues (on l. 13) with registration of the remaining templates, that is the ones with identifiers (id) template-avatar-cbot, dirty-location and parking-location.

6.4   Interaction with Safety Critical Controller

The VE is completed and what remains is to implement the rest of the controller, its interaction with the VE and to include the SCC.

The B-KPI compiler translates implementations from Sect. 5.1 into separate JavaScript modules, containing classes with the same functionality as the implementations and same names as the B-machines they implement. The top-level component of the B specification is CBotController and its implementation CBotController_i (Fig. 9). So, to have the controller in the VE, we need an instance of the corresponding class CBotController. Such instance is created by the script in Listing 3. The script names the instance SCC and assigns it to the browser's window object to make it easily accessible (l. 3). The instance has maxMsrblDst set to 30 m and both safeDstCl and safeDstMov to 5 m. As the locations to clean are relatively small, the cleanRange is 0.1 m only.

Listing 3. initCBotController.js.

```
1 import CBotController
2    from "./CBotScController/CBotController.js";
3 window.SCC = new CBotController(30000,5000,5000,100);
```

The rest of the controller and the interaction with the VE are implemented in two components

- cbot-dirty-location (Listing 4), which is responsible for the representation of the cleaning process and
- cbot-master-controller (Listing 5), which implements the overall control procedure as described in Sect. 5 and manipulates the *cBot* entity accordingly.

Each component is described below the corresponding listing.

Listing 4. cbot-dirty-location.component.js.

```
1  AFRAME.registerComponent("cbot-dirty-location", {
2    schema: {
3      dirtiness: {type: "int", default: 250},
4      MAX_DIRTINESS : {type: "int", default: 250}
5    },
6
7    init: function () {
8      this.el.setAttribute("material", "color", "brown");
9      this.el.setAttribute("material", "transparent", true);
10   },
11
12   update: function () {
13    this.el.setAttribute("material", "opacity",
14      this.data.dirtiness / this.data.MAX_DIRTINESS);
15   },
16
17   doCleaningStep: function (dirtinessDecrease) {
18      let newDirtiness=this.data.dirtiness-dirtinessDecrease;
19      if(newDirtiness<0) {newDirtiness=0;}
20      this.el.setAttribute("cbot-dirty-location",
21        "dirtiness", newDirtiness);
22   },
```

```
23
24   isClean: function () {
25     if(this.data.dirtiness >0) {return false;}
26     else {return true;}
27   },
28
29   makeDirty: function () {
30       this.el.setAttribute("cbot-dirty-location",
31         "dirtiness", this.data.MAX_DIRTINESS);
32   }
33
34 });
```

Every A-Frame component needs to be registered by calling the register-Component function (l. 1 in Listing 4). Its first argument is the name of the component. The name is used in entities to include the component (see l. 72 in Listing 1). The second argument is an object with the component itself.

The first item of cbot-dirty-location is schema (l. 2–5 in Listing 4). It defines the state of the component and in this case consists of two properties. The first one, dirtiness, determines how dirty the location is and the second one, MAX_DIRTINESS, defines the value of total dirtiness. It is written in capital letters to indicate that it should be handled as a constant. The location is considered clean when dirtiness = 0.

The functions init (l. 7–10) and update (l. 12–15) define the life cycle of the component. The init function is called only once, when the component is loaded. In our case it sets the color of the plate to brown and indicates that it can be transparent. The update function is called when the component is initialized and after updating any of its properties. Here, it sets the opacity of the plate according to the value of dirtiness.

The last three functions are intended to be called from outside of the component, in our case from the cbot-master-controller component. The doCleaningStep function (l. 17–22) decreases the dirtiness by the amount set by its input parameter dirtinessDecrease, isClean (l. 24–27) returns true if dirtiness = 0 and makeDirty (l. 29–32) resets the location to its original state of total dirtiness.

Listing 5. A part of cbot-master-controller.component.js with the life cycle items.

```
 1 AFRAME.registerComponent("cbot-master-controller", {
 2   schema: {
 3     cBotState: {type: "int", default: 0},
 4     MOVEMENT_SPEED: {type: "number", default: 0.02},
 5     PARKING_LOCATION:{type: "vec2", default: {x:0, y:0}},
 6     TICKS_TO_RESTART: {type: "int", default: 500}
 7   },
 8
 9   init: function () {
10     this.actualTicks2Restart = this.data.TICKS_TO_RESTART;
11     this.citizens =
12       document.getElementsByClassName("avatarCitizen");
```

```
13    this.head = document.querySelector(".cbot-head");
14    this.dirtyLocations =
15     document.querySelectorAll("[cbot-dirty-location]");
16    this.noOfDirtyLocations=this.dirtyLocations.length;
17    this.nextDirtyLocationIndex=0;
18    this.loadNextLocation2Clean();
19    this.PARKING_LOCATION=
20     new THREE.Vector3(this.data.PARKING_LOCATION.x, 1,
21      this.data.PARKING_LOCATION.y);
22    this.el.object3D.position.set(this.PARKING_LOCATION.x,
23     this.PARKING_LOCATION.y, this.PARKING_LOCATION.z);
24    this.el.object3D.rotation.set(0, 0, 0);
25    this.instr4CBot=[false, false, 0, 0];
26    this.vector2Target = new THREE.Vector3();
27    this.targetAngleInMin=0;
28    this.targetDistanceInMm=0;
29    this.sensorArrayReadings=
30     [30000,30000,30000,30000,30000,30000,30000,30000];
31    this.newCBotState=this.data.cBotState;
32  },
33
34  tick: function () {
35   if(window.SCC){
36    if(this.location2Clean){
37     this.readSensorArray();
38     this.updateDistanceAndAngleToTarget(
39      this.location2Clean.object3D.position);
40     this.instr4CBot =
41      window.SCC.updateAndEvaluate(1,
42      this.targetAngleInMin,this.targetDistanceInMm,
43      ...this.sensorArrayReadings);
44     this.newCBotState=this.calculateCBotState(
45      this.instr4CBot[0],this.instr4CBot[1]);
46     if(this.newCBotState===2){
47      this.moveCBotTowardsTarget();
48     }else if(this.newCBotState===1){
49      this.location2CleanAFrameComponent.doCleaningStep(1);
50      if(this.location2CleanAFrameComponent.isClean())
51       { this.loadNextLocation2Clean(); }
52     }
53    }else{
54     if(this.data.cBotState===0){
55      if(this.actualTicks2Restart >0){
56       this.actualTicks2Restart --;
57      }else{
58       this.actualTicks2Restart=this.data.TICKS_TO_RESTART;
59       this.makeLocationsDirty();
60       this.loadNextLocation2Clean();
61      }
62     }else{
```

```
63      this.readSensorArray();
64      this.updateDistanceAndAngleToTarget(
65       this.PARKING_LOCATION);
66      this.instr4CBot = window.SCC.updateAndEvaluate(3,
67       this.targetAngleInMin, this.targetDistanceInMm,
68       ...this.sensorArrayReadings);
69      this.newCBotState= this.calculateCBotState(
70       this.instr4CBot[0], this.instr4CBot[1]);
71      if(this.newCBotState===2)
72       { this.moveCBotTowardsTarget(); }
73     }
74    }
75    if(this.newCBotState!=this.data.cBotState){
76     this.el.setAttribute("cbot-master-controller",
77      "cBotState", this.newCBotState);
78    }
79   }
80  },
81
82  update: function () {
83   if(this.head){
84    let cBotheadColor="grey";
85    if (this.data.cBotState===1){ cBotheadColor="red"; }
86    else if (this.data.cBotState===2){
87     cBotheadColor="darkorange";
88    }
89    this.head.setAttribute("material", "color",
90     cBotheadColor);
91   }
92  },
93
94 });
```

The component cbot-master-controller works in a similar way as cbot-dirty-location. It is more than 300 lines long, so Listing 5 presents only the schema and functions directly related to its life cycle. This component is not intended to be accessed from outside and the functions left out implement auxiliary procedures, used in the life-cycle-related ones. The code of items shown in Listing 5 is complete and "..." is a JavaScript operator.

The schema of cbot-master-controller (l. 2–7) has four properties. The first one is cBotState. The meaning of its values and relation to the output of updateAndEvaluate are given in Table 2. The MOVEMENT_SPEED defines the speed of the *cBot* in meters per tick. One tick is a time between the rendering of two subsequent frames. For example, with the rendering speed of 30 frames per second, one tick takes 33.3 ms. This component is designed in such a way that after the *cBot* cleans all the locations, the whole cleaning process restarts with certain delay. The delay is defined in ticks by the TICKS_TO_RESTART property.

Table 2. Values of cBotState and their relation to botOn and botCleaning.

cBotState		Output of updateAndEvaluate	
Value	cBot *status*	botOn *value*	botCleaning *value*
0	standby (grey)	false	true/false
1	on and cleaning (red)	true	true
2	on and not cleaning (orange)	true	false

The first function in cbot-master-controller is init (l. 9–32). It defines and initializes additional state variables of the component and puts the *cBot* to the parking location. The state variable actualTicks2Restart is used later (l. 55–61) to count down the delay before restarting the cleaning process. Here (l. 10), it is set to its maximum allowed value. The variable citizens (l. 11–12) is a collection of entities representing all citizens in the shared VE. It is a live collection, so it is automatically updated when a citizen enters or leaves the VE (i.e. when in a web browser a user opens or closes index.html, hosted on the same server as cbot.html with this component). The following variables store a link to the entity with the head of the *cBot* (l. 13) and locations to clean (l. 14–17). The next location, in this case the first one, is loaded to the location2Clean state variable on l. 18. The *cBot* is placed at the parking location on l. 19–24. The rest (l. 25–31) initializes variables used for interaction with the SCC.

The second function is tick (l. 34–80). It is called before rendering each frame and its primary purpose is to update the state of the component. The previous component, cbot-dirty-location, does not need such function as its state is always changed from outside, by calling the corresponding functions. In cbot-master-controller, tick is the function implementing the control procedure from Sect. 5. It can only work when the SCC instance is already loaded, that's why the condition on l. 35.

If there still is a location to clean (l. 36–53), the *cBot* checks its surroundings (l. 37) by reading the sensor array (sensorArrayReadings), updates the distance (targetDistanceInMm) and compass angle (targetAngleInMin) to the desired destination according to its actual position (l. 38–39) and calls the SCC to determine its next activity (l. 40–43) and state (l. 44–45). Next, the activity is executed (l. 46–52).

There are two situations possible when all the locations are cleaned (l. 53–74). If the *cBot* is at the parking location and on standby (l. 54–62), the countdown to the cleaning process restart happens (l. 55-57). After that, all four locations are made dirty again (l. 59) and the first of them is selected to be cleaned (l. 60). The second one is that the *cBot* is still on and moves to the parking location while trying to avoid citizens (l. 62–73).

The final action in tick is an update of the cBotState property (l. 75–78).

The last life-cycle-related function, update (l. 82–92), is responsible for changing the color of the head of the *cBot* when cBotState changes.

7   Conclusion

The case study presented in this paper shows how a formally developed software prototype can be embedded into a virtual environment for validation and evaluation purposes. The case study is minimalistic, as it is intended to be used in short courses. At the time of publishing (2024), it is available online[8], including the complete source code[9].

The two topics covered by the case study, verified software development and web-based virtual environments implementation, are quite distinct in nature. Therefore, for courses using the study, it is recommended to focus only on one of the topics in detail, with respect to the interests of the target audience. The less interesting topic can be covered by a lecture while a hands-on experience is provided for the more interesting one.

An aspect not dealt with here is the validation and evaluation itself. For this to work with the case study, its JavaScript code needs to be extended with recording the *cBot* activity. The validation should be about checking whether the safety properties SP.1 and SP.2, defined in Sect. 3 and formally specified in the implementation component CBotControllerCore_i (Fig. 7), are the correct ones. Provided that the SCC fulfills the properties, this can be determined from counting the collisions of the *cBot* with the citizens when moving and too close encounters when cleaning. Both should be zero. Of course, this is only an indirect indicator, because when we remove the formalized SP.1 and SP.2 from CBotControllerCore_i, the SCC will work in exactly the same way. The critical parameters affecting both the safety of the citizens and efficiency of the *cBot* are, primarily, safeDstCl and safeDstMov. When set too low, the collisions and close encounters may happen. When set too high, the efficiency may decrease as the *cBot* will try to evade or stop cleaning unnecessarily often. The efficiency can be evaluated by recording the time (ticks) the *cBot* needs to clean all the dirty locations. Besides tweaking the parameters, the particular form of the evasive action and whether it should be implemented at all can be determined.

References

1. Abrial, J.R.: The B-book: Assigning Programs to Meanings. Cambridge University Press, New York (1996)
2. Abrial, J.R.: Modeling in Event-B: System and Software Engineering, 1st edn. Cambridge University Press, New York (2010)
3. Abrial, J.R.: On b and event-b: Principles, success and challenges. In: Butler, M., Raschke, A., Hoang, T.S., Reichl, K. (eds.) Abstract State Machines, Alloy, B, TLA, VDM, and Z, pp. 31–35. Springer, Cham (2018)

---

[8] To open the VE instance with citizen, please, visit https://lirkis-cbot-simple.glitch. me/. For the *cBot* instance, visit https://lirkis-cbot-simple.glitch.me/cbot.html.

[9] If you wish to examine the source code and create your own copy of the case study, please, use the link https://glitch.com/edit/#!/lirkis-cbot-simple.

4. Butler, M., et al.: The first twenty-five years of industrial use of the b-method. In: International Conference on Formal Methods for Industrial Critical Systems, pp. 189–209. Springer (2020)
5. Chen, X., Gong, L., Berce, A., Johansson, B., Despeisse, M.: Implications of virtual reality on environmental sustainability in manufacturing industry: a case study. Procedia CIRP **104**, 464–469 (2021). https://doi.org/10.1016/j.procir.2021.11.078
6. Crocker, D.: Safe object-oriented software: the verified design-by-contract paradigm. In: Redmill, F., Anderson, T. (eds.) Practical Elements of Safety, pp. 19–41. Springer, London (2004)
7. Dijkstra, E.W.: A Discipline of Programming. Prentice-Hall (1976)
8. Fitzgerald, J., Larsen, P.G., Mukherjee, P., Plat, N., Verhoef, M.: Validated Designs for Object-oriented Systems. Springer, New York (2005)
9. Goodland, R.: Environmental sustainability and the power sector. Impact assessment **12**(4), 409–470 (1994)
10. Goodland, R., Daly, H.: Environmental sustainability: universal and non-negotiable. Ecol. Appl. **6**(4), 1002–1017 (1996)
11. Hamid, N.S.S., Aziz, F.A., Azizi, A.: Virtual reality applications in manufacturing system. In: 2014 Science and Information Conference, pp. 1034–1037. IEEE (2014)
12. Hudák, M., Korečko, Š, Sobota, B.: Lirkis global collaborative virtual environments: current state and utilization perspective. Open Comput. Sci. **11**(1), 99–106 (2021). https://doi.org/10.1515/comp-2020-0124
13. Korečko, Š, Dancák, M.: Some aspects of bkpi b language compiler design. Egyptian Comput. Sci. J. **35**(3), 33–43 (2011)
14. Korečko, Š., Hudák, M., Sobota, B., Sivý, M., Pleva, M., Steingartner, W.: Experimental performance evaluation of enhanced user interaction components for web-based collaborative extended reality. Appl. Sci. **11**(9) (2021) https://doi.org/10.3390/app11093811, https://www.mdpi.com/2076-3417/11/9/3811
15. Lecomte, T., Deharbe, D., Prun, E., Mottin, E.: Applying a formal method in industry: a 25-year trajectory. In: Cavalheiro, S., Fiadeiro, J. (eds.) Formal Methods: Foundations and Applications, pp. 70–87. Springer, Cham (2017)
16. Letan, T., Régis-Gianas, Y.: Freespec: specifying, verifying, and executing impure computations in coq. In: Proceedings of the 9th ACM SIGPLAN International Conference on Certified Programs and Proofs, pp. 32 46. CPP 2020. Association for Computing Machinery, New York (2020). https://doi.org/10.1145/3372885.3373812, https://doi.org/10.1145/3372885.3373812
17. Letan, T., Régis-Gianas, Y., Chifflier, P., Hiet, G.: Modular verification of programs with effects and effect handlers in coq. In: Formal Methods: 22nd International Symposium, FM 2018, Held as Part of the Federated Logic Conference, FloC 2018, Oxford, UK, July 15-17, 2018, Proceedings 22, pp. 338–354. Springer (2018)
18. Leuschel, M.: Spot the Difference: A Detailed Comparison Between B and Event-B, pp. 147–172. Springer, Cham (2021). https://doi.org/10.1007/978-3-030-76020-5_9
19. Rocca, R., Rosa, P., Sassanelli, C., Fumagalli, L., Terzi, S.: Integrating virtual reality and digital twin in circular economy practices: a laboratory application case. Sustainability **12**(6) (2020). https://doi.org/10.3390/su12062286, https://www.mdpi.com/2071-1050/12/6/2286
20. Schneider, S.: The B-method: An Introduction. Cornerstones of computing, Palgrave (2001)
21. A-Frame homepage (2023). https://aframe.io/
22. Networked-Aframe homepage (2023). https://github.com/networked-aframe
23. three.js homepage (2023). https://threejs.org/

Sustainable Programming in the Julia Language

Anikó Kopacz$^{(\boxtimes)}$ ⓘ, Zoltán Tasnádi ⓘ, and Lehel Csató ⓘ

Faculty of Mathematics and Computer Science, Babeş-Bolyai University,
Cluj-Napoca, Romania
`{aniko.kopacz,zoltan.tasnadi,lehel.csato}@ubbcluj.ro`

Abstract. We highlight the sustainable aspects of the newly developed Julia language: which is an excellent tool that can be used as a tool developing for mathematical modelling applications. Julia builds on modern notational- and language design principles that can lead to more compact and readable source code, close to mathematical notations. Numerical modellers can – in principle – use Julia and its programming/coding environment for implementing, understanding, and testing new algorithms. The benefit of the language is that – if necessary – the algorithms can be scaled up and run on a massive scale without code-level changes.

In the article we present the distinctive features of Julia; the presentation is based on the Python language and on novel programming concepts, followed by a few examples that highlight Julia features that *can* make this language a popular teaching and research tool.

Keywords: Technical computing · Sustainable software development · Two-language dilemma

1 Introduction

The rapid spread of computers in the 20th century foreshadowed the growth of information and communication technologies. The development of these technologies adheres to Moore's law.[1] As a consequence of the continuous hardware

This work received financial support through the Erasmus+ Strategic Partnership for Higher Education *SusTrainable—Promoting Sustainability as a Fundamental Driver in Software Development Training and Education* (project number 2020-1-PT01-KA203-078646), funded by the European Union and coordinated by the University of Coimbra, Portugal.

The information and views set out in this publication are those of the authors and do not necessarily reflect the official opinion of the European Union. Neither the European Union institutions and bodies nor any person acting on their behalf may be held responsible for the use which may be made of the information contained therein.

[1] Moore predicted that the number of units on a chip would double every 2 years, https://www.britannica.com/technology/Moores-law. (acc: 02.11.2022).

C. Grelck et al. (eds.), *Promoting Sustainability as a Fundamental Driver in Software Development Training and Education*, LNCS 15670, pp. 55–73, 2026.
https://doi.org/10.1007/978-3-032-22278-7_3

improvements, a significant amount of computational resources are easily accessible for solving complex optimizations – such as market analysis, cyber-security, and industrial manufacturing and transportation –, as well as for the average user. Some of the recent technological break-troughs are centered on improving the efficiency of utilizing the available computational resources. Several programming paradigms facilitated the design and maintenance of code, e.g. structured and object-oriented programming. Later, the focus shifted towards efficiency via e.g. parallel programming, interpreted languages, or functional programming.

Advances in the quality and accessibility of software design facilitated the widespread application of coding practices in multiple areas, such as desktop publishing, and developer tools for complex functionalities, e.g. Visual Studio, which is a well-known *integrated development environment* (IDE) for object-oriented languages. This work focuses on making mathematical - and numerical - programming accessible and transparent. Several development tools exist for modelling and implementing numerical algorithms, and we highlight Mathematica[2] and Matlab[3] development environments and languages; these were the pioneers in creating high-level mathematical concepts for programming.

Recent trends indicate that *sustainability* improves the ability to adapt to the evolving requiremental and societal conditions of software development. In [17], the authors stress the importance of software architectures to facilitate the maintenance, evolution, and readability of software systems. In [11] it is claimed that sustainability is important in all stages of software development: the development process itself, the maintenance, system production, and system usage and it correlates sustainability with the shift towards mathematical and functional style of programming, an example of which is Julia. Sustainable software development leverages financial, ecological, and human resources, as opposed to vanilla software maintenance, the management of resources for up-keeping software quality, repairs, and modifications in line with requirement changes [11].

We highlight the following aspects of programming languages to endorse mathematical programming: (1) obtaining a development environment suitable for implementing and manipulating mathematical models from algebra, and numerical analysis; (2) delivering integrated tools for testing code segments during the software development process and partial evaluation of independent components of the code base – packages with functionalities to visualize input data efficiently and intuitively. (3) platform-independence, the ability of the code to run on all popular operating systems. Due to the technological advancements in the information technology sector, various hardware and computational limitations and rules appear that should be considered during code testing or evaluation. Besides differences in time efficiency and energy consumption, the instruction set of computational units might differ depending on the manufacturer and device model. The heterogeneous market of computational units is responsible for adapting the hardware-specific requirements of the programmer and the applied

---

[2] https://www.wolfram.com/mathematica (acc. 15.05.2024).
[3] https://www.mathworks.com/products/matlab.html (acc. 15.05.2024).

development tools. The programmer might have to develop device-specific code considering the capacity and guidelines of the specific computational unit.

While compiled languages must first be converted into executable code (the *compilation*), interpreted languages are compiled "when run". Interpreted languages have advantages over compiled languages, e.g. dynamic typing and the opportunity to write high-level code; e.g. code snippets written in Mathematica or Matlab can be executed within an IDE, meaning that the implemented model can be tested during the development.

We focus on two relatively new programming languages: Python[4] and Julia[5]. Python is a *Swiss-knife* language, very popular in deep learning, mainly due to the vast number of libraries delivering machine learning algorithms. Julia – started in 2008 – is built on the principles of clean and high-level programming; the program can run "optimally" on modern computing systems. Whilst being performant, Julia provides abstraction tools that bring the program codes as close as possible to a mathematical – abstract – formalism. This high-level programming encourages those interested in data mining to use Julia to implement various methods and algorithms. Julia outperforms Python in means of CPU run-time while maintaining a similar level of abstraction of the code.

1.1 Julia and the Two-Language Dilemma

In industry research-and-development practice, it is common to divide the algorithm design process into two stages:(1) defining a model addressing the research problem, (2) improving the efficiency of the model. In the first stage, we implement the "conceptual" details of the model, e.g. model architecture, hyperparameters; the implementation is often done in a high-level language, like Mathematica, Matlab, or Python. Several decisions are taken that impact the performance and complexity, such as: deciding which elements and features from data are beneficial to use; estimating the run-time and the complexities, followed by design decisions on parameters.

In the second stage – with the algorithm and parameter details fixed – the model training and deployment process is optimized to reduce run-time or associated costs. A possible cost-reduction practice is shifting from a high-level programming language to a more efficient programming language – typically C/C++/Fortran – and migrating the final algorithm to an efficient language. As a consequence, program codes are written twice; this separation not only doubles the development cost, but also the development time and other resources.

With the adoption of Julia, it is anticipated that the previously described issue of code duplication can be resolved, and code written for the conceptual model can be used in production with minimal changes: recent research using Bayesian inference on *petascale* for cataloging images from the telescopes [14], where the code was written entirely in Julia. The research highlighted the feasibility of the Julia language for large-scale mathematical computing. Previous

---

[4] https://www.python.org/ (acc. 15.05.2024).
[5] https://julialang.org/ (acc.15.05.2024).

```julia
# defining a function
r((x₁,x₂)) = (1-x₁)^2+10(x₂-x₁^2)^2

# computing the gradient
∇r(z) = gradient(r,z)[1]

# alternative function definition
function grad_d(Dtr, Φ, ∇loss, w;
    η = 0.1, T = 100)

    for t ∈ 1:T
        w -= η*mean( ∇loss(x,y,w,Φ)
                    for (x,y) ∈ Dtr )
    end
    return w
end

x  = range(-2, 6, length=801)
ss = [ r((t,u)) for t ∈ x, u ∈ x ]
```

Fig. 1. Example of Julia code: one-line function definition (line 2); gradient computation (line 5), gradient optimisation (**grad_d**), and a double loop.

studies examining the execution times and energy efficiency of programming languages have demonstrated that low-level programming languages – such as C, C++ – outperform Python [12,13]. In this paper, we explore the role of Julia, a novel programming language, in building efficient conceptual models to experiment with various machine learning techniques. Python offers numerous libraries for developing machine learning models; thus, serves as a reliable option in industrial applications. However, several high-level programming languages are adequate for demonstrating state-of-the-art methods and testing different adaptations. Hereby, we emphasize the fact that the reduction in programming time and the efficient use of computational resources make the systems developed in Julia more sustainable.

The Structure of the Paper: Section 2 introduces the Julia language, emphasizing functional characteristics, Sect. 3 provides solved exercises, Sect. 4 compares – using code-snippets written in Python and in Julia – the implementation of more complex tasks; and Sect. 5 lists our conclusions.

2   The Julia Language

The programming language Julia is a *dynamically typed* programming language, similar to Python, Matlab, or Javascript, where variables can "just" be declared, but the type is assigned to a variable when value is assigned to it [3,16]. One can write Julia code without type declarations or data structure definitions; however, "typed" versions of program codes certainly increase code safety and make it more intuitive and more readable. Code readability can be enhanced by code formatting: we can use the full UTF-8 character set – as seen in Fig. 1 –

making the functions we write more like the "mathematical" description of an algorithm. Figure 1 illustrates (1) the *short* definition of a function – on line 2 – (2) the usage of the *mathematical* formalism – on lines 2,5 – by using subscripts and the gradient sign, and (3) and the usage of intuitive iterators within functions that build vectors (see the `mean` function on line 12).

Being close to Matlab, Julia uses the backslash operator – "\" – for solving linear systems: it is the short form of the linear system $A \cdot x = b$, where the unknown is the vector x, and one assumes that the quantities A and b are such that the equation is meaningful. To make programs fast, in Julia, multiple functions implement the backslash operator: essentially *one for every special case*; and the special cases are identified by separate types for operator arguments A and b. The choice of the best fitting implementation – in Julia slang – is called *dynamic dispatch* and it is akin to the operator/function overloading known e.g. from C/C++. The different cases are identified by the *types of the arguments*, and the design of a type hierarchy helps the system in choosing the implementation that will be the most accurate and the fastest.[6]

2.1   Types and Type Hierarchy

By default Julia supports most commonly used types; and one can create new – custom – types using the `struct` keyword – similarly to the structures in e.g. C/C++. Added to the existing type list using the `struct` construct, we can declare new *abstract* types using the `abstract type` construct and we can declare that a concrete type as a subtype of an abstract type using the `subtype` keyword. All declared types form a type hierarchy and in Fig. 2 an example is provided for built-in types: the types `UInt8`, `UInt32`, and `UInt64` store positive numbers, all derived from the `Unsigned` parent-type. The `Unsigned` type is a direct child of the `Integer` type; a sub-type of `Number`.

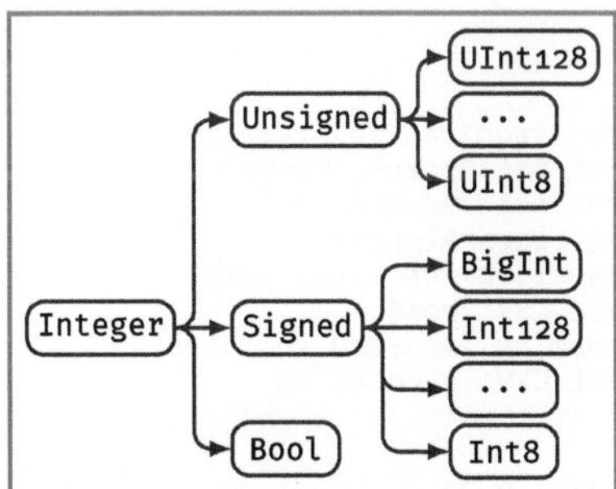

Fig. 2. Julia type hierarchy: the sub-graph of the `Integer` abstract type.

---

[6] In Julia – with some extra packages installed – there are more, than 100 implementations behind the backslash operator – using `methods(\)`.

In Julia there are tools for checking types at run-time: we can ask, for example, whether one type is more general than another type: `Number >: Integer` – true –; we can also ask for the subtypes of a given type – `subtypes` – or for the ancestor – `supertype`. Similarly to Haskell, types can be used in function declarations to specify parameters with specific types upon calling. If abstract types are used in the definitions, the compiler will select the function – as it should happen – if the actual parameter is a subtype of the declared abstract type. Julia is not an object-oriented language and in general, types are not permitted to have methods *except* for the *constructor*, which allows setting the default values for the structures. Using control structures in constructors, we can constrain the fields, disallow certain configurations, or check for correctness.

The classical function and operator overload in Julia is heavily used: different functions with the same name, but different parameter lists can be declared, and the choice of the "best" match is found using "multiple dispatch". If a function is called, from the set of functions with the same name, the instance best matching the actual parameter list is going to be executed. As consequence, functions with abstract types are selected only if the type of actual parameter is a descendant of that abstract type. For example, if we have a function defined for `Integer`, `Unsigned` and `UInt8`, then for `UInt8(123)` it will use the last definition, while for -13 it will use the first one (see also the notion of dynamic dispatch).

2.2   Packages

As with other languages (such as Python, and Haskell), Julia structures the code into modules. Thus, frequently used data structures and methods may be organized by their functionality and managed as distinct units. Julia has several built-in modules – packages – implementing common functionalities for different applications. Using functions and types from packages can be imported using the mechanism described in Fig. 3. In what follows, we describe packages that one needs when implementing data-processing algorithms.

```
# name required for reference
import DataFrames, Plots
# name not required
using LinearAlgebra
# importing functions only
import CSV: read
# renaming imported components
import CSV: read as rd
```

Fig. 3. Package handling in Julia.

One of the strengths of the Julia language is implementing effective data manipulation, including input and output operations from the file system. Data import and export are possible with the *CSV*, *DataFrames* packages. The *CSV* package is a "pure-Julia" package – written entirely in Julia – and can handle delimited text streams, be it in the comma-separated value (CSV) format, tab-separated value (TSV) format, or any other texts that have specific delimiters consisting of a single- or multiple characters.

The functions `CSV.read` and `CSV.File` read the whole file; the difference is that the former requires the programmer to specify the type of the data within the file, while the latter returns `CSV.File` object, a handler that can be used for tailored data reading. Using the functions from the *CSV.jl* is ergonomic – there are sensible defaults for the format configurations and the return values are also straightforward. Multiple dispatch mechanism is again to the advantage since the same syntax is used for reading different sources, like files, vectors of bytes, IO types, compressed .gz files, or even internet links.

These functions do not return vectors or matrices, but rather either a table or a table row, where the table is loosely defined as a 2D structure with named columns. The exact interface is defined in the *Tables* package, and one implementation of this interface is the `DataFrame` type from the *DataFrames* package. It is similar to the *pandas* library in Python, or tables in Matlab. Specific rows/columns are chosen similarly to Matlab or Python: rows can be selected with a number or a range of numbers; for columns one can also use names, making the code more readable. Selections are automatically typed – if possible – as `Vector` if only one column was selected. We also have selectors, for example `Not`, `Between`, or regular expressions – using the `Regex` type and strings prefixed with the 'r' character. An efficient solution in Julia is the use of "views": logical indexing of arrays without the need to copy, achieved by the `view` function or the `@view` macro. Since this is not a copy, operations on views change the original data.

The visualisation – *Plots* – package is also a nice feature of Julia: one can use the package for data visualisation.[7] There is a significant effort in making it sufficiently generic, providing visualisation for the computer screen and scientific papers with the same code; achieved with different "backends": we mention the *GR* backend for print-quality plots and the *Plotly* or *PlotlyJS* packages for producing an interactive web- and browser-based data visualisation.

Last but not least, the *Flux.ml* library for *machine learning* and deep learning, similar to the PyTorch framework. *Flux.ml* is implemented in Julia, thus, benefits from implicit code optimisation and more efficient running on different architectures allowing – with minimal change – the use of GPU-s to speed up computations. It builds on the mathematical optimisation library – *Optim.jl* – and the automatic source-code level differentiation – the *Zygote.jl* package – and the code written is small, easy to maintain, contributing to the sustainability of systems.

---

[7] An extensive documentation can be found at https://juliaplots.org (acc: 02.11.2022).

3   Solved Exercises in Julia

In this section, we provide solutions for several independent exercises demonstrating the features of Julia that facilitate mathematical programming. We recommend Jupyter notebooks when testing the Julia language.[8]

We build our introductory coding practices presented in this section on the use of arrays, as here we have only weak assumptions about the structure, thus a wide area of application is possible. Multiple types are feasible for processing arrays, such as `Vector`, `Matrix` and `Array`. There are many ways to initialize arrays: one can enumerate explicitly, enclosing the values, like in `a = [1,2,3]`; one can fill with specific values: `zeros`, `ones` or `rand`. *List comprehension* can also be used for initializing arrays, this is a common practice in functional languages, which makes the code more transparent and more compact.

Exercise 1. If given an $n \times m$ array, where each row represents a sample consisting of m elements and a label associated with each vector, select the vectors of the first 1000 that are labeled 6 or 7.

Array indexing is defined in accordance with well-known programming languages: elements of vectors can be accessed individually by using integer indexes. In contrast to several programming languages, Julia arrays are 1-indexed by default. Similarly to Python or Matlab, sub-arrays can be specified using *slice-index*, which, although similar to integer-indexing, enables the user to select multiple elements from the array. To specify which elements to choose, one can use the " : " operator between the index of the first and last elements of the selection. Omitting the indices means that every element of that dimension is requested. Retrieving elements on custom indices can be achieved using a `BitVector` as an index. If the provided array index is a `BitVector` of the same length as the indexed array, all elements corresponding to 1-s will be retrieved, and positions that have 0-s in the index array will be omitted.

Figure 4 shows the usage of indexing in Julia through selecting a sub-set of given inputs. In the code, `data` is an $n \times m$ matrix where n is the number of samples and m is the number of data attributes. The `labels` is an n-length vector which contains the associated labels for each sample (row) of the `data`. The first 1000 samples are selected utilizing slice-indexes for vectors and matrices. Then, the indexes those samples are selected from the data set that are labelled as 6 or 7: the indexes of the labels 6 and 7 are identified in the `labels` vector in order to retrieve the elements sought from the `data` matrix.

Exercise 2. Use list comprehension to preprocess the data. Transform the labels of 7 to +1 and the labels of 6 to −1. For each input, keep the first two elements and the product of the first two elements.

---

[8] The benefits of using Jupyter include (1) creating portable code, (2) lightweight development environment and (3) support for creating high quality info-graphics, text formatting for documentation, accessible editing of mathematical formulas.

```
# slice index
labels = labels[1:1000]
# slice (2) - 1000 rows
data = data[1:1000, :]

# using BitVectors
bv6 = labels .== 6
bv7 = labels .== 7
bv  = bv6 .| bv7

labels = labels[bv]
# Rows with 6 or 7
data = data[bv, :]
```

Fig. 4. Indexing vectors and matrices with slices and BitVectors.

```
# List comprehension
# Preparing labels
y_train = [ x == 7 ? 1 : -1
    for x in labels ]

# Features functions
features = [
    x -> x[1]
    x -> x[2],
    x -> x[1]*x[2]   ]

# Bi-dimensional comprehension
# Extracting features
x_train = [ f(x)
    for x in eachrow(data),
    f in features ]
```

Fig. 5. One and bi-dimensional list comprehension examples in Julia.

In data science, arrays are often created during data transformation. The task of array construction can be accomplished with for loops, but it may become tedious, as it involves duplicating similar lines of code repeatedly with only slight modifications. To improve code readability and performance, some programming languages, including Julia, offer the concept of list comprehension. List comprehension consists of building a new data structure by iterating over existing collections and evaluating an expression for every element. The size of the new data structure depends on the size of the iterated collections.

Figure 5 shows a possible solution for a well-known preprocessing task in which we extract *features* from the data. First, the training labels are based on the original labels; the label 7 becomes the positive class marked with +1, while labels different from 7 are re-tagged as negative using the −1 value. Feature extraction can be applied to the original input to decrease the dimensions of the training data set.[9]

---

[9] A well-known method applied for feature extraction is Principal Component Analysis.

The vector **features** contains functions to extract features from the data-rows – treated as abstract points. The **eachrow** function provides a view of a matrix as if the matrix was a vector of rows, and feature extraction is called for every row. In this example, we highlighted that the matrix **x_train** is built with bi-dimensional list comprehension, and each row contains a vector of the extracted features for the original data-points of the same index.

Exercise 3. Calculate a prediction for any $(1, m)$ dimensional array x as: $x \times theta + bias$, where *theta* is a $(m, 1)$ dimensional array and *bias* is a number. Calculate a prediction for each row of a 2-dimensional array using a broadcast operation.

Another powerful tool provided by Julia to simplify array manipulation is the broadcast operation. Similarly to Matlab, we can use the .* operator to calculate a dot-product. In addition, . can be appended to other operators, indicating that the original operator should be applied element-wise. Possible applications of the . syntax include addition or subtraction given a number and an array of numbers as operands. Almost all operators and some of the one-parameter functions may be broadcast over arrays. The functionality of broadcasting is defined for comparison operators: the result is a **BitArray** of the same length as the operand-array marking those indices where the array element satisfies the condition. That can be seen in Fig. 4 where we chose only those elements, which had 6 or 7 as labels.

```julia
function predict_all(xs, theta, bias)
    # Broadcast for dot product
    vecs = [x .* theta for x in xs]
    # Broadcasting a function
    # over a vector
    x_pred = sum.(vecs)
    # Broadcasting an operator
    # over a vector
    x_pred .- bias
end
```

Fig. 6. Prediction using broadcast operator.

Other examples of the broadcast operator are shown in Fig. 6, where broadcasting is applied to make a prediction using a linear regression model. First, we implement the dot-product of two vectors by broadcasting the multiplication operator: each element of the vector x is multiplied by the element of **theta** with the same index (Fig. 6, line 3). On line 6, the **sum** built-in function is broadcast over the dot-product computed for each element of the input: each row of the vector the **sum** function is applied and the results are stored in a new vector, similarly to list comprehension. Lastly, we broadcast a subtraction operation with a floating point value to the entire `x_pred` array.[10]

Exercise 4. If given a log-stream and a function `is_error` that returns if the parameter object is an error, retrieve the messages associated with errors. Select elements of a list in a functional programming style.

Depending on the length of the available data stream and the size of the memory available for computations, it may not be efficient (or possible) to allocate space and store the raw data in the memory section of the code. The problem of manipulating arbitrarily large lists is addressed in functional languages by implementing the notion of "lazy evaluation": choosing to apply operations on the items of the series as soon as the element in question becomes available, rather than retrieving the whole vector before processing.

```
# Filtering out non-error messages
# using a predefined function
errs = filter(is_error, logs)
# Extracting text information
# using a lambda function
messages = map(x -> x.message, errs)
```

Fig. 7. `map` and `filter` for information extraction.

In Julia, vectors and matrices are expandable after creation using the `push!` and `append!` methods in the former case and the `hcat` and `vcat` methods in the latter case. However, these are costly operations and should be avoided when possible. Typical examples for this are the transformation and filtering of arrays. Building a new array with a for loop and the `push!` method comes with a significant performance penalty.

A preferable alternative to populating arrays with `for` loops is to use the `map` and `filter` functions. The `map` function can be used to transform an array into another of the same length, but not necessarily of the same type. The `map` function essentially replaces a for loop iterating over the original array that applies a function to each element and saves them in a new array. The `map` function usually follows the "lazy evaluation" paradigm – outperforming a conventional for loop. Choosing elements of a given property from a collection can be achieved using the `filter` function. This replaces the common pattern of iterating over an array and copying over only those elements which fit a given criterion. Both must be provided with a function as a parameter that processes individual elements of the collection from the parameter list, as presented in Fig. 7. In this exercise, the goal is to identify every error message from a list of logs. Not all elements in the logs are error messages; moreover, logs also have meta-information, which we are

---

[10] Functions written in Julia return by default the last expression computed, thus, the function `predict_all` in Fig. 6 returns the result of the `x_pred .- bias` expression.

not interested in the context of this exercise. As a consequence, the first step is to filter out irrelevant logs, and then we have to extract the message attribute from the relevant hits.

Exercise 5. Implement the median filter to vizualize the COVID-19 data of patients from multiple countries. The data-set is available in CSV format[11].

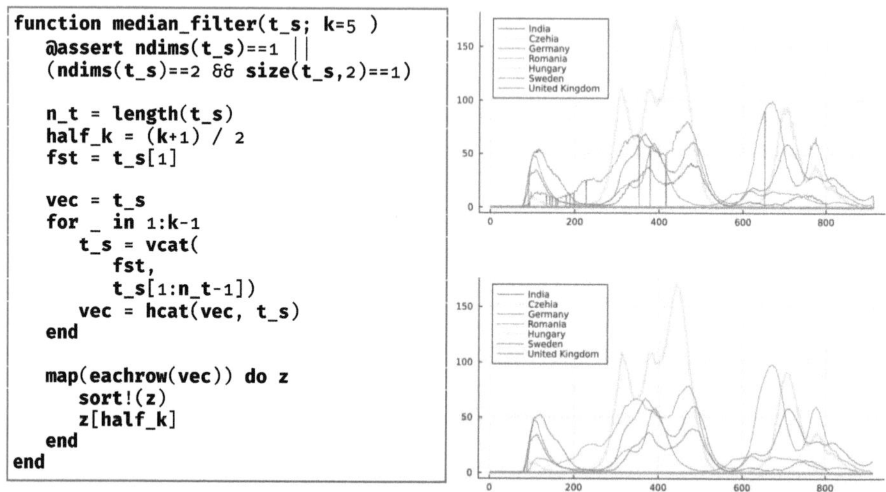

```
function median_filter(t_s; k=5 )
    @assert ndims(t_s)==1 ||
    (ndims(t_s)==2 && size(t_s,2)==1)

    n_t = length(t_s)
    half_k = (k+1) / 2
    fst = t_s[1]

    vec = t_s
    for _ in 1:k-1
        t_s = vcat(
            fst,
            t_s[1:n_t-1])
        vec = hcat(vec, t_s)
    end

    map(eachrow(vec)) do z
        sort!(z)
        z[half_k]
    end
end
```


Fig. 8. Median filtering example. Left: code for the median filtering function of a one-dimensional data stream. Right: Covid-19 data showing ICU occupancy per million people; (top) raw data, and (bottom) a clearer image after median filtering (k = 7).

In the data set, the patients reported per day are split into multiple categories, e.g. total number of patients, number of patients (per million), the mortality rate (per million), and number of patients in the ICU. The numbers of patients in a given category for each country correspond to time-dependent data streams.

First, we visualized the raw data containing the number of patients occupying beds in the ICU for a few selected countries, as seen in Fig. 8, top-right figure. Although raw data provides a comprehensive report of changes in the number of patients, to better communicate our insights to the audience, post-processing is beneficial. A possible improvement is smoothing of the number of new patients curve by applying a median filter to the raw values, as seen in Fig. 8 – the bottom right part.

We implement the median filter (Fig. 8, left) applying various operations on matrices and higher-order functions. The original `t_s` one-dimensional vector is

---

[11] Covid-19 data source https://github.com/owid/covid-19-data (acc: 03.12.2022).

copied into the columns of a two-dimensional data structure, where each subsequent column is shifted with 1 position compared to the previous row. The matrix is constructed using the `hcat` and `vcat` built-in functions to"stack"vectors. Constructing the $n\_t \times k$ matrix in the memory provides an easier understanding of the method, as well as ensures that the result may be computed iterating over the rows of the matrix exactly once. We use the `map` higher-order function to select the median from k the rows using Julia's built-in `sort` method.

```python
def eig_vv(X):
    x_ = X - np.mean(X, axis=0)
    C = np.cov(x_, rowvar=False)
    vals, vecs = np.linalg.eigh(C)
    vals = np.fliplr([vals])[0]
    vecs = np.fliplr(vecs)
    return vals, vecs

def PCA_k(data, prec = 0.98):
    eig_vals, _ = eig_vv(data)
    pr = np.cumsum(eig_vals) /
            np.sum(eig_vals)
    K = np.argmax( pr >= prec)+1
    return K
```

```julia
eig_vv(X) = eigen(
    cov(X);
    sortby = x-> -x
);

function PCA_k(data, prec = 0.98)
    values = eig_vv(data).values
    pr = cumsum(values)/sum(values)
    findfirst(pr .>= prec)
end
```

Fig. 9. Implementation of PCA in (left) Python using *NumPy* and (right) in Julia using *LinearAlgebra* libraries. It is worth observing the shorter and more functional style of the Julia code.

4   A Comparison of Julia and Python

In this section, the solution of two compound tasks is addressed demonstrating coding practices in Julia and in Python. In industrial applications the main drivers for the programming language selection process include resource consumption, scalability, and having support for a wide variety of methods – several data analysis approaches are implemented in Python and available through libraries, such as PyTorch[12], Tensorflow[13], scikit-learn[14] and others. However, in cases where the understanding and analysis of methods is the main goal, the main drivers for selecting a programming language shift towards readability and providing support for mathematical operations. In this article, we demonstrate the usage of two high-level programming languages to present the building blocks of well-known data analysis approaches. The implementations of the principal component analysis and determining modularity in both languages are com-

---

[12] https://pytorch.org/ (acc. 15.05.2024).
[13] https://www.tensorflow.org/ (acc. 15.05.2024).
[14] https://scikit-learn.org/stable/ (acc. 15.05.2024).

pared[15]. The CPU run times of written code snippets are reported with the scope of highlighting energy-efficient implementations.

4.1   Principal Component Analysis (PCA)

Principal component analysis (PCA) is a well-known approach to analyzing data and performing dimensionality reduction for data with a large number of attributes. PCA is a numerical method that requires the computation of covariance matrices and the finding of eigenvalues or eigenvectors; therefore, the PCA problem was chosen to demonstrate mathematical programming and the applicability of the Julia language for data analysis.

Figure 9 shows the implementation of the PCA with code similar to the mathematical notation. The *PCA_k* function returns the number of eigenvalues for data transformation so that the eigenvalues determined describe the original data set up to an arbitrary percentage specified by the *precision* parameter. For comparing the two implementations of the PCA, we opted for the – MNIST [2] – data-set, that contains representations of handwritten digits. With Python, we used the *NumPy*[16] implementation, a set of functions for processing multi-dimensional matrices. For Julia, the built-in data type `Vector` is suitable for representing the inputs. The *Statistics*[17] and *LinearAlgebra*[18] Julia frameworks are imported for testing with optimized versions of the covariance matrix generation and the eigenvector computation tasks.

4.2   Modularity

One of the experiments to compare Python and Julia consists of calculating the modularity [1,10] of the community structures within graphs. Given a graph and a presumed community for each node, we can calculate the modularity of that community structure. Modularity measures the divisive nature of a given partition of the network into groups – communities – taking into account the underlying network structure. Partitions yielding high modularity feature a large number of connections within communities and relatively few edges connecting distinct groups. Higher modularity is associated with more fitting community structures given the assumption that in a randomly generated graph, no community structure is expected. If there are more edges inside a presumed community than expected then our assumptions of those nodes belonging to the same community that is computed as:

$$Q = \frac{1}{2m} \sum_{v_i, v_j \in V} \left[A_{ij} - \frac{k_i k_j}{2m} \right] \delta(c_i, c_j)$$

---

[15] The provided code snippets are possible implementations, where we prioritized including the same steps to solve the tasks with the goal of demonstrating and teaching well-known data analysis approaches.

[16] https://numpy.org/ (acc: 24.03.2023).

[17] https://docs.julialang.org/en/v1/stdlib/Statistics/ (acc: 24.03.2023).

[18] https://docs.julialang.org/en/v1/stdlib/LinearAlgebra/ (acc: 24.03.2023).

where δ is the Kronecker delta function.

We conducted our experiments on modularity, evaluating networks with various structures to evaluate the Python and Julia custom implementations of calculating modularity: the [15] network with 24 vertices, sub-sample of the epinions.com social network [7] and the LFR benchmark [6]. The sample network was generated in an iterative manner, selecting nodes from the original network to be added to the sample graph. The starting node of the sample graph is selected randomly following a uniform distribution. In consecutive steps, candidates to be added to the sample are drawn from a *neighbor-pool*, which contains the nodes connected to the vertices of the sample graph, but not in the sample graph. The sampling method stops when the neighbor pool is exhausted, or the sample size of the network reaches a threshold.

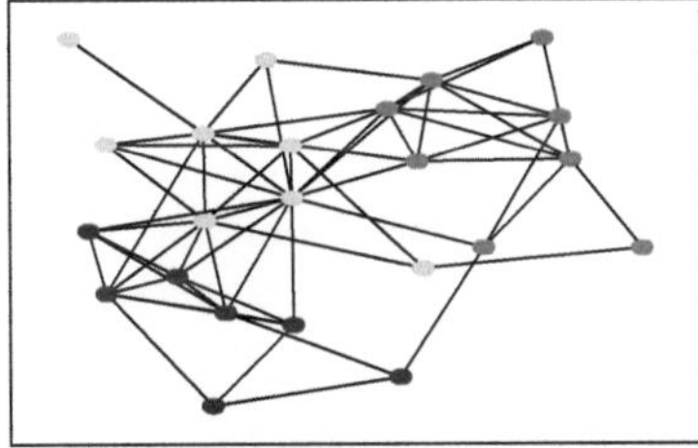 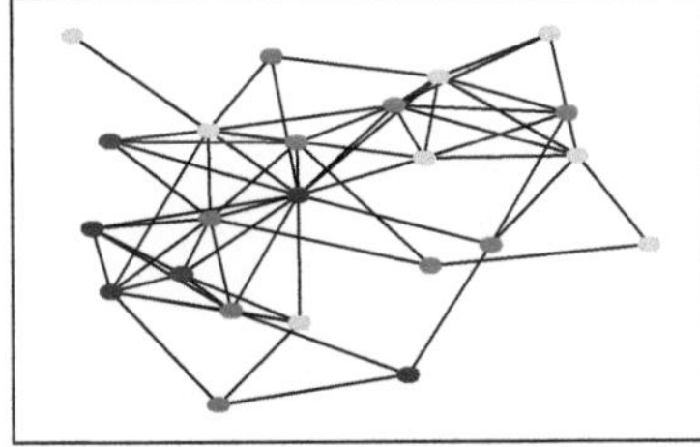

Fig. 10. Graph nodes are assigned to 3 different groups; the modularities are (left) 0.360 and (right) −0.003.

The LFR benchmark consists of several graphs with a given amount of communities where each community has the same number of nodes. These graphs are generated with a mixing parameter, which sets the probability of edges starting from a node pointing outside of the community. The graph we have used for testing is generated with 4 communities, each community consisting of 32 nodes[19] with the mixing parameter set to 0.3 meaning that 70% of the edges connect nodes from the same community.

Figure 10 shows the input graph from [15] utilized for evaluating computational differences between Julia and Python. The two possible community assignments of nodes shown in Fig. 10 marked by different colors are arbitrary partitions respectively. With the scope of measuring the effectiveness of node distributions, modularity is employed as a performance metric. Applying modularity for community detection problems calls for an efficient and straightforward implementation to decrease the computational overhead (Fig. 11).

4.3 Comparative Experiments

To compare the impact of programming in different coding languages, the running time of code snippets is reported. For evaluating the Python functions, the

---

[19] 128 nodes in total.

```python
def modul_nodes(graph, i, j):
    m = graph.number_of_edges()
    A_ij = int(graph.has_edge(i, j))
    k_i = (len(graph[i])
            if i in graph else 0)
    k_j = (len(graph[j])
            if j in graph else 0)
    return A_ij - (k_i*k_j) / 2 / m

def modul_gr(graph, c):
    m = graph.number_of_edges()
    n = graph.number_of_nodes()
    nodes = tuple(graph.nodes)
    s = sum( sum(
        modul_nodes(
            graph,nodes[i],nodes[j])
            for j in np.where(c==c[i])[0]
        )
        for i in range(n)
        )
    return s / 2 / m
```

```julia
function modul_nodes(graph, i, j)
    # Number of edge
    m = ne(graph)
    A_ij = Int(has_edge(graph,i, j))
    k_i = length(all_neigh(graph, i))
    k_j = length(all_neigh(graph, j))
    A_ij - (k_i * k_j) / 2m
end

function modul_gr(graph, c)
    # Edge and vertex number
    m = ne(graph)
    n = nv(graph)
    numbers = collect(1:n)
    s = sum( sum.(
        [ [modul_nodes(graph, i, j)
            for j in numbers[c .== c[i]]
        ] for i in  1:n
        ]
    )
    s / 2m
end
```

Fig. 11. Python (left) and Julia (right) implementations of the modularity score. In Python, *NetworkX* and *NumPy* libraries are used for managing graph and vector operations, respectively. In Julia, the *Graphs* and *SimpleWeightedGraphs* packages are utilized for the same purpose.

timeit library[20] is used. The duration of code snippets is calculated as an average difference of the wall-clock time[21] for 1000 method calls. In order to mitigate the impact of background processes on our experiments, 5 timing experiments were run, and the lowest run-times are shown in Table 1.

The *Benchmarktools* package provides a user-friendly interface for measuring the performance of code snippets written in Julia 1.8.2.[22] The *btime* function provided by the *Benchmarktools* framework lists run-time statistics of arbitrary function calls. The evaluated functions are called 1000 times for the same parameters as for the Python part, the best run-times are reported in Table 1.

According to our experiments, in the custom implementations of PCA, the code written in Julia proves to be approximately 2× faster than the Python implementation. The performance metrics reported for calculating the modularity on undirected graphs of various sizes demonstrate that Julia outperforms Python in terms of CPU run-time. For the smallest graph input, the Julia implementation of modularity is 400× faster than the Python implementation included in this article. As the size of the input graph increases, the difference between the CPU run-times of the modularity functions in Julia and Python decreases: Julia displayed an 180× decrease in CPU run-time for the input graph with 128 nodes and an 290× decrease for the graph with 340 nodes.

---

[20] https://docs.python.org/3/library/timeit.html (acc. 15.05.2024).
[21] The wall-clock time is obtained using the *perf_counter_ns* function from the `time` library, see https://docs.python.org/3/library/time.html (acc: 21.03.2023).
[22] Julia benchmark tools: https://juliaci.github.io/BenchmarkTools.jl (acc: 21.03.2023).

In practical applications of mathematical computations, the overall efficiency of code is crucial, including custom function implementations, as well as functions provided by various frameworks, libraries, and modules. Since the coding efficiency is affected by the choice of frameworks, we reported the overall CPU usage, which includes the resource consumption of functions imported from libraries. The PCA method based on matrix operations implemented using Python's *NumPy* is significantly closer to the performance seen for Julia than the modularity method featuring the application of graph-based *NetworkX* and *Graphs* libraries.

Table 1. Mean CPU run-times reported for Python and Julia code snippets

Function	Input	Python	Julia
PCA	256 vectors [2]	67.65 ms	27.486 ms
Modularity	24 nodes [15]	2.72 ms	47.72 μs
	128 nodes [6]	197.56 ms	1.13 ms
	340 nodes [7]	2865.20 ms	9.92 ms

The implementations of *PCA* and *modularity* functions prioritize code readability and mathematical programming style. Higher-order functions and generators were applied to optimize memory and time consumption. The usage of well-known frameworks is demonstrated for Python and Julia. We employed Python's *NumPy* and Julia's *Statistics* libraries for efficient implementations of various mathematical notions and methods.

4.4   Code Quality

Quality metrics take into account various factors regarding the code bases. [8] highlights reliability, program correctness, and maintainability as essential techniques to asses the code quality. This section addresses code readability as an instrument to ensure code maintainability and focuses on coding practices that facilitate program correctness in Julia.

A benefit of code readability is the reduction in overhead time and capital imposed by modifying and maintaining existing code bases. Julia enhances the readability of code-snippets by supporting syntax close to well-known functional programming languages, which minimize "boilerplate" code[23] in comparison with most lower-level programming languages. Generic programming [9] is a suitable approach to avoid code sequence repetition [4] by using a functional language – *Haskell* – for development. The works [4,5] illustrate minimizing boilerplate code and direct toward pragmatic snippets building on the type-safe cast and higher-rank types that use generic functions and generic *zip* like functions.

---

[23] Highly similar code sections re-appearing in the code-basis multiple times are coined as "boilerplate" (https://en.wikipedia.org/wiki/Boilerplate_code, acc: 13.04.2023).

List comprehensions and higher-level functions (such as `map, filter`) not only improve run-time, but also simplify writing code that would require `for` loops which could introduce extra complexity. Julia includes several numerical functions in the standard library to simplify the usage of prevalent methods excluding the need to reference modules and reducing the risk of name collisions.

5   Conclusions

In this article, we presented the main characteristics of the Julia programming language and argued that Julia is suitable for implementing and testing high-level code. We focus on sustainable aspects of the Julia programming language, which facilitate the development of portable, robust, and readable code. By prioritizing technical computing, Julia can manage high-level concepts while utilizing the underlying hardware efficiently. The functional aspects of Julia, such as the support of list comprehensions, facilitate writing a clean and transparent code base. Code transparency, which aims to avoid possible inefficiencies, paired with built-in high-performance code translation to the hardware, brings closer programming languages that are efficient by design.

Our experiments show that high-level codes for various data analysis methods can be implemented using Julia while maintaining a fast CPU run-time. Our future efforts regarding the analysis of sustainable aspects of programming languages - especially Julia - include investigating the run-time and energy usage of code snippets utilizing tools running on the operating system level; e.g. *perf* for the Linux operating system family.[24]

References

1. Girvan, M., Newman, M.E.J.: Community structure in social and biological networks. Proce. Natl. Acad. Sci. **99**(12), 7821–7826 (2002)
2. Hull, J.: A database for handwritten text recognition research. IEEE Trans. Pattern Anal. Mach. Intell. **16**(5), 550–554 (1994)
3. Kwong, T., Karpinski, S.: Hands-On Design Patterns and Best Practices with Julia: Proven Solutions to Common Problems in Software Design for Julia. Packt Publishing (2020)
4. Lämmel, R., Jones, S.P.: Scrap your boilerplate: a practical design pattern for generic programming. In: Proceedings of the 2003 ACM SIGPLAN International Workshop on Types in Languages Design and Implementation. TLDI '03, pp. 26–37. Association for Computing Machinery, New York, NY, USA (2003). https://doi.org/10.1145/604174.604179
5. Lämmel, R., Jones, S.P.: Scrap more boilerplate: reflection, zips, and generalised casts. In: Proceedings of the Ninth ACM SIGPLAN International Conference on Functional Programming. ICFP '04, pp. 244 255. Association for Computing Machinery, New York, NY, USA (2004). https://doi.org/10.1145/1016850.1016883
6. Lancichinetti, A., Fortunato, S., Radicchi, F.: Benchmark graphs for testing community detection algorithms. Phys. Rev. E **78**, 046110 (2008)

---

[24] For documentation: https://perf.wiki.kernel.org (acc: 22.03.2023).

7. Leskovec, J., Huttenlocher, D.P., Kleinberg, J.M.: Signed networks in social media. In: Proceedings of the SIGCHI Conference on Human Factors in Computing Systems (2010)
8. Mills, E.E.: Metrics in the software engineering curriculum. Ann. Softw. Eng. **6**(1), 181–200 (1988)
9. Musser, D.R., Stepanov, A.A.: Generic programming. In: Gianni, P. (ed.) ISSAC 1988. LNCS, vol. 358, pp. 13–25. Springer, Heidelberg (1989). https://doi.org/10.1007/3-540-51084-2_2
10. Newman, M.E.J., Girvan, M.: Finding and evaluating community structure in networks. Phys. Rev. E **69**, 026113 (2004)
11. Penzenstadler, B.: Towards a definition of sustainability in and for software engineering. In: Proceedings of the 28th Annual ACM Symposium on Applied Computing, pp. 1183–1185 (2013)
12. Pereira, R., et al.: Ranking programming languages by energy efficiency. Sci. Comput. Program. **205**, 102609 (2021)
13. Pereira, R., et al.: Energy efficiency across programming languages: how do energy, time, and memory relate? pp. 256–267 (2017). https://doi.org/10.1145/3136014.3136031
14. Regier, J., et al.: Prabhat: cataloging the visible universe through Bayesian inference in Julia at petascale. J. Parallel Distrib. Comput. **127**, 89–104 (2019). https://doi.org/10.1016/j.jpdc.2018.12.008
15. Reynolds, J.J.H., Hirsch, B.T., Gehrt, S.D., Craft, M.E.: Raccoon contact networks predict seasonal susceptibility to rabies outbreaks and limitations of vaccination. J. Anim. Ecol. **84**(6), 1720–1731 (2015)
16. Sherrington, M.: Mastering Julia. Packt Publishing (2015)
17. Venters, C.C., et al.: Software sustainability: research and practice from a software architecture viewpoint. J. Syst. Softw. **138**, 174–188 (2018)

Algorithms for Sustainable System Topologies

Tihana Grbac Galinac$^{(\boxtimes)}$, Emili Puh, and Neven Grbac

Juraj Dobrila University of Pula, Zagrebačka 30, HR-52100 Pula, Croatia
`{tgalinac,epuh,neven.grbac}@unipu.hr`

Abstract. These are the follow-up lecture notes of the lectures presented at the SusTrainable Summer School 2022 held at the University of Rijeka, Croatia, in July 2022. The main goal of the lectures is to provide a gentle introduction to topological data analysis because of its possible applications to sustainable software engineering. When applying topological data analysis to software system structures, one can observe additional structural metrics from the topological space which may be useful as a complement analysis technique in understanding complex software system behavior while aiming to address required quality attributes and sustainable goals.

Specific challenges on addressing sustainability as software quality attribute, but also as an aspect of engineering software, are discussed. A running example is presented and how the topological data analysis can be used to analyse software structures is explained, related implementation details are discussed, and these observations are related to sustainable software and software engineering practice. The topological notions and ideas are introduced and explained on a toy example, in which the basic topological invariants are calculated explicitly, and then an exercise is provided in which the students can test their understanding and check the efficiency of their algorithms on examples of complex software structures.

Keywords: Software systems · Software structure · Topological data analysis · Computational topology · Algebraic topology

---

This work received financial support from the Croatian Science Foundation under the projects HRZZ-IP-2019-04-4216 and HRZZ-IP-2022-10-4615.

This work received financial support through the Erasmus+ Strategic Partnership for Higher Education *SusTrainable—Promoting Sustainability as a Fundamental Driver in Software Development Training and Education* (project number 2020-1-PT01-KA203-078646), funded by the European Union and coordinated by the University of Coimbra, Portugal.

The information and views set out in this publication are those of the authors and do not necessarily reflect the official opinion of the European Union. Neither the European Union institutions and bodies nor any person acting on their behalf may be held responsible for the use which may be made of the information contained therein.

1   Introduction

Digitalization is nowadays recognized as a strategy for driving business success and a key strategy for implementing global sustainable goals in various domains. The software is the main tool used to abstract and model the physical world driving digitalization in all aspects of human life and as such it becomes a crucial target in addressing global sustainability goals. Therefore, sustainability has become one of the global goals for further technological evolution [18,19].

Adequate modeling of software logic and algorithms is important to address sustainability goals from different perspectives. **Engineering sustainable software** targets optimization of computation in big data analysis in applications such as medicine and environmental modeling cases, optimization of physical resource usage targeting performance or energy efficiency constraints supporting decision-making processes in various applications such as business processes, traffic management, communal infrastructure management, etc. and addressing reliability, security and responsibility aspects of software operation in various mission-critical domains. To support these tasks, the software implements various advanced machine learning and artificial intelligence algorithms that may operate on big data, may be distributed across the entire telecommunication network, and may require extensive processing power and related resources. Here the main problem is the effective and efficient use of computational power and resources. In the software engineering domain, **sustainable software engineering** is a twofold act, it considers theory and practice that support software engineering tasks developing software that would secure sustainable software operation within the target application domain as mentioned above, but also to support sustainable business goals for the software engineering industry like for example to support long-term software evolution and responsible profession towards all stakeholders involved in the software lifecycle value chain. Sustainability goals significantly reflect on the software engineering profession, and all traditional software engineering methods need additional reassessment from this new perspective. Some examples are code optimization, code analysis, detecting anomalies, bug tracking and resolution, support in communication and social networking among stakeholders engaged in the software lifecycle, support in predictive maintenance, and software modeling oriented to long-term software evolution.

In all these software applications the key problem is the management of complexity. In our previous lecture notes [13,14], we define **complex systems** as systems in which it is hard or impossible to derive simple rules of global software behavior from local system behaviors. Global system behavior may be characterized through the system properties measured as a consequence of the software system execution. Examples of global system properties are defined by ISO Quality model [20]. For example, in software systems, we may measure reliability as the software system quality property of being able to work or operate for long periods without breaking down or needing attention. Furthermore, there is work in progress to revise standard software quality attributes following global sustainability goal [23,24,32], in which we may measure sustainability as the global system property of the use of natural resources and energy in a way that does not harm the environment. Both global properties are sensitive to decisions we undertake during the system design phase while modeling software logic and algorithms. On the other hand, local software behavior may be characterized by dimensions of the local system properties. Examples of local system properties are various metrics that can be measured during the software and system engineering phases on the product, process, or even project level, like for instance, various cyclomatic complexity metrics on system components, component defectiveness, or effort spent on component development, people expertise, etc. As is defined in the above-stated definition of a complex system, the main problem in engineering such a complex system is the absence of simple models that can guide design decisions, e.g. how to develop system structure and its components' complexities that would lead to desired global system behavior during its execution and use in the real environments.

The *system structure* is the main artifact we model in any complex system design. The possible solution space in which we design components of complex systems and their interactions is extremely large. Proper design decisions can select system structures whose execution can achieve the best global properties during system execution. We may understand this concept of connecting two levels of system abstraction through the case of music. The music is designed through music composition and the composed music is felt as sonority during music reproduction, see Fig. 1 reused from [12]. Note that these two levels of music abstraction provide completely different views on music.

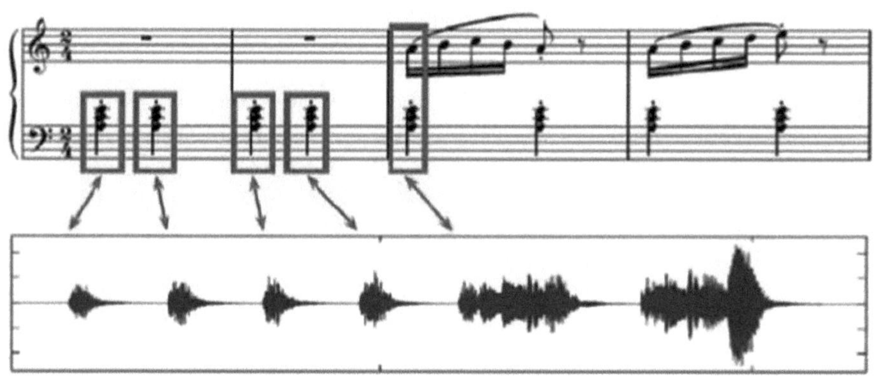

Fig. 1. Design time music composition view (top) versus sonority during music reproduction view (bottom). Image reused from [12].

Graphs are an essential and frequently used tool in various fields of engineering complex systems structures due to their ability to model complex systems by connecting global behaviors to local properties and their internal relationships in a comprehensible and manipulable manner. Numerous graph-based algorithms can be used to analyze and model system behavior effectively. However, the main problem of the majority of graph-based algorithms is computational efficiency and lack of structural property metrics. Topological Data Analysis (TDA) is a useful analysis method because of its ability to analyze shapes and structures within the topological space. Graphs used in the computing discipline may capture topological space of data where data can be any analyzed artifact from big data in various applications to the data coming from software structure. The use of TDA provides a unique perspective on the structural and dynamic properties of analyzed data, offering insights that complement traditional analysis methods. The main benefit of TDA is that it brings a new qualitative perspective on the analyzed system structure (or data structure) that may be related to other global and local system properties [25].

These lecture notes explain how we may apply TDA to engineering sustainable software. In our running case, we will show how TDA may be used in the analysis of software structures with the help of TDA aiming to address sustainable software evolution. The results of our case of using TDA to model software structures in evolution are presented in [37].

The main learning objectives of these lecture notes are as follows:

- To understand specific challenges of modeling and management of sustainable software systems
- To understand the needs and benefits of generalized approaches to software modeling and software management
- To introduce students with the key concepts from topological data analysis
- to learn how to implement abstract topological notions,
- to understand the difficulties in terms of computational power and sustainability of the algorithm by testing the code on different types of graphs,
- to understand how we can impact on sustainable software structures

The lecture notes are structured as follows. This introductory section is followed by Sect. 2 in which we define a playground for the application of TDA within software engineering context. Here the application of complex system definition is discussed in the context of a software system, the complex software system modeling and design problems and solutions are pointed out, and the landscape of sustainable software structures is defined. In Sect. 3, TDA is motivated and introduced. In Sect. 4, the examples of algorithms in TDA are presented. Section 5 explains the principles of TDA algorithms on a toy example. In Sect. 6, an exercise for students to practice the TDA principles on software structures is given. Finally in Sect. 7 we conclude the paper.

2   Sustainable Software Structures

There are several specific challenges of the modeling and management of sustainable software systems.

Software engineering discipline addresses problems of system reuse, independent evolution of system parts, and development of generic software functions that are getting harder and more expensive as the complexity of software increases. The consequences of badly designed software solutions may be severe and affect operational software behavior, making software product operations and their development ineffective in achieving sustainable goals. Advanced software management tools are needed to enable smarter software creation thus fulfilling not only functional goals but also operational sustainability of resource management.

The specific challenge of designing complex systems is that it is hard to understand, manage, maintain, and evolve, and especially several people with diverse knowledge is needed to develop them. So, the responsible software engineering profession has to enable the independence of people working together on the same complex system, to enable its understanding for its easier maintenance and long-term evolution. The design principles used in the software engineering discipline that we introduced in [13,14] are modular system designs aiming to develop the system as a set of loosely coupled components, and with a number of levels of abstraction with a layered design and clear hierarchy. Therefore, any (non-trivial) software system has modular structures and consists of modules (components) and their interactions. The term module refers to a component with the standard and loosely coupled interfaces that are used by other modules within its environment. Depending on the context and the purpose of the study, the modules may be subroutines, functions, units, classes, objects,. . .

Another challenge of modern software systems is that they usually operate in a globally interconnected Internet environment, modules may be distributed over the Internet network, modules may be invoked at runtime, and modules may be replaced at runtime. The system structure is then composed not only of many components that are run within isolated computer nodes, but the software structure is a set of software components that are interconnected on a geographically distributed Internet network. Decisions on software structure may and will also influence the sustainability of network operations. One of the key challenges that 6G networks should address is how to lower energy consumption [1]. This work is an attempt in that direction, since the majority of software-based resource management solutions are deployed via telecommunication networks and smart designs of software structures may lead to significant savings. These are specific challenges of modeling and management of sustainable software systems for distributed software deployment environments.

To address the aforespecified challenges the software engineering discipline has developed various generalized approaches to software modeling and management. One widely used abstract artifact is software structure. The software structure is defined as a set of modules and their interactions that are used to achieve global system functionality. The software structure is an abstract artifact and there are numerous ways to represent software structure. One of the most common software structure representations is by using call graphs [4] (Fig. 2).

An example of representing Java code as a call graph is presented in [34] and the concept is reused from [38]. In this example, the graph is used for abstract software structure representation, in which program classes (A, B, C) are represented as graph nodes (vertices) and method calls (methodB(), methodC()) are

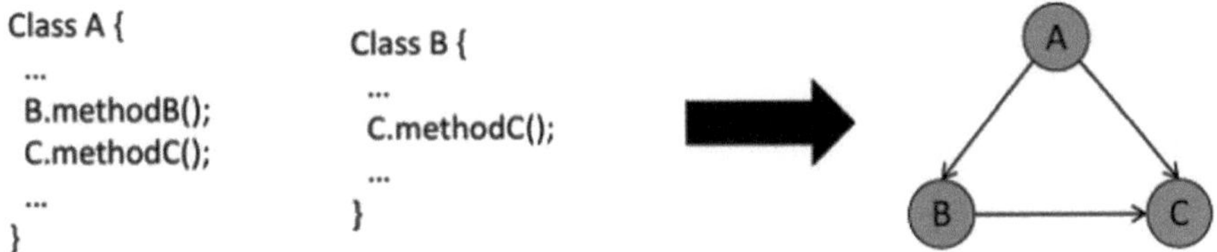

Fig. 2. Call graph as a representation of software structure [34].

represented as directed graph edges. The same principle we reused to develop a graph extraction tool *rFind* to extract call-graphs from Java code [34], which were analyzed with the help of motifs. Although, this analysis has provided us with useful insights into software structure evolution the main problem was in its computational inefficiency. Within this lecture, the students were encouraged to use the same call-graph structures to exercise with TDA described in Sect. 6. The results of the TDA analysis on the same call graphs were presented in [37] where the experiences of using motifs versus TDA were explained in detail.

Our previous studies analyzed software system behavior with the help of this abstract representation of software communication structure as a graph network, in which we experiment with network science models on such software graphs [33,34,39]. With the help of various structural characterization approaches, the aim is to understand the system behavior such as reliability as a global system property. These works are an extension of our previous studies in *Software Defect Prediction* [27] that aim to predict high-risk defective modules based on the models trained on historical data of local software metrics. The main challenge in these works is dimensionality reduction and feature selection procedures, [16,31]. In the related survey studies, it has been identified that an improvement in SDP models could be gained with the additional characterization of the problem, not represented within the standard software metrics. These models suffer from a lack of system structure knowledge that may be incorporated into the data analysis and provide a more powerful basis for data clustering. TDA may be used for clustering of the graph structure by determining the extent to which nodes in a graph tend to form local clusters or groups that may improve the effectiveness and efficiency of decision-making algorithms. In our previous work, we have already demonstrated that software structure may contain useful information to model software defect prediction [34,39]. Some hidden structural characterization obtained from topological information and higher dimensional structure can be obtained from the TDA models such as Betti numbers (which we explain in Sect. 4). These may help to better distinguish between differences and similarities during the data clustering phase. This topological information has been already identified as powerful tool in various data analysis domains [6,26], e.g. in healthcare and disease detection [22], neuroscience [21], image analysis and computer vision [2,3,8], protein classification [5], gene expression data classification [9], epidemiology [35], software failure dynamics [36], etc.

3   Topological Data Analysis

Topology is a branch of mathematics that deals with qualitative geometric information. This discipline studies geometric properties in a way that is much less sensitive to the actual choice of metric than simple geometric methods, which involve quantitative geometric properties such as distance, angle, area, or volume. More precisely, topology ignores the quantitative values of distance functions and replaces them with the notion of infinite proximity of a point to a subset in space. Roughly speaking, topology studies the shape and form of a geometric object and ignores its size and position.

This metric insensitivity is useful in studying situations in which one only understands the metric roughly. Topological data analysis refers to the study of data using topological methods. It is particularly useful to describe and compare the shape, discover trends, and search for hidden patterns in the data.

The standard reference for the basic general topology is the classic book by Munkres [30]. For the algebraic topology, the recommended references are [17,29]. There are several books and survey articles explaining the various applications of topological data analysis, such as [6,7,11,28]. For an account of open problems in computational topology see [15].

3.1   Topology as a Qualitative Model of Reality

Sometimes in the study of real-world behavior, it is not important to have precise quantitative models. In such cases, the interest is in the shape, trends or form, and not in size, scale or time frame. The models of reality are then qualitative, not quantitative. The qualitative models come in two types: discrete and continuous. Common examples of discrete models are graphs, which may be viewed as a set of vertices (also called nodes, points) together with a set of edges (also called branches). Edges are pairs of vertices, usually represented as a line connecting two vertices. Graphs may be used to model connections between objects, but without actual scale or size of them. Weighted graphs may be used if more quantitative information is required.

Topology is a part of geometry that deals with the qualitative properties of geometrical objects. It provides a model of the shape of a geometrical object, without referring to its size. Very roughly speaking, two objects are topologically equivalent if they can be transformed into each other by stretching and modeling but without gluing and cutting. A folklore example for topological equivalence is the example of a doughnut and a coffee mug in Fig. 3, because if they were made of clay one could model one to another without cutting and gluing. However, the pot in the figure is not topologically equivalent to the doughnut and the coffee mug. The reason is that the pot has two handles with a hole in each, and the doughnut and a coffee mug have just one hole. Hence, in order to get equivalence we must either glue one of the holes in the handles of the pot or cut a hole in the coffee mug, but both gluing and cutting is forbidden.

Topological space is a set of points for which a certain notion of "closeness" is defined. This is achieved by defining for each point a set of (open) neighborhoods.

(a) Doughtnut

(b) Coffeemug

(c) Pot

Fig. 3. Example of topological equivalence.

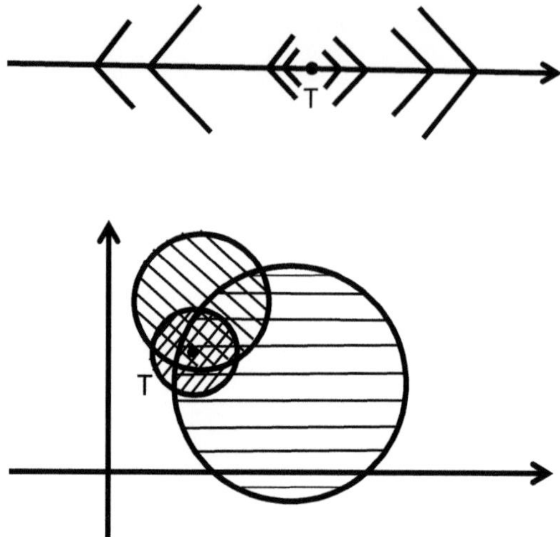

Fig. 4. Neighborhoods on a line and in the plane as examples of topological spaces.

For example, the real line is a topological space in which the neighborhoods of each point are open intervals containing that point. Another example would be the plane. It is a topological space in which the neighborhoods of each point are open discs containing that point. These examples are shown in Fig. 4.

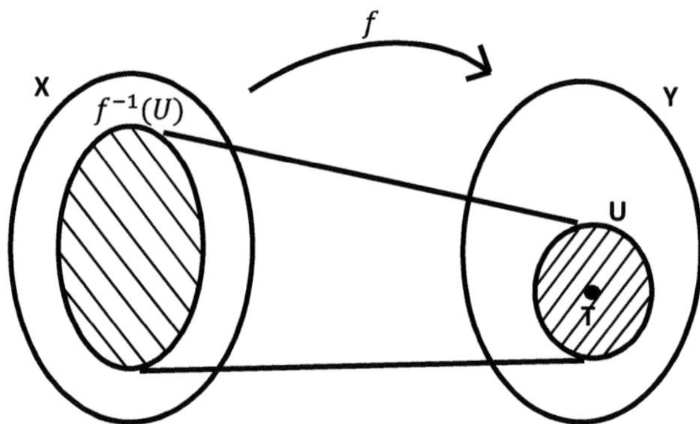

Fig. 5. Continuous functions between topological spaces preserve neighborhoods.

The topological spaces are compared using continuous maps, which in some sense preserve "closeness", which is encoded in neighborhoods. Under a continuous map, the preimage of an open neighborhood of each point is again an open neighborhood. A schematic image of this condition is given in Fig. 5. Homeomorphism is a continuous bijection between two topological spaces such that its inverse is also continuous. Two topological spaces are topologically equivalent if there is a homeomorphism between them. It is extremely difficult to check that two topological spaces are homeomorphic or, in other words, topologically equivalent. Often one cannot find or construct a homeomorphism between topological spaces, but also cannot exclude the possibility of its existence, so that the question of their equivalence cannot be settled.

Considering and comparing topological spaces is important in topological data analysis, as the topological space itself contains information on the underlying data, its structure and properties.

3.2   Topological Invariants as Approximations

Topological invariants are certain objects attached to a topological space, which remain unchanged under homeomorphisms, i.e., topologically equivalent spaces have the same topological invariants. Thus, if one finds any topological invariant of two topological spaces which is not the same, they are certainly not topologically equivalent, but still, if all topological invariants that can be explicitly computed or described of two topological spaces are the same, it is still not sufficient to decide whether they are topologically equivalent or not. There are many topological invariants, most of them of algebraic nature. Hence, the study of topological invariants is often referred to as algebraic topology. The most common invariants are the fundamental group, homotopy, homology and cohomology, and persistent homology.

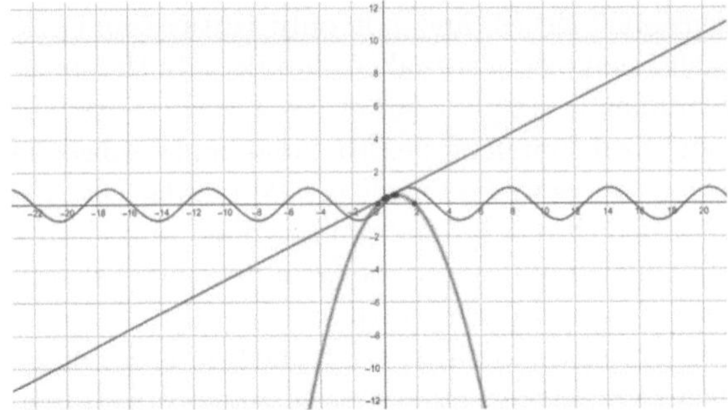

Fig. 6. Approximations of the sine function with a straight line and a parabola around the point $x = \frac{\pi}{3}$.

We draw now an analogy to, hopefully, more familiar subject of mathematical analysis. In mathematical analysis, which is a very quantitative field of mathematics, the approximations of functions is one of the most useful tools in applications. Given, for instance, a real function on some interval, one may use derivatives to obtain the best approximation with a straight line of that function near a point (the so-called tangent line). Using higher order derivatives, the Taylor polynomials provide approximations with polynomials of higher degree. The example[1] of approximations of the sine function with a straight line (the Taylor polynomial of degree one) and a parabola (the Taylor polynomial of degree two) aroung point $x = \frac{\pi}{3}$ is given in Fig. 6. It is often in applications that only such approximation of the observed phenomenon is available (interpolation polynomials). Polynomials are elementary functions, which are more suitable for calculations than arbitrary functions, while at the same time provide a reasonable approximations of the real world phenomena.

Unlike mathematical analysis, topological spaces are not at all quantitative. The approximations of topological spaces should somehow approximate the shape and play the same role as polynomials do in mathematical analysis. (Algebraic) topological invariants may be viewed precisely as algebraic objects that approximate a topological space. They are simpler than the original topological space, but still resemble its shape often in highly sophisticated ways. Algebra is in some sense easier than topology.

3.3   From Topological Space to Its Invariants

The passage from a topological space to its algebraic topological invariants usually requires some kind of discrete model of a continuous object such as a topological space. This is achieved using the so-called triangulation. The topological space is described using a skeleton, which is similar to a graph describing the space, but contains also higher dimensional pieces (not only vertices and edges). Such skeleton is called a simplicial complex and it is a generalization of a graph.

A triangulation by a simplicial complex is a generalization of graphs in which also higher dimensional pieces, and not only vertices and edges, are present. These higher dimensional pieces are called simplices. In the spirit of graph theory, a d-dimensional simplex is simply a set of $d + 1$ vertices (points), together with all lower-dimensional subsimplices, i.e., the non-empty subsets. Observe that an edge of a graph may be viewed as a set of two points. Thus, together with its non-empty subsets, which are its end-points, it is an 1-simplex. Vertices of a graph are 0-dimensional simplices, as they can be viewed as sets with one point. More generally, a triangle, together with its sides and vertices, may be viewed as a 2-dimensional simplex. Similarly, a tetrahedron is a 3-dimensional simplex.

Any (finite) set of simplices, viewed as a single object, is a simplicial complex. The point is that any topological space may be triangulated by a simplicial

---

[1] More precisely, the function $f(x) = \sin x$ is approximated around the point $x = \frac{\pi}{3}$ by the straight line

$$T_1(x) = \frac{1}{2}x + \left(\frac{\sqrt{3}}{2} - \frac{\pi}{6} \right)$$

using the first derivative of f at $x = \frac{\pi}{3}$, and by the parabola

$$T_2(x) = -\frac{\sqrt{3}}{4}x^2 + \left(\frac{\pi\sqrt{3}}{6} + \frac{1}{2} \right) x + \left(\frac{\sqrt{3}}{2} - \frac{\pi}{6} - \frac{\pi^2\sqrt{3}}{36} \right)$$

using also the second derivative of f at $x = \frac{\pi}{3}$.

complex, which may be viewed as a discretization of a topological space. The triangulation by a simplicial complex is not unique, but different triangulations lead to certain topological invariants which are equal. Hence, assignment of certain topological invariants to a topological space using a convenient discretization in terms of the simplicial complex really makes sense.

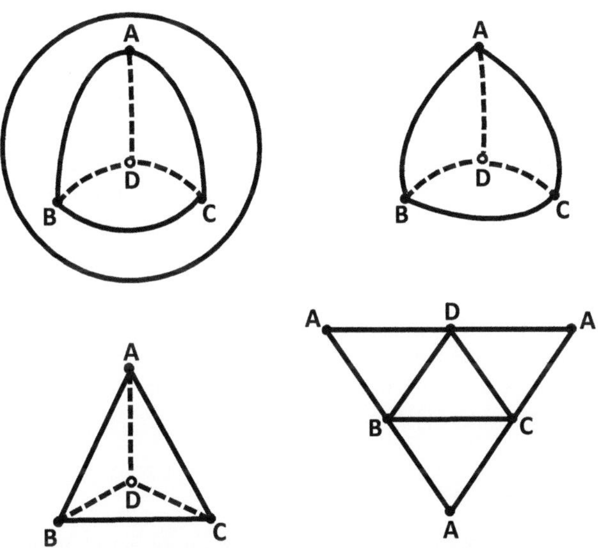

Fig. 7. A triangulation of a two-dimensional sphere.

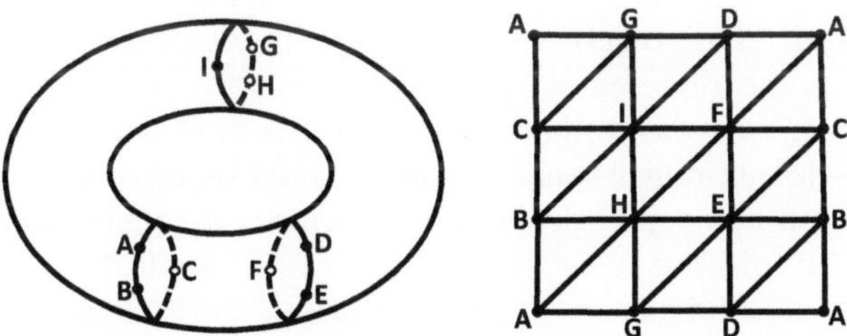

Fig. 8. A triangulation of a two-dimensional torus.

Examples of triangulations are given in Fig. 7 and Fig. 8. The former shows a triangulation of a 2-sphere, and the latter the trinagulation of a 2-dimensional torus. A triangulation of a 2-sphere is presented on the sphere itself, extracted out of a sphere and with straight edges, but also as a plane diagram in which the points and segments with the same names are identified. In the case of a 2-dimensional torus, a triangulation is more complicated, so that only the points of the triangulation are shown in the figure, and not the full triangulation of the torus. It is given as a plane diagram instead, in which the points and segments with the same names are identified.

The relationship between the study of topological spaces and the study of their invariants can be nicely pointed out by comparing zoology and paleontology. In paleontology the study of dinosaurs relies mostly on the animal skeletons, which can be compared to the study of simplicial complices and the associated algebraic objects. On the other hand, zoology can study animals as a whole, but also consider only the skeletons, which can be compared to the study of a topological space as it is given, but also the algebraic invariants can be used. However, it is quite often the case in topology that the topological spaces are so complicated that we are more in the skin of a paleontologist than a zoologist. Or, in analogy with the music example provided in Sect. 1 the software structure is like a music sheet structure where we explore its properties and try to understand its consequences on sonority in music waveforms during its reproduction. Again, the relation among various structural differences in sheet music and its audio representations are so complicated and we need advanced methods for their better understanding, [12].

3.4   Topological Analysis of Software

The study of software in terms of a graph is an old and widespread idea. It turns out to be very useful in all stages of the software life-cycle from the design, testing, and verification to maintenance.

Any (non-trivial) software system consists of components. Depending on the context and the purpose of the study, the components may be modules, software units, classes, objects, functions, and so on. The components may be viewed as vertices of a graph and the connections between components may be viewed as edges of a graph. These together form a graph representing a software system. However, the graph representation of software captures only one-dimensional relations between software components. To get a finer description of the software structure higher-dimensional relations should be considered. Here, we will provide details on how topological space in terms of its higher-dimensional relations in complex software systems structures can be applied.

Hence, one step further into the structure of the software system and its components is to find groups of components tied together. Such groups may be viewed as simplices, just like two connected software components were viewed as edges (1-dimensional simplices). This gives the simplicial complex of a software system. It can be viewed as a higher-dimensional version of a graph. The motifs, discussed in the previous sections can also be viewed as such higher-dimensional objects in the software structure.

The simplicial complex of a software system is a discretization of some topological space. The topological invariants of that space may be viewed as topological invariants of the software. These invariants can be computed directly from the simplicial complex of software. They may be viewed as approximations of the software structure. The key point of this approach to software system structure is that one should approximate the structure by topological invariants obtained from the simplicial complex of the software system. The topological approach ignores the scale of the system and concentrates only on its shape. The topological analysis can detect hidden similarities and differences in software system structures.

In the analogy to the zoology and paleontology elaborated above, the topological study of software is entirely within the paleontology side. The software is like a dinosaur skeleton waiting for us to explore and extrapolate and try to figure out the properties of the whole animal. We will explain the first steps in that endeavour in the following sections.

4   Topological Algorithms

This section is devoted to the description of the topological algorithm which will be applied to the software structure. It is very well known algorithm and can be viewed as the "hello world" of topological analysis of software. The reference for this material could be any textbook devoted to computational topology, such as [10].

The algorithm presented here is the most basic algorithm of computational topology, and meant as an invitation to dive into the field of topological data analysis. It is the basic step in the study of computational algebraic topology, in particular, the homology, cohomology, persistant homology, among other.

The section should be read in combination with Sect. 5, because the latter contains the running example. The notions and operations described here are performed and explained in detail on the running example in Sect. 5.

The main concepts introduced here are the basic notions of homological algebra. The ultimate goal is to introduce the Betti numbers, i.e., the ranks of homology groups, associated to a simplicial complex. For simplicity of exposition, the underlying field is the field $\mathbb{Z}_2 = \{0, 1\}$ of two elements with addition and multiplication modulo two. The notions required and defined in this section are summarized in Table 1 for convenience of the reader.

The importance of homology groups, and the associated Betti numbers, lies in the fact that they encode certain topological information about the space, or in our case the software graph. These are mostly related to connectedness properties of the topological space whose "skeleton" is the software graph in question. Although the homology groups over $\mathbb{Z}_2$ considered in these lectures capture only limited topological information, they provide a nice introduction to the subject of topological data analysis. In particular, they are computationally accessible and the students can make their one code for computing the Betti numbers.

Table 1. Summary of basic notions

Notion	Symbol	Definition
Simplicial complex	Calligraphic letters $\mathcal{K}$	A family of non-empty sets which contains all non-empty subsets of its members
Simplex	σ, τ,...	Members of a simplicial complex
p-simplex		Simplex of dimension p, i.e., containing $p+1$ elements
p-chain	c, d,...	Formal sums of simplices of the same dimension p
Boundary operator	∂	The linear operator defined on chains as the chain of lower dimension given as the formal sum of the boundary faces
p-boundary operator	∂_p	The boundary operator acting on p-chains
Cycles	z, x, y,...	Chains with zero boundary
Boundaries	b, a,...	Chains obtained as boundaries of chains of higher dimension
Rank of p-cycles	z_p	The basis two logarithm of the number of p-cycles
Rank of p-boundaries	b_p	The basis two logarithm of the number of p-boundaries
Betti number	β_p	The non-negative integer obtained as $\beta_p = z_p - b_p$

4.1   Simplicial Complex

The definition of a simplicial complex is essentially very simple. The simplicial complex is just a bunch of sets, but whenever some set is in the bunch, then all its non-empty subsets are in the same bunch. It is as simple as that.

Formally, a finite family $\mathcal{K}$ of non-empty finite sets is called a simplicial complex if every non-empty subset τ of any set σ in the family $\mathcal{K}$ is also a member of the family $\mathcal{K}$ (called a face of σ), i.e.,

$$\text{if } \sigma \in \mathcal{K} \text{ and } \emptyset \neq \tau \subseteq \sigma, \text{ then } \tau \in \mathcal{K}.$$

Note that here Greek letters denote sets. This is usual in the study of simplicial complices.

The sets in a simplicial complex are called simplices. The elements of sets in a simplicial complex are usually called points. In other words, every simplex consists of points. The number of points in a simplex determines its dimension, sometimes also called the degree of a simplex. More precisely, a simplex with $p+1$ points is of dimension p and usually referred to as a p-simplex. Thus, a p-simplex is a set in $\mathcal{K}$ containing exactly $p+1$ points, so that

0-simplices are just the points,

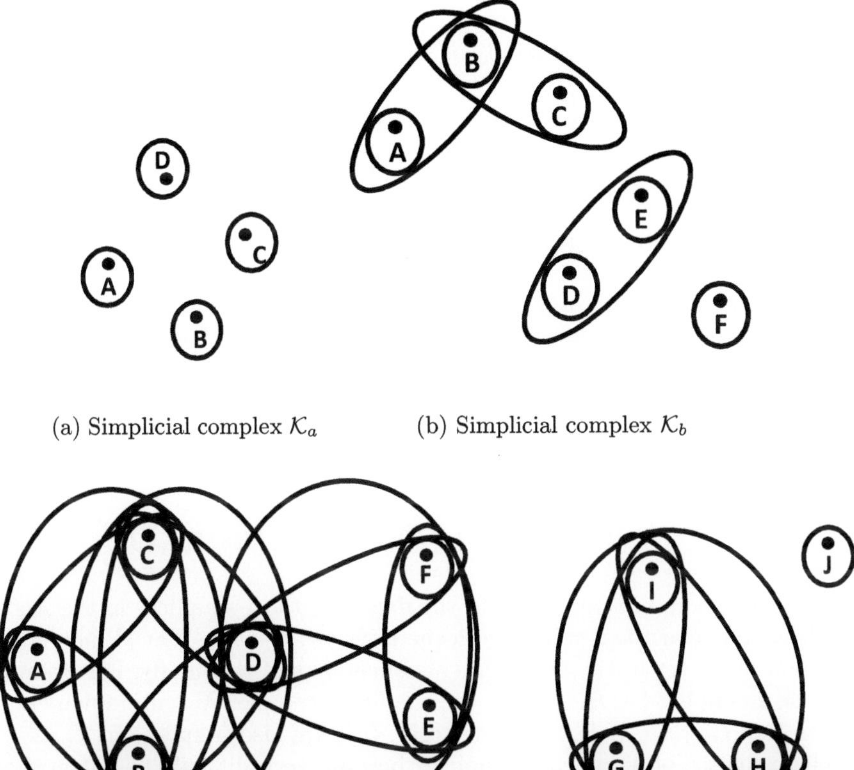

(a) Simplicial complex $\mathcal{K}_a$ (b) Simplicial complex $\mathcal{K}_b$

(c) Simplicial complex $\mathcal{K}_c$

Fig. 9. Examples of simplicial complices viewed as families of sets.

1-simplices contain two points (sometimes called segments),
2-simplices contain three points (sometimes called triangles),
...
p-simplices contain $p+1$ points.

Examples of simplicial complices, viewed as families of sets, are given in Fig. 9.

The first simplicial complex $\mathcal{K}_a$, given in Fig. 9a, consists of four sets with one element in each of these sets, that is, $\mathcal{K}_a$ is the family of sets

$$\mathcal{K}_a = \Big\{\{A\}, \{B\}, \{C\}, \{D\}\Big\}.$$

Thus, it contains four 0-simplices (points), and no higher-dimensional simplices.

The second simplicial complex $\mathcal{K}_b$, given in Fig. 9b, contains 1-simplices. As shown in the figure, it is the family of sets

$$\mathcal{K}_b = \Big\{\{A,B\},\{B,C\},\{D,E\},$$
$$\{A\},\{B\},\{C\},\{D\},\{E\},\{F\}\Big\},$$

where the 1-simplices (segments) are listed in the first row and the 0-simplices (points) in the second row.

The third simplicial complex $\mathcal{K}_c$, given in Fig. 9c, is more complicated. It contains the 2-simplices. As shown in the figure, $\mathcal{K}_c$ is the family of sets

$$\mathcal{K}_c = \Big\{\{A,B,C\},\{B,C,D\},\{D,E,F\},\{G,H,I\},$$
$$\{A,B\},\{A,C\},\{B,C\},\{B,C\},\{C,D\},\{D,E\},$$
$$\{D,F\},\{E,F\},\{G,H\},\{G,I\},\{H,I\},$$
$$\{A\},\{B\},\{C\},\{D\},\{E\},\{F\},\{G\},\{H\},\{I\},\{J\}\Big\}$$

where is the 2-simplices are listed in the first row, the 1-simplices in the second and third row, and the 0-simplices in the last row. Observe that all non-empty subsets of every set in the family are also members of the family. For example, since the 2-simplex $\{A,B,C\}$ is in the simplicial complex $\mathcal{K}_c$, all its non-empty subsets $\{A,B\}$, $\{A,C\}$, $\{B,C\}$, $\{A\}$, $\{B\}$, $\{C\}$ are also simplices in $\mathcal{K}_c$.

These figures of simplicial complices become very difficult to follow if the simplical complex contains higher-dimensional simplices. Already Fig. 9c is quite messy. Therefore, we will develop data structures for handling the simplicial complices. These are introduced in Sect. 5. The simplicial complex and its simplices in the running example of Sect. 5 are constructed inductively in Sect. 5.4, starting with the lowest dimensions zero and one in Sect. 5.3.

4.2   Geometric Viewpoint

Simplicial complices arise from geometry (or more precisely topology), and may be interpreted as the family of polyhedra in a sufficiently high-dimensional space. That is where the terminology using points, faces, boundaries etc., comes from. The geometric viewpoint is a way of visualizing the set-based definition of simplicial complex, introduced in Sect. 4.1, but not a suitable form of representation for computational purposes.

The simplicial complices of Fig. 9 are represented from the geometric viewpoint in Fig. 10. The sets of two points are represented by real geometric segments, the sets of three points by triangles, and those of four points by tetrahedra. The problem of representing the higher-dimensional simplices remains, as it requires more than three dimensions.

Observe how the simplicial complex $\mathcal{K}_a$, from the geometric point of view, is represented in Fig. 10a by four points A, B, C, D in space. The simplicial

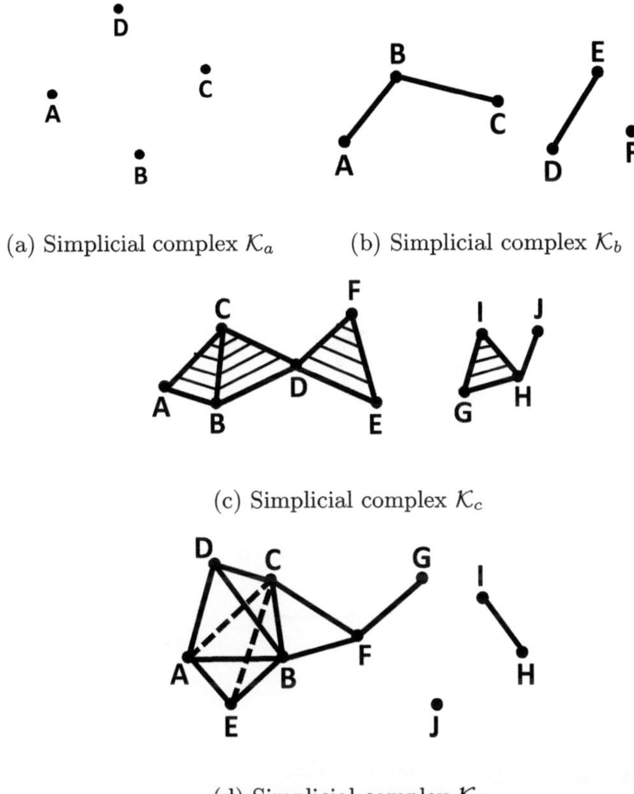

Fig. 10. Examples of simplicial complices from the geometric viewpoint.

complex $\mathcal{K}_b$ is presented in Fig. 10b. Its 1-simplices are represented by line segments $\overline{AB}$, $\overline{BC}$, $\overline{DE}$, while 0-simplices are points. The simplicial complex $\mathcal{K}_c$ is presented in Fig. 10c. Its 2-simplices are represented by the triangles $\triangle ABC$, $\triangle BCD$, $\triangle DEF$, $\triangle GHI$, its 1-simplices are the segments in the figure, and the 0-simplices the points. The figures of simplicial complices from the geometric point of view seem simpler than the figures of the same simplicial complices as families of sets above. However, the limitation of representing the higher-dimensional simplices remain.

The last simplicial complex $\mathcal{K}_d$, given in Fig. 10d, contains 3-simplices. These are the tetrahedra $ABCD$ and $ABCE$. The 2-simplices are triangles $\triangle ABC$, $\triangle ABD$, $\triangle ABE$, $\triangle ACD$, $\triangle ACE$, $\triangle BCD$, $\triangle BCE$. Observe that the triangle $\triangle BCF$ is not a 2-simplex as it is not shaded, although all of its sides are 1-simplices. The 1-simplices are all segments in the figure, and 0-simplices are all points in the figure.

As another example, consider a tetrahedron $ABCD$ given in Fig. 11. It represents a 3-simplex. In the abstract approach taken above, this 3-simplex is just

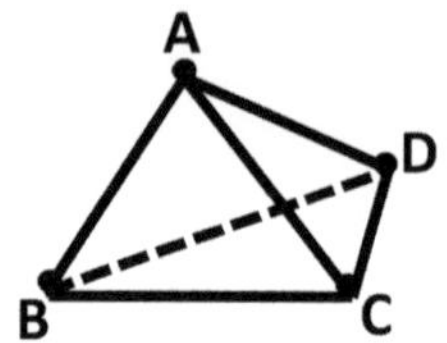

Fig. 11. The tetrahedron $ABCD$ representing a simplicial complex.

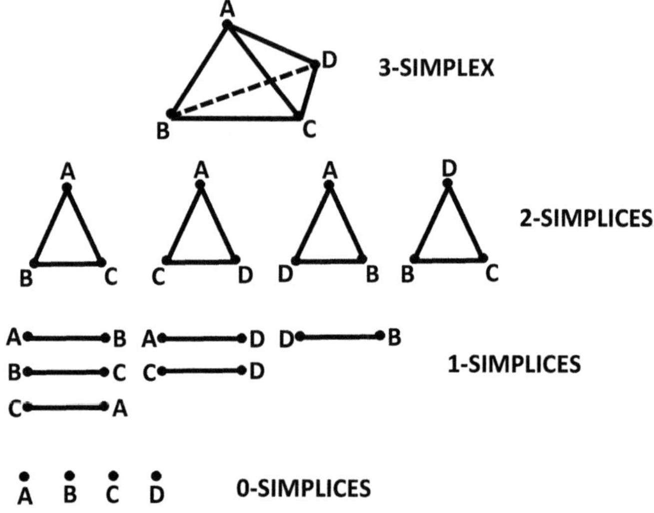

Fig. 12. The simplices of the simplicial complex represented by the tetrahedron $ABCD$.

the set

$$\sigma = \{A, B, C, D\}$$

of vertices of the tetrahedron. Notice that the geometrical dimension of a tetrahedron coincides with the simplicial dimension. That is the reason why simplices with $p+1$ points are considered to be p-dimensional.

The faces of the tetrahedron are the four triangles $\triangle ABC$, $\triangle ABD$, $\triangle BCD$ and $\triangle CAD$, as shown in Fig. 12. These are the 2-simplices that are subsimplices of the tertrahedron. In the abstract setting, these are the sets

$$\{A, B, C\}, \{A, B, D\}, \{B, C, D\}, \{C, A, D\},$$

of the vertices of the triangles. Observe that these are all subsets with three elements of the 3-simplex $\sigma = \{A, B, C, D\}$. In geometric language, these faces form the boundary of the tetrahedron. The notion of boundary will play a prominent role in the abstract simplicial complices as well.

Going one step further, the six edges $\overline{AB}$, $\overline{BC}$, $\overline{CA}$, $\overline{AD}$, $\overline{BD}$, $\overline{CD}$ of the tetrahedron, shown in Fig. 12, represent 1-simplices. These edges are at the same time the sides of the triangle faces of the tetrahedron. In the abstract setting, the 1-simplices are just the sets of endpoint of these edges, i.e.,

$$\{A, B\}, \{B, C\}, \{C, A\}, \{A, D\}, \{B, D\}, \{C, D\}.$$

Observe again that these are all subsets with two elements of the 3-simplex $\sigma = \{A, B, C, D\}$, but they are also subsets of some of the 2-simplices arising from triangle faces of the tetrahedron.

Finally, the four vertices A, B, C, D of the tetrahedron represent 0-simplices. They are endpoints of the edges and also vertices of the triangle faces of the tetrahedron. In the abstract setting, they form the sets

$$\{A\}, \{B\}, \{C\}, \{D\}.$$

These are again all subsets with one element of the 3-simplex $\sigma = \{A, B, C, D\}$, but they are also subsets of some of the 2-simplices represented by edges and 3-simplices represented by triangle faces of the tetrahedron.

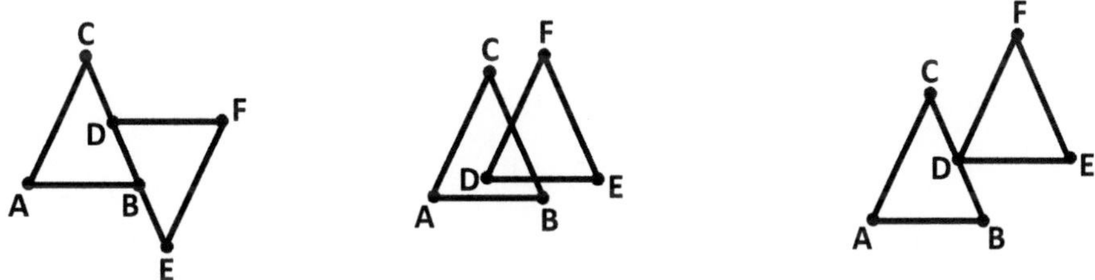

Fig. 13. Non-examples of simplicial complices.

Although one may think of a simplex as a polyhedron, and its subsimplices as its faces of all dimensions (vertices, edges, faces, higher-dimensional faces), it is easier, and computationally more convenient, to work with an abstract data structure such as our definition of the simplicial complex. In that way, we avoid geometric concerns. For example, there could be several intersecting polyhedra in a simplicial complex, and they must intersect only in their faces. More precisely, the intersection of any two such polyhedra must be a face of both. But in geometry, polyhedra may intersect in different geometric objects. See, for example, the intersections of two triangles in Fig. 13. Only if the polyhedra intersect in a face or an edge or a vertex, they represent a simplicial complex. This is easier to handle by considering a bunch of sets, than looking at the geometric picture and properties of polyhedra.

4.3   Simplicial Complex of a (software) Graph

There are several ways to assign a simplicial complex to a graph. We explain here the simplest way as an example, although this is too naive for serious applications. Our goal here is to explain a topological algorithm, so this simple approach will do. Be aware that our approach could result in computationally highly demanding algorithm in the case of an arbitrary graph. However, the nature of software graphs is such that the computation will be feasible. As already mentioned in Sect. 4.1, the construction is illustrated in the running example of Sect. 5, more precisely, in Sect. 5.3 and Sect. 5.4.

Let $G = (V, E)$ be a graph with the set of vertices V and the set of edges E. We consider only simple graphs, i.e., G is undirected, unweighted, has no loops and no multiple edges. Recall that a loop is an edge with equal endpoints, and a multiple edge refers to the existence of several edges with the same endpoints. An example is given in Fig. 17 and explained in Sect. 5.

A p-simplex in the graph G is defined as any subgraph containing $p + 1$ vertices such that each pair of vertices is connected by an edge. Recall that such subgraph is called a complete graph with $p + 1$ vertices. In applications, such subgraphs are often referred to as cliques in a graph.

More precisely, in our approach, the simplices of the simplicial complex associated to a given graph are the following:

0-simplices are just vertices (complete subgraphs with 1 vertex),
1-simplices are just edges (complete subgraphs with 2 vertices),
2-simplices are triangles (complete subgraphs with 3 vertices),
3-simplices are tetrahedra (complete subgraphs with 4 vertices),
...
p-simplices are complete subgraphs with $p + 1$ vertices.

Observe that all subgraphs of a complete graph are also complete. Therefore, we really obtained a simplicial complex, because the faces of a simplex are indeed simplices.

4.4   Chains

We proceed with an arbitrary simplicial complex $\mathcal{K}$. In applications, it will be the simplicial complex assigned to a software graph. Examples of all the notions introduced here are given in the running example of Sect. 5. However, since the notions in the running example are all expressed in an appropriate basis given by simplices, the chains do not appear explicitly in Sect. 5. They are hidden in the linear algebra formalism.

The motivation for introducing chains and their addition is to have a linear algebra formalism which allows the consideration of several simplices of the same dimension as a single object. This will allow the study of higher-dimensional topological structures, and in particular the Betti numbers, using linear algebra.

Let S_p denote the family of all p-simplices in the simplicial complex $\mathcal{K}$. Denote by n_p the number of p-simplices, i.e., $n_p = |S_p|$ is the cardinality of S_p. Write

$$S_p = \{\sigma_1, \sigma_2, \ldots, \sigma_{n_p}\}$$

for the p-simplices in $\mathcal{K}$.

A p-chain in $\mathcal{K}$ is any subfamily of S_p. In other words, any choice of any number of p-simplices produces a p-chain. Even the empty choice, in which none of the p-simplices is chosen, is allowed.

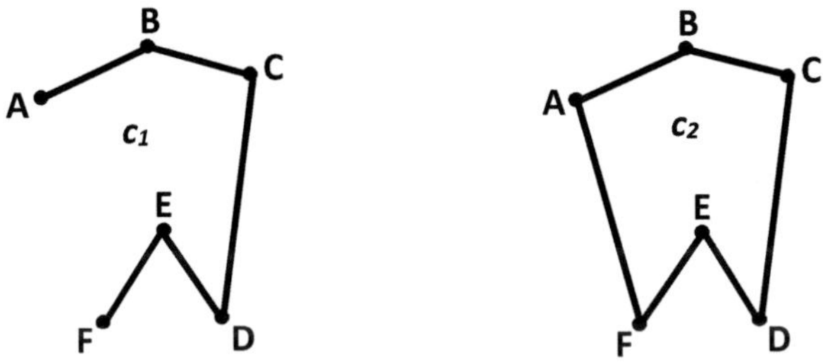

Fig. 14. Examples of 1-chains c_1 and c_2.

Examples of 1-chains are given in Fig. 14. The 1-chain c_1 on the left-hand side in the figure consists of 1-simplices $\overline{AB}$, $\overline{BC}$, $\overline{CD}$, $\overline{DE}$ and $\overline{EF}$, while the 1-chain c_2 on the right-hand side in the figure consists of the same 1-simplices with $\overline{FA}$ as an extra 1-simplex.

There is a convenient way to write p-chains using formal sums of p-simplices with coefficients in $\mathbb{Z}_2 = \{0, 1\}$. The coefficients encode our choice of p-simplices in a p-chain. If the coefficient of a p-simplex is zero, then it is not chosen in a p-chain, while if the coefficient is one, then it is chosen to be in a p-chain.

More precisely, a p-chain c in $\mathcal{K}$ can be written as a formal sum

$$c = \sum_{i=1}^{n_p} a_i \sigma_i,$$

where $a_i \in \mathbb{Z}_2 = \{0, 1\}$. If the coefficient $a_i = 0$, then σ_i is not in the p-chain c, while if $a_i = 1$, then σ_i is in the p-chain c. The 1-chains in Fig. 14 can be written as

$$c_1 = \overline{AB} + \overline{BC} + \overline{CD} + \overline{DE} + \overline{EF}$$
$$c_2 = \overline{AB} + \overline{BC} + \overline{CD} + \overline{DE} + \overline{EF} + \overline{FA}$$

in the formal sum notation.

Another point of view, perhaps more familiar, is to interpret a p-chain c in $\mathcal{K}$ as a sequence of coefficients

$$c = (a_1, a_2, \ldots, a_{n_p}),$$

where $a_i \in \mathbb{Z}_2 = \{0, 1\}$, so that a p-chain is any n_p-bit word. The coefficient a_i again indicates whether the p-simplex σ_i is in the chain c or not.

Let C_p denote the set of all p-chains in the simplicial complex $\mathcal{K}$. The addition of p-chains is defined, either as the mod 2 addition of formal sums, or as the exclusive or operation on n_p-bit words. Let

$$c = \sum_{i=1}^{n_p} a_i \sigma_i = (a_1, a_2, \ldots, a_{n_p}),$$

$$c' = \sum_{i=1}^{n_p} b_i \sigma_i = (b_1, b_2, \ldots, b_{n_p})$$

be two p-chains in C_p. Then, their sum is given as

$$c + c' = \sum_{i=1}^{n_p} (a_i +_2 b_i)\sigma_i$$
$$= (a_1 \oplus b_1, a_2 \oplus b_2, \ldots, a_{n_p} \oplus b_{n_p}),$$

where $+_2$ stands for the mod 2 addition and $\oplus$ stands for the exclusive or operation. The sum of the two 1-chains in Fig. 14 is

$$c_1 + c_2 = \overline{FA},$$

because the other 1-simplices appear in both 1-chains, so that they cancel in mod 2 addition.[2]

Observe that the cardinality $|C_p| = 2^{n_p}$, because it is the number of n_p-bit words. The exponent n_p is called the rank of C_p. The reason for exclusive-or, i.e., mod 2, addition of chains is used here, because we restrict the scope of this lecture notes to the simplest case of the field $\mathbb{Z}_2$ of two elements. The analogous theory can be developed over other fields and rings, such as the ring of integers $\mathbb{Z}$. However, in that case the boundary operator, introduced in Sect. 4.5 below, must be defined in a different way, using certain alternating sums of boundary faces. This is necessary in order to satisfy the fundamental property of boundary operators, explained in Sect. 4.7.

4.5 Boundary Operator

The boundary operator is the glue that fits together all the simplices of different dimensions in a simplicial complex. The p-boundary operator ∂_p assigns to each p-simplex its boundary, i.e., the family of its $(p-1)$-simplex faces. But such family is a $(p-1)$-chain. Given a p-simplex $\sigma_i \in S_p$, the action of the p-boundary operator can be expressed as

$$\partial_p \sigma_i = \sum_{\substack{\tau \in S_{p-1} \\ \tau \subseteq \sigma_i}} \tau \in C_{p-1},$$

which is a sum of $(p-1)$-simplices τ, and thus a $(p-1)$-chain in C_{p-1}. In the running example of Sect. 5, the boundary operators are determined in their matrix representation as part of the construction of the simplicial complex in Sect. 5.3 and Sect. 5.4.

---

[2] We mention, for the record, that the set C_p of all p-chains in the simplicial complex $\mathcal{K}$ forms an Abelian group with addition. However, this fact and the algebraic notion of groups will not be necessary for understanding the rest of the paper.

For example, the 3-boundary ∂_3 of the 3-simplex given by the tetrahedron $ABCD$ in Fig. 11 is given as

$$\partial_3(ABCD) = \triangle ABC + \triangle ACD + \triangle ADB + \triangle BCD,$$

which is the 2-chain given as the sum of 2-simplex faces in Fig. 12 of the tetrahedron $ABCD$.

More generally, given a p-chain

$$c = \sum_{i=1}^{n_p} a_i \sigma_i \in C_p,$$

the p-boundary operator ∂_p acts as

$$\partial_p c = \partial_p \left(\sum_{i=1}^{n_p} a_i \sigma_i \right)$$
$$= \sum_{i=1}^{n_p} a_i \partial_p \sigma_i \in C_{p-1},$$

where

$$\partial_p \sigma_i = \sum_{\substack{\tau \in S_{p-1} \\ \tau \subseteq \sigma_i}} \tau \in C_{p-1},$$

as above. Hence, ∂_p is a map from C_p to C_{p-1}. It is now clear from the definition how the boundary operator glues together information on simplices and chains of different dimensions.

For example, the 1-boundary operator ∂_1 applied to the 1-chains c_1 and c_2 in Fig. 14 is computed as follows

$$\partial_1 c_1 = \partial_1 \left(\overline{AB} + \overline{BC} + \overline{CD} + \overline{DE} + \overline{EF} \right)$$
$$= \partial_1 \overline{AB} + \partial_1 \overline{BC} + \partial_1 \overline{CD} + \partial_1 \overline{DE} + \partial_1 \overline{EF}$$
$$= (A + B) + (B + C) + (C + D) + (D + E) + (E + F)$$
$$= A + F$$

which is the 0-chain given as the sum of 0-simplices A and F, while

$$\partial_1 c_2 = \partial_1 \left(\overline{AB} + \overline{BC} + \overline{CD} + \overline{DE} + \overline{EF} + \overline{FA} \right)$$
$$= \partial_1 \overline{AB} + \partial_1 \overline{BC} + \partial_1 \overline{CD} + \partial_1 \overline{DE} + \partial_1 \overline{EF} + \partial_1 \overline{FA}$$
$$= (A + B) + (B + C) + (C + D) + (D + E) + (E + F) + (F + A)$$
$$= 0$$

which is the empty 0-simplex denoted by zero. The boundary operator computes the boundary of the simplices in a chain expressed as a sum over $\mathbb{Z}_2$, i.e., with mod 2 addition. The boundary operator gives 0 if the simplices form a cycle, as in the example of c_2.

At this point, it becomes clear why the formal sum notation for p-chains is much more convenient than the n_p bit words approach. The reason is that, in the formal sum notation, it is not necessary to maintain the order of p-simplices. Given a p-chain c in C_p, one may write

$$c = \sum_{\sigma \in S_p} a_\sigma \sigma,$$

where $a_\sigma \in \mathbb{Z}_2 = \{0, 1\}$ is the coefficient of the p-simplex σ. As above $a_\sigma = 1$ if σ belongs to c, and $a_\sigma = 0$ if σ does not belong to c. The order of σ is irrelevant.

This advantage of the formal sum notation is useful in the definition of the p-boundary operator. It is not necessary to specify which simplices τ are in the boundary when defining and using the p-boundary operator. However, as we will see below, for explicit calculation of topological invariants, it is convenient to have the order of p-simplices fixed.

4.6   Matrix of the Boundary Operator

The convenient way to view the boundary operator is as the linear operator on the vector spaces of chains. In this context, the matrix of the boundary operators as a linear operator can be introduced. That is the subject of this section. In the running example of Sect. 5, the matrix of the boundary operators are determined inductively in Sect. 5.4, starting with the lowest dimension in Sect. 5.3.

The p-chains in C_p may be viewed as linear combinations with coefficients in $\mathbb{Z}_2 = \{0, 1\}$ of simplices in S_p. Hence, they form a vector space over the field $\mathbb{Z}_2$ of two elements. The basis of C_p as a vector space over $\mathbb{Z}_2$ is the set S_p of all p-simplices. Thus, the dimension of C_p over $\mathbb{Z}_2$ is the number $n_p = |S_p|$ of p-simplices.

By the very definition of the p-boundary operator, it is a linear operator from C_p to C_{p-1} as vector spaces over $\mathbb{Z}_p$, because it respects the linear combinations

$$\partial_p \left(\sum_{i=1}^{n_p} a_i \sigma_i \right) = \sum_{i=1}^{n_p} a_i \partial_p \sigma_i,$$

for any $a_i \in \mathbb{Z}_2$.

As any linear operator, the p-boundary operator ∂_p can be represented by a matrix. The matrix of the p-boundary operator ∂_p has n_{p-1} rows, because it is the dimension of C_{p-1}, and n_p columns, because it is the dimension of C_p. It contains only 0's and 1's, because the vector spaces are over $\mathbb{Z}_2 = \{0, 1\}$. The rows represent $(p-1)$-simplices in S_{p-1}, and the columns represent p-simplices in S_p, in a fixed order.

The j-th column represents the p-simplex $\sigma_j \in S_p$. It contains 1 in the rows representing its $(p-1)$-simplex faces in S_{p-1}, and 0 in the rows which represent $(p-1)$-simplices in S_{p-1} which are not its faces. Every p-simplex has exactly $p+1$ faces, so that each column contains precisely $p+1$ ones. More precisely, the element at the crossing of the i-th row and the j-th column equals 1 if the

i-th simplex in S_{p-1} is a $(p-1)$-simplex face of the j-th simplex in S_p, and it equals 0 otherwise. More precisely,

$$
\partial_p = \begin{bmatrix} d_{1,1} & \cdots & & & d_{1,n_p} \\ & \ddots & & & \\ \vdots & & d_{i,j} & & \vdots \\ & & & \ddots & \\ d_{n_{p-1},1} & \cdots & & & d_{n_{p-1},n_p} \end{bmatrix},
$$

where

$$
d_{i,j} = \begin{cases} 1, \text{ if the } i\text{-th } (p-1)\text{-simplex } \tau_i \in S_{p-1} \text{ is a face} \\ \quad \text{ of the } j\text{-th } p\text{-simplex } \sigma_j \in S_p, \\ 0, \text{ otherwise.} \end{cases}
$$

4.7   Boundary Operators as Differentials

In this section, we explain the most fundamental property of boundary operators. It is the fact that the boundary of a boundary is zero.

The boundary operators in different dimensions fit into this picture

$$
\cdots \xrightarrow{\partial_{p+2}} C_{p+1} \xrightarrow{\partial_{p+1}} C_p \xrightarrow{\partial_p} C_{p-1} \xrightarrow{\partial_{p-1}} \cdots \xrightarrow{\partial_2} C_1 \xrightarrow{\partial_1} C_0.
$$

The fundamental property of the boundary operators is that the composition of consecutive boundary operators is the zero operator, i.e.,

$$
\partial_p \circ \partial_{p+1} = 0
$$

for all p. In other words, the boundary of the boundary of any chain is 0.

This fundamental property of the boundary operators is the defining property of differentials of a simplicial complex. Hence, by the definition of differentials, this property of the boundary operators means that they are differentials of the simplicial complex.

The proof of the fundamental property of boundary operators follows from the fact that every $(p-1)$-simplex face of a $(p+1)$-simplex is at the same time a face of exactly two of its p-simplex faces, and since $2 = 0 \bmod 2$, the resulting chain is the zero chain. This can be easily checked in the abstract view of the simplicial complex. Given a $(p+1)$-simplex as a set of $p+1$ points, its $(p-1)$-simplex face is a subset of $p-1$ points. In other words, the $(p-1)$-simplex face is obtained by removing two points from the $(p+1)$-simplex. But then, adding each of the removed points to the $(p-1)$-simplex face produces exactly two p-simplex faces of the $(p+1)$-simplex which share the $(p-1)$-simplex face. Hence, the number of appearances of any $(p-1)$-simplex face in the boundary of the boundary of the $(p+1)$-simplex is even, which equals 0 in mod 2 arithmetic of $\mathbb{Z}_2$.

4.8   Cycles and Boundaries

The cycles and boundaries in a simplicial complex are special types of chains. They are defined in terms of the boundary operators. In the running example of Sect. 5, the cycles and boundaries are not determined explicitly. Instead, only their numbers are determined in Sect. 5.5 from the matrices of boundary operators using the linear algebra notions of rank and defect of a linear operator.

A p-cycle z is a p-chain with zero boundary, i.e.,

$$\partial_p z = 0.$$

The set of all p-cycles is denoted Z_p. It is a subset of the set C_p of all p-chains.[3]

An example of 1-cycle is the 1-chain c_2 in Fig. 14, because we already calculated in Sect. 4.5 its 1-boundary and obtained $\partial_1 c_2 = 0$ is the empty 0-chain. The other 1-chain c_1 in Fig. 14 is not an 1-cycle, because its 1-boundary $\partial_1 c_1 = A + F$ is not the empty 0-chain. The figure also explains where the name of the cycles comes from. The 1-cycle c_2 is represented by a closed loop, while the 1-chain c_1, which is not an 1-cycle, is not.

The number $|Z_p|$ of p-chains is a power of two

$$|Z_p| = 2^{z_p},$$

where the exponent z_p is called the rank of Z_p. In the special case of $p = 0$, there is no boundary operator ∂_0, but we make the convention that all 0-chains are 0-cycles, i.e., $Z_0 = C_0$, and the rank of Z_0 is $z_0 = n_0$.

A p-boundary b is a p-chain which is a boundary of some $(p+1)$-chain, i.e.,

$$b = \partial_{p+1} c$$

for some $c \in C_{p+1}$. The set of all p-boundaries is denoted B_p. It is a subset of the set C_p of all p-chains.[4]

The number $|B_p|$ of p-boundaries is also a power of two

$$|B_p| = 2^{b_p},$$

where the exponent b_p is called the rank of B_p.

If t is the top dimension of a simplex in a simplicial complex, i.e., there are no p-simplices of dimension $p > t$, then there are no t-boundaries, because there is no $(t+1)$-simplex in the simplicial complex. Hence, B_t is trivial and the rank $b_t = 0$ for the top dimension t.

The fundamental property of the boundary operators implies that every p-boundary is a p-cycle, because the boundary of the boundary is always zero. More precisely, if $b = \partial_{p+1} c$ is any p-boundary, then

$$\partial_p b = \partial_p(\partial_{p+1} c) = (\partial_p \circ \partial_{p+1}) c = 0,$$

---

[3] For the record, the set Z_p of p-cycles forms an Abelian group under addition of chains.

[4] For the record, the set B_p of p-boundaries forms an Abelian group under addition of chains.

which means that b is a p-cycle. Thus,

$$B_p \subseteq Z_p$$

for all p.

4.9 Homology

The p-th homology group H_p of a simplicial complex counts the p-cycles which are not obtained as p-boundaries, i.e., the "true" cycles among p-chains.[5] For our purposes, it is sufficient to know that the number $|H_p|$ of elements in the p-th homology group is also a power of two, and that its rank is the difference between the ranks of of Z_p and B_p. In the running example of Sect. 5, the Betti numbers are determined in the final step of Sect. 5.6. If we write

$$|H_p| = 2^{\beta_p},$$

then the rank β_p of H_p is called the p-th Betti number. It is obtained by the formula

$$\begin{aligned}
\beta_0 &= z_0 - b_0 = n_0 - b_0, \\
\beta_p &= z_p - b_p, &\quad 1 \le p \le t - 1, \\
\beta_t &= z_t - b_t = z_t, \\
\beta_q &= 0, &\quad q > t,
\end{aligned}$$

where t is the top dimension of a simplex in the simplicial complex. All these values can be read off from the matrix of ∂_p, as we shall see in the examples below in Sect. 5. However, in order to provide intuition for the chains representing elements of the homology groups over $\mathbb{Z}_2$, we begin here with a simple illustrative example.

Let $\mathcal{K}$ be the simplicial complex given in Fig. 15. The shaded triangle in the figure is a 2-simplex in $\mathcal{K}$, while the unshaded one is not in $\mathcal{K}$. Hence, the simplicial complex $\mathcal{K}$ consists of the following simplices

$$\begin{aligned}
S_2 &= \{\triangle ABC\}, \\
S_1 &= \{\overline{AB}, \overline{AC}, \overline{BC}, \overline{BD}, \overline{CD}\}, \\
S_0 &= \{A, B, C, D\}.
\end{aligned}$$

The chain representing an element in homology group is best exhibited in dimension one. Hence, we first determine 1-cycles.

The 1-boundary operator acts on an arbitrary 1-chain as

$$\begin{aligned}
\partial_1 (x&\overline{AB} + y\overline{AC} + z\overline{BC} + u\overline{BD} + v\overline{CD}) \\
&= x(A + B) + y(A + C) + z(B + C) + u(B + D) + v(C + D) \\
&= (x + y)A + (x + z + u)B + (y + z + v)C + (u + v)D,
\end{aligned}$$

---

[5] Formally, the p-th homology group is defined as the quotient group $H_p = Z_p/B_p$, which is again an Abelian group.

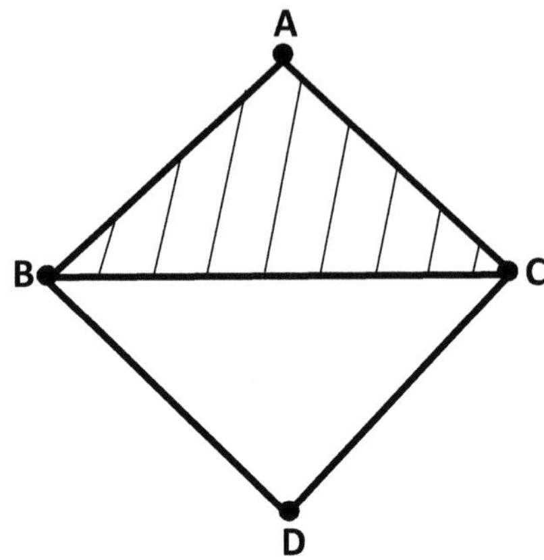

Fig. 15. Simplicial complex $\mathcal{K}$ in the example of Sect. 4.9.

where $x, y, z, u, v \in \mathbb{Z}_2$ are arbitrary coefficients. The condition for 1-cycles is that the obtained 0-chain is trivial, i.e., all the coefficients must be zero. Solving the system of equations over $\mathbb{Z}_2$ gives

$$y = x$$
$$v = u$$
$$z = x + u,$$

where $x, u \in \mathbb{Z}_2$ are arbitrary. Hence, there are four 1-cycles, and the rank of Z_1 is $z_1 = 2$. More precisely, the 1-cycles are all 1-chains of the form

$$x\overline{AB} + x\overline{AC} + (x + u)\overline{BC} + u\overline{BD} + u\overline{CD},$$

which can be rearranged as

$$x(\overline{AB} + \overline{AC} + \overline{BC}) + u(\overline{BC} + \overline{BD} + \overline{CD}),$$

where $x, u \in \mathbb{Z}_2$ are arbitrary.

The next task is to determine the 1-boundaries. The only 2-simplex is the triangle $\triangle ABC$, hence the 2-boundary operator acts on an arbitrary 2-chain as

$$\partial_2(k\triangle ABC) = k(\overline{AB} + \overline{AC} + \overline{BC}),$$

so that the only non-trivial 1-boundary is the 1-chain

$$\overline{AB} + \overline{AC} + \overline{BC},$$

and the rank of B_1 is $b_1 = 1$.

From the above, we can easily compute the Betti number in dimension one, which is

$$\beta_1 = z_1 - b_1 = 1.$$

However, in this example we would like to provide an exemplary 1-chain representing the non-trivial element of the homology group H_1. Comparing the obtained 1-boundary with the description of 1-cycles, we see that 1-cycle in the first bracket, the one with coefficient x, is actually an 1-boundary, and thus not a non-trivial element of H_1. This implies that the non-trivial element in H_1 is represented by the other bracket in the description of 1-cycles above. It is the 1-cycle

$$\overline{BC} + \overline{BD} + \overline{CD}.$$

In this way we determined an example of a cycle, which is not a boundary.

There is another subtlety arising from the quotient group definition of H_1. Namely, there could be more than one 1-cycle representing the same element in H_1. In our example, this happens with the 1-cycle

$$(\overline{AB} + \overline{AC} + \overline{BC}) + (\overline{BC} + \overline{BD} + \overline{CD}) = \overline{AB} + \overline{AC} + \overline{BD} + \overline{CD},$$

which is the mod 2 sum of the two 1-cycles considered above. This 1-cycle is not an 1-boundary by itself, so that it should also represent a non-trivial element of H_1. However, the mod 2 difference between the two representatives is precisely the 1-boundary $\overline{AB} + \overline{AC} + \overline{BC}$, so that, up to boundary, the two 1-cycles are equal, and therefore represent the same element in H_1.

4.10   Example 1: Homology of a Tetrahedron

We consider now a simple geometric example, which can be done directly by hand and can be observed in pictures. It is the example of a tetrahedron, which is already a 3-simplex, so that the simplicial complex of a tetrahedron consists of a single 3-simplex with all its subsimplices. This example is already described in Sect. 4.2 and depicted in Fig. 11 and Fig. 12. The sets S_p of p-simplices are already listed there, and their geometric presentation is discussed. For determining the homology of a tetrahedron and its Betti numbers, we only require the description of the boundary operators.

The 3-boundary operator ∂_3 acts on the set C_3 of 3-chains. Since there is only one 3-simplex $\{A, B, C, D\}$, i.e., the whole tetrahedron $ABCD$, all the 3-chains are of the form

$$c_3 = k \cdot ABCD,$$

where $ABCD = \{A, B, C, D\}$ denotes the tetrahedron as a 3-simplex and $k \in \mathbb{Z}_2$ is the coefficient. In other words, there are only two 3-chains: the empty chain and the chain $ABCD$. The action of the 3-boundary operator on the 3-chain c_3 is given by the formula

$$\partial_3 c_3 = k \cdot (\triangle BCD + \triangle ACD + \triangle ABD + \triangle ABC),$$

where the triangles in the brackets are 2-simplices that form the boundary of the tetrahedron ABCD.

From this formula we can read off the 3-cycles and 2-boundaries. There are no non-trivial 3-cycles, because $\partial_3 c_3 = 0$ only if the coefficient $k = 0$. Thus,

c_3 is a 3-cycle if and only if it is trivial, i.e., the empty 3-chain. Since 3 is the top dimension of the simplicial complex of the tetrahedron, it follows that the homology group

$$H_3 = Z_3 = \{0\}$$

is trivial, and the Betti number $\beta_3 = 0$.

The 2-boundaries are the 2-chains obtained as the image of the 3-boundary operator. Hence, B_2 consists of

$$B_2 = \Big\{ k \cdot (\triangle BCD + \triangle ACD + \triangle ABD + \triangle ABC), \quad \text{where } k \in \mathbb{Z}_2 \Big\}.$$

In other words, there are only two 2-boundaries: the empty 2-chain and the 2-chain

$$\triangle BCD + \triangle ACD + \triangle ABD + \triangle ABC.$$

Thus, the rank of B_2 is $b_2 = 1$.

We should consider next the 2-boundary operator ∂_2, which acts on 2-chains. Since there are four 2-simplices, i.e., four triangles, in the simplicial complex, any 2-chain is of the form

$$c_2 = a \cdot \triangle BCD + b \cdot \triangle ACD + c \cdot \triangle ABD + d \cdot \triangle ABC,$$

where the triangles represent 2-simplices and $a, b, c, d \in \mathbb{Z}_2$ are the coefficients. Note that the number of 2-chains is $2^{n_2} = 2^4 = 16$.

The action of the 2-boundary operator on the 2-chain c_2 is given by the formula

$$\begin{aligned}
\partial_2 c_2 =\ & a \cdot \left(\overline{BC} + \overline{CD} + \overline{DB} \right) + b \cdot \left(\overline{AC} + \overline{CD} + \overline{DA} \right) \\
& + c \cdot \left(\overline{AB} + \overline{BD} + \overline{DA} \right) + d \cdot \left(\overline{AB} + \overline{BC} + \overline{CA} \right) \\
=\ & (c + d) \cdot \overline{AB} + (b + d) \cdot \overline{AC} + (b + c) \cdot \overline{AD} \\
& + (a + d) \cdot \overline{BC} + (a + c) \cdot \overline{BD} + (a + b) \cdot \overline{CD}.
\end{aligned}$$

Here in the first line we compute the boundary of each triangle separately, and then in the second line sum up according to segments. i.e., 1-simplices.

We can obtain from this formula the 2-cycles and 1-boundaries. The 2-cycles are those 2-chains c_2 for which $\partial_2 c_2 = 0$. In mod 2 arithmetic, this happens if and only if all the coefficients are zero, i.e.,

$$a = b = c = d = e = f.$$

Therefore, the set of 2-cycles is

$$Z_2 = \Big\{ a \cdot (\triangle BCD + \triangle ACD + \triangle ABD + \triangle ABC), \quad \text{where } a \in \mathbb{Z}_2 \Big\}.$$

But this is equal to the set B_2 of 2-boundaries obtained above. Hence, the homology group

$$H_2 = \{0\}$$

is also trivial, and the Betti number $\beta_2 = 0$.

Finding all 1-boundaries by hand is a bit more involved. The trick is to determine first the 1-cycles, because every 1-boundary is at the same time an 1-cycle, by the fundamental property of differentials in a simplicial complex. Finding 1-boundaries among 1-cycles turns out to be easier.

Therefore, consider now the 1-boundary operator ∂_1 which acts on 1-chains. There are six 1-simplices in the simplicial complex, given by the edges of the tetrahedron. Thus, any 1-chain is of the form

$$c_1 = x \cdot \overline{AB} + y \cdot \overline{AC} + z \cdot \overline{AD} + u \cdot \overline{BC} + v \cdot \overline{BD} + w \cdot \overline{CD},$$

where the segments represent the 1-simplices and $x, y, z, u, v, w \in \mathbb{Z}_2$ are coefficients. The number of 1-chains is $2^{n_1} = 2^6 = 64$.

The action of the 1-boundary operator ∂_1 on c_1 is given by the formula

$$\begin{aligned}
\partial_1 c_1 =& x \cdot (A + B) + y \cdot (A + C) + z \cdot (A + D) \\
& + u \cdot (B + C) + v \cdot (B + D) + w \cdot (C + D) \\
=& (x + y + z) \cdot A + (x + u + v) \cdot B + (y + u + w) \cdot C + (z + v + w) \cdot D,
\end{aligned}$$

where in the first line we just determined the boundaries of the segments as the sum of their endpoints, and in the second line we summed up according to points, i.e., 0-simplices.

The 1-cycles are those 1-chains for which $\partial_1 c_1 = 0$, so that they are determined by the solutions of the system of linear equations

$$\begin{aligned}
x + y + z &= 0 \\
x + u + v &= 0 \\
y + u + w &= 0 \\
z + v + w &= 0
\end{aligned}$$

in $\mathbb{Z}_2$. Since, in mod 2 arithmetic, the last equation is the sum of the first three equations, it can be erased. The remaining three equation can be solved in terms of x, y and u as

$$\begin{aligned}
z &= x + y \\
v &= x + u \\
w &= y + u
\end{aligned}$$

where $x, y, u \in \mathbb{Z}_2$ are arbitrary. Hence, the set of 1-cycles consists of

$$Z_1 = \Big\{ x \cdot \left(\overline{AB} + \overline{AD} + \overline{BD}\right) + y \cdot \left(\overline{AC} + \overline{AD} + \overline{CD}\right) + u \cdot \left(\overline{BC} + \overline{BD} + \overline{CD}\right)$$

$$\text{where } x, y, u \in \mathbb{Z}_2 \Big\}.$$

We are now ready to find the 1-boundaries among the 1-cycles described above in the set Z_1. Observe that the three brackets in the expression for an

1-cycle are all boundaries of one of the triangle faces of the tetrahedron. More precisely,

$$\partial_2 \left(\triangle ABD\right) = \overline{AB} + \overline{AD} + \overline{BD}$$
$$\partial_2 \left(\triangle ACD\right) = \overline{AC} + \overline{AD} + \overline{CD}$$
$$\partial_2 \left(\triangle BCD\right) = \overline{BC} + \overline{BD} + \overline{CD}.$$

Therefore, for any $x, y, u \in \mathbb{Z}_2$, we have

$$\partial_2 \left(x \cdot \triangle ABD + y \cdot \triangle ACD + u \cdot \triangle BCD\right) = x \cdot \left(\overline{AB} + \overline{AD} + \overline{BD}\right)$$
$$+ y \cdot \left(\overline{AC} + \overline{AD} + \overline{CD}\right)$$
$$+ u \cdot \left(\overline{BC} + \overline{BD} + \overline{CD}\right),$$

which shows that every 1-cycle is at the same time an 1-boundary. In other words, $Z_1 = B_1$, so that the homology group

$$H_1 = \{0\}$$

is trivial, and the Betti number $\beta_1 = 0$.

It remains to determine homology in dimension zero. Since all 0-chains are 0-cycles, we have

$$Z_1 = C_1 = \left\{q \cdot A + r \cdot B + s \cdot C + t \cdot D \quad \text{where } q, r, s, t \in \mathbb{Z}_2\right\}.$$

The number of 0-cycles is $2^{n_0} = 2^4 = 16$. It is sufficient to determine which of these 0-cycles are 0-boundaries. Since the 1-boundary operator acts on the edges of the tetrahedron, and each of them has two end-points, it turns out that in the boundary of any 1-chain, there is an even number of points in total. This means that among elements in Z_1, the 1-boundaries are determined by the condition that

$$q + r + s + t = 0$$

in mod 2 arithmetic. This means that the number of 0-boundaries is $2^{4-1} = 8$, as the last coefficient is always determined by the other three according to the condition above. Therefore, the Betti number equals $\beta_0 = 4 - 3 = 1$.

In conclusion, we have determined the Betti numbers

$$\beta_0 = 1,$$
$$\beta_q = 0, \text{ for } q > 0,$$

for homology over $\mathbb{Z}_2$.

More generally, the homology over $\mathbb{Z}_2$ of any p-simplex, viewed as a simplicial complex with all of its subsimplices, exhibits a similar pattern of Betti numbers. The only non-zero Betti number is $\beta_0 = 1$. There is a geometric reason underlying this result. Any p-simplex, as for instance the tetrahedron, can be contracted into a point, without gluing or cutting. Hence, the topological invariants of a p-simplex (or any other convex set), are those of a single point. But for a single point, which is a 0-simplex, the top dimension is zero. And in dimension zero, the point represents a cycle, which is not a boundary. Thus, $\beta_0 = 1$, and $\beta_q = 0$ for all $q > 0$, for homology over $\mathbb{Z}_2$ of any p-simplex.

4.11   Example 2: Homology of a 2-Sphere

The homology of a 2-sphere is calculated using its triangulation. The result is independent of the choice of triangulation and we use the one already given in Fig. 7. This triangulation is a simplicial complex very close to the simplicial complex of the tetrahedron in the previous example. The only difference is that the 3-simplex given by the whole tetrahedron is removed from the simplicial complex.

Hence, all the calculations in the previous example apply here, except for the homology groups in the top dimension. In the case of 2-sphere the top dimension is two, because there are no 3-simplices in the simplicial complex. However, the 2-cycles remain the same as in the case of the tetrahedron. Thus, the number of 2-cycles is 2. But since there are no 3-simplices, there are no 2-boundaries, so that the homology group is

$$H_2 = Z_2,$$

and the Betti number is $\beta_2 = 1$. The other homology groups and the Betti numbers are exactly the same as in the example of the tetrahedron. Thus, the Betti numbers of the homology over $\mathbb{Z}_2$ of the 2-sphere are

$$\beta_0 = \beta_2 = 1,$$

$$\beta_p = 0 \text{ for all } p \neq 0, 2.$$

The same argument, relying on the homology of an $(n + 1)$-simplex, implies that the Betti numbers for the n-sphere, i.e., the n-dimensional sphere in the $(n + 1)$-dimensional space, are given by

$$\beta_0 = \beta_n = 1,$$

$$\beta_p = 0 \text{ for all } p \neq 0, n.$$

for homology over $\mathbb{Z}_2$.

5   A Running Example

In the real world, a software graph will never be a simple graph. The vertices of a software graph may represent different things: classes, objects, functions, modules, components, software units etc. The edges of a software graph represent communication or calls between these parts of the software. In any case, the software graph is always directed, as the calls between parts of the software are directed from one part to the other. The software graph is often weighted or with multiple edges, because some calls between parts of the software are more frequent than others. This is encapsulated in the graph by assigning weights to the edges or allowing multiple edges for multiple calls. Even the loops may appear in a software graph, as they would represent recursive calls to the same part of software.

On the other hand, the topological algorithm presented in Sect. 4 requires a simple graph as input, works over the field $\mathbb{Z}_2 = \{0, 1\}$ of two elements, and defines simplices as complete subgraphs. These simplifications are necessary to make the presentation of the basic concepts more accessible to students. Nevertheless, it still captures certain amount of topological insight in higher dimensional structure of software. The topological techniques of similar nature can be applied to more general settings, such as the cases of directed and weighted graphs and working over other field and rings than $\mathbb{Z}_2$. The reference for these techniques could be any textbook on algebraic topology such as [29].

In the real world software is certainly a limitation, although there is still some topological insight gained from the presented algorithm.

5.1   Input to the Algorithm

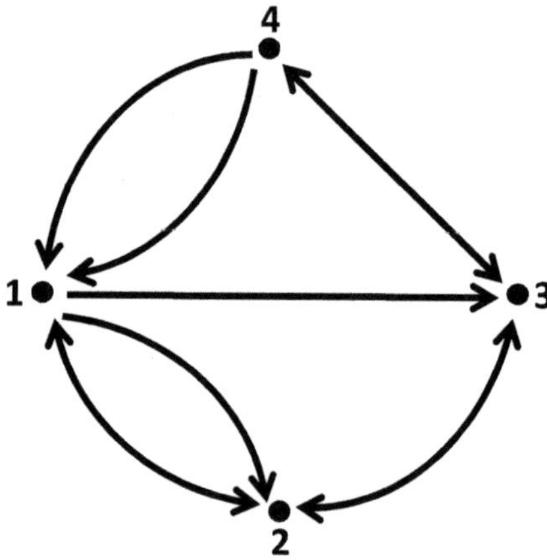

Fig. 16. The graph of the running example.

The running example considered now is given by the graph given in Fig. 16. It is a small graph consisting of only four vertices, denoted in the figure by numbers 1, 2, 3 and 4, but it is not a simple graph. There are multiple edges and it is directed. Hence, the input to the running example is just a list of edges of a software graph. The first two numbers in each line represent an edge between the vertices labeled by these numbers. The direction of the edge is the order of numbers. It is also possible that the edges are weighted, which is indicated in the list by the third number in each row. In the running example, all the weights are set to 1. For the running example in Fig. 16, the list of edges is given by

$$
\begin{array}{ccc}
1 & 2 & 1 \\
2 & 3 & 1 \\
1 & 3 & 1 \\
2 & 1 & 1 \\
1 & 2 & 1 \\
3 & 2 & 1 \\
3 & 4 & 1 \\
4 & 1 & 1 \\
4 & 1 & 1 \\
4 & 3 & 1
\end{array}
$$

However, since the topological algorithm that we introduced in Sect. 4 requires a simple graph as input, the list of edges must be cleaned up. The multiple edges must be replaced by a single undirected copy of an edge, and the weights must be erased. Hence, the input to the topological algorithm for the running example is the list

$$
S_1 = \begin{bmatrix} 2 & 3 \\ 1 & 3 \\ 2 & 1 \\ 4 & 1 \\ 4 & 3 \end{bmatrix}.
$$

where multiple edges and weights are removed. The simple graph obtained by this procedure is given in Fig. 17.

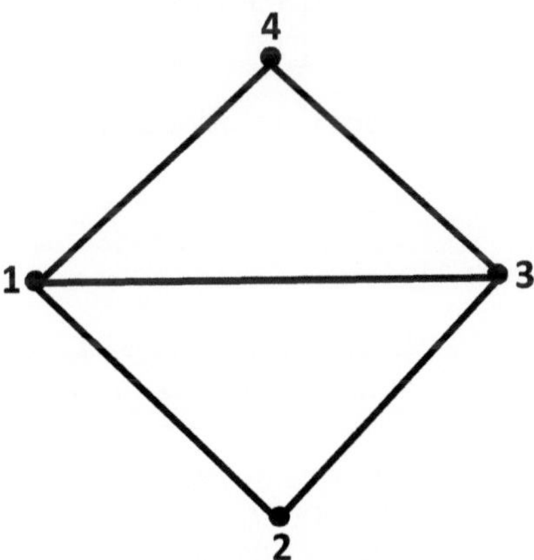

Fig. 17. The simple graph obtained in the running example.

5.2 Step 0 - Adjacency Matrix

The step 0 of the algorithm is to write down the adjacency matrix of the given simple graph. It will be used repeatedly in the following steps as it encodes information about neighboring vertices. This information is required in constructing the simplicial complex and boundary operators introduced in Sect. 4, more precisely, in Sect. 4.1, Sect. 4.3 and Sect. 4.5.

Recall that the adjacency matrix of a graph is a square matrix with the number of rows and columns equal to the number of vertices of the graph. Let n_0 denote the number of vertices in the graph, so that the adjacency matrix is an $n_0 \times n_0$ matrix. The order of vertices must be chosen and fixed. Then, the elements of the adjacency matrix are determined as follows.

There is 1 at the crossing of the i-th row and the j-th column, if the i-th and the j-th vertex are connected by an edge,
There is 0 at the crossing of the i-th row and the j-th column, otherwise.

Observe that we consider a simple graph. Hence, the adjacency matrix is symmetric, because the edges are undirected. It contains only ones and zeroes, because there are neither multiple edges nor weights. The diagonal entries are zero, because there are no loops.

In the running example $n_0 = 4$, and the order of vertices is already fixed by the labels 1, 2, 3, 4. The adjacency matrix is

$$A = \begin{bmatrix} 0 & 1 & 1 & 1 \\ 1 & 0 & 1 & 0 \\ 1 & 1 & 0 & 1 \\ 1 & 0 & 1 & 0 \end{bmatrix}.$$

Observe that A has all the properties mentioned for the adjacency matrix of a simple graph.

5.3 Step 1 - Incidence Matrix

The next step is to determine the incidence matrix of a given graph. It is nothing else than the matrix of the boundary operator in dimension one, as defined in Sect. 4.6.

Recall that the incidence matrix of a graph is a matrix with the number of rows equal to the number n_0 of vertices, and the number of columns equal to the number n_1 of edges. Thus, the incidence matrix is an $n_0 \times n_1$ matrix. The order of vertices and edges must be fixed once and for all. Then, the entries of the incidence matrix are determined as follows.

There is 1 at the crossing of the i-th row and the j-th column, if the i-th vertex belongs to the j-th edge,
There is 0 at the crossing of the i-th row and the j-th column, otherwise.

Since every edge has two vertices as its end-points, each column of the incidence matrix contains exactly two entries equal to 1, and the rest are 0.

In the running example, there are $n_0 = 4$ vertices, ordered by their labels, and $n_1 = 5$ edges, ordered as in the given list S_1 of edges. The order of edges may be chosen differently. The most canonical choice would be to order the edges in the lexicographical order with respect to the order of edges, but this is not so important in this small example. With the fixed order of vertices and edges as above in S_1, the incidence matrix in the running example is

$$\partial_1 = \begin{bmatrix} 0 & 1 & 1 & 1 & 0 \\ 1 & 0 & 1 & 0 & 0 \\ 1 & 1 & 0 & 0 & 1 \\ 0 & 0 & 0 & 1 & 1 \end{bmatrix}.$$

Observe that the incidence matrix is denoted by ∂_1, as the 1-boundary operator. This is not a coincidence. The incidence matrix is precisely the matrix of the 1-boundary operator written with the fixed order of vertices, which are 0-simplices, and edges, which are 1-simplices of the simplicial complex assigned to the graph. The point is that the j-th column, which represents the j-th edge, i.e., j-th 1-simplex, contains 1 precisely in the rows representing its end-points, i.e., its 0-simplex faces. Thus, the incidence matrix is indeed equal to the matrix of the 1-boundary operator.

For convenience of the reader, following the suggestion of the referee, we draw here the table from which the matrix of the boundary operator ∂_1 is determined

$\partial_1 \sim$

	$2,3$	$1,3$	$2,1$	$4,1$	$4,3$
1	×	√	√	√	×
2	√	×	√	×	×
3	√	√	×	×	√
4	×	×	×	√	√

The rows are labeled with 0-simplices (i.e., vertices) and columns are labeled with 1-simplices (i.e., edges). The signs √ and × in the table represent whether the vertex of a given row is the end-point of the edge of a given column, or not. In the matrix, these signs become ones and zeroes.

Similar table for the boundary operator ∂_2 is as follows.

$\partial_2 \sim$

	$2,3,1$	$1,3,4$
$2,3$	√	×
$1,3$	√	√
$2,1$	√	×
$4,1$	×	√
$4,3$	×	√

In this table, the rows are labeled with 1-simplices (i.e., edges) and columns are labeled with 2-simplices (i.e., triangles), and signs √ and × represent whether a given edge is the side of a given triangle.

In this way, it becomes clear how the matrices of the boundary operator are determined from the knowledge of boundary faces of every simplex in a simplicial complex.

5.4   Step 2 - Higher Differentials

The algorithm proceeds iteratively by computing step-by-step the higher differentials of the simplicial complex, that is, the matrices of the p-boundary operators for $p \geq 2$.

There are a few sloppy places in applying this procedure. First of all, one should make sure that the list S_p of p-simplices does not contain several copies of the same p-simplex. Hence, it is desirable to fix an order of vertex labels (increasing or decreasing) when making the list in order to avoid double simplices in S_p.

Another issue is that adding a column in the matrix ∂_p of p-boundary operator requires the knowledge of the row numbers corresponding to all of its faces. Perhaps a good idea is to make the list S_p in such a way that the order of simplices is easy to handle and allows fast search.

Recall from Sect. 4.6 that the matrix of the p-boundary operator ∂_p is a matrix with n_{p-1} rows and n_p columns, where n_k is the number of k-simplices, i.e., the number of complete subgraphs with $k + 1$ vertices. The entries in the matrix are determined as follows.

> There is 1 at the crossing of the i-th row and the j-th column, if the i-th $(p - 1)$-simplex, i.e., the i-th complete subgraph with p vertices, is a $(p - 1)$-simplex face of the j-th p-simplex, i.e., a subgraph of the the j-th complete subgraph with $p + 1$ vertices,
>
> There is 0 at the crossing of the i-th row and the j-th column, otherwise.

Observe that this requires a fixed ordering of complete subgraphs with a given number of vertices. Such complete subgraphs are determined iteratively, and their order is fixed along the way. Once it is fixed, the order should not be changed. In particular, the numbering of complete subgraphs with p vertices should be the same in the construction of matrices of boundary operators ∂_p and ∂_{p-1}. In this way, we obtain at the end the list of all simplices in the simplicial complex, as in Sect. 4.1.

The problem of finding all complete subgraphs in an arbitrary simple graph is computationally highly demanding. However, the software graphs are not, and should not, be arbitrary. Larger complete subgraphs in a software graph would indicate a high level of communication between many different parts of software, which is not appropriate for design, testing, verification and maintenance of software. Software design principles specify that such communication should be avoided. Therefore, in the special case of software graphs, the search for complete subgraphs is feasible.

The list S_1 of edges may and will serve as the list of 1-simplices with the fixed order, as above in the construction of the incidence matrix ∂_1. The iterative computation of differentials requires to keep track of the p-simplices in each dimension p, and their fixed order. The list S_p of p-simplices is a matrix with n_p rows and $p+1$ columns, where n_p is the number of complete subgraphs with $p+1$ vertices (i.e. p-simplices) such that each row contains $p+1$ integers which are labels of points that form a p-simplex.

As the first iterative step, one should determine the matrix ∂_2 and the list S_2 of complete subgraphs with 3 vertices (i.e. 2-simplices) from the matrix ∂_1 and the list S_1 of edges (i.e. 1-simplices) using the adjacency matrix A. This is done as follows.

1. Consider the first previously unused row in the list S_1 of edges (1-simplices)
2. Check in the adjacency matrix A if all the vertices of the 1-simplex defined by that row have a vertex as a common neighbor, that is, if all the vertices of the 1-simplex are connected by an edge to the same vertex of a graph.
3. For each common neighbor determined in Step 2, adding such a common neighbor to the 1-simplex gives a 2-simplex (a triangle in the graph), which is stored in the list S_2 of 2-simplices.
4. For each 2-simplex determined in Step 2, add a column on the right-hand side of the 2-boundary operator matrix ∂_2 which contains ones in the rows of edges of that 2-simplex.
5. Go back to Step 1, unless there are no more unused rows in S_1.

In this way, adding all possible neighbors to the 1-simplices, we obtain the list S_2, and the order of the list is fixed once and for all. At the same time, for each 2-simplex, a column on the right-hand side of the 2-boundary operator matrix ∂_2 is added. The outcome of the algorithm is the ordered list S_2 and the matrix ∂_2 obtained simultaneously.

In the running example the list of 2-simplices (i.e., triangles in the graph) are the following

$$S_2 = \begin{bmatrix} 2 & 3 & 1 \\ 1 & 3 & 4 \end{bmatrix}$$

and the 2-boundary operator matrix is

$$\partial_2 = \begin{bmatrix} 1 & 0 \\ 1 & 1 \\ 1 & 0 \\ 0 & 1 \\ 0 & 1 \end{bmatrix}.$$

There are only two triangles in the graph, so that the list S_2 has two rows and the matrix ∂_2 has two columns.

In general, how to determine the p-boundary operator matrix ∂_p and the list S_p of p-simplices from the $(p-1)$-boundary operator matrix ∂_{p-1} and the list S_{p-1} of $(p-1)$-simplices using the adjacency matrix A? In precisely the same way!

Go through the rows of the list S_{p-1} of $(p-1)$-simplices, and for each row of S_{p-1}, check in the adjacency matrix A if all the vertices of the $(p-1)$-simplex defined by that row have a common neighbor. Adding such common neighbor to the $(p-1)$-simplex gives a p-simplex, which should be stored in the list S_p of p-simplices. When a p-simplex is detected, at the same time, add a column in the p-boundary operator matrix ∂_p. This added column contains ones in the rows of $(p-1)$-simplex faces of that p-simplex.

In the running example, as small as it is, there are no 3-simplices (i.e. complete subgraphs with four vertices). Hence, the list S_3 is empty, and there is no matrix ∂_3.

5.5   Step 3 - Rank of a Matrix over $\mathbb{Z}_2$

All the information required for the calculation of homology groups and their ranks, introduced in Sect. 4.9, are encoded in the matrix ∂_p of the p-boundary operator for all p, which are defined in Sect. 4.6. In particular, the rank of the matrices ∂_p contains the information regarding the rank of the cycle and boundary groups, as defined in Sect. 4.8, which are required for computing the Betti numbers, i.e., the rank of homology groups, as in Sect. 4.9.

This information can be read off from a normal form of a matrix. The reduction of a matrix to a normal form can be made using the well-known Gauss elimination method. The students learn this algorithm in the basic linear algebra course, and over $\mathbb{Z}_2$ it is even simpler.

The Gauss elimination over $\mathbb{Z}_2$ consists of the following steps, where the input is any matrix with entries in $\mathbb{Z}_2 = \{0,1\}$.

1. Set the considered part of the matrix to be the whole matrix.
2. Find a place in the considered part of the matrix containing 1. If it does not exist, the matrix is already in a normal form, exit the algorithm.
3. Exchange two rows and two columns, so that this 1 appears in the upper-left corner of the considered part of the matrix.
4. Find 1's in the first column of the considered part of the matrix.
5. Add the first row of the considered part of the matrix to those rows that contain 1's in the first column of the considered part of the matrix. The addition is made mod 2, so that these 1's are canceled.
6. Change the considered part of the matrix by removing its first row and its first column. Go through steps 2–5 with this new considered part of the matrix.

The outcome of the Gauss elimination is the matrix of the form

$$
M = \begin{bmatrix}
1 & * & \dots & * & * & \dots & * \\
0 & 1 & \ddots & \vdots & \vdots & & \vdots \\
\vdots & \ddots & \ddots & * & \vdots & & \vdots \\
0 & \dots & 0 & 1 & * & \dots & * \\
0 & \dots & \dots & 0 & 0 & \dots & 0 \\
\vdots & & & & & & \vdots \\
0 & \dots & \dots & 0 & 0 & \dots & 0
\end{bmatrix}.
$$

The rank of the matrix M is defined as the number of 1's along the diagonal. The defect of the matrix M is the number of remaining columns. Although we did not define the rank and defect in general, note that the rank and defect of any matrix is equal to the rank and defect of its normal form as defined for the matrix M.

It turns out that in the case of the matrix ∂_p of the p-boundary operator

the rank of ∂_p equals the rank b_{p-1} of the boundary group B_{p-1}, as defined in Sect. 4.8,
the defect of ∂_p equals the rank z_p of the cycle group Z_p, as defined in Sect. 4.8.

Thus, the rank and defect of ∂_p for all p provides all the necessary information to compute the Betti numbers of the given graph, that is, the Betti numbers of the considered software.

In the running example, the Gauss elimination of the matrix ∂_1 is performed as follows. In the first step we move the 1 at the crossing of the second row and first column to the upper-left corner. This is achieved by replacing the first two rows of the matrix. In the second step, the first row is added to the third row to cancel the 1 in the first column.

$$\partial_1 = \begin{bmatrix} 0&1&1&1&0 \\ 1&0&1&0&0 \\ 1&1&0&0&1 \\ 0&0&0&1&1 \end{bmatrix} \sim \begin{bmatrix} 1&0&1&0&0 \\ 0&1&1&1&0 \\ 1&1&0&0&1 \\ 0&0&0&1&1 \end{bmatrix} \sim \begin{bmatrix} 1&0&1&0&0 \\ 0&1&1&1&0 \\ 0&1&1&0&1 \\ 0&0&0&1&1 \end{bmatrix}.$$

Now the considered part of the matrix is obtained by removing the first row and the first column. In the considered part of the matrix, there is 1 already in the upper-left corner (which is the crossing of the second row and the second column of the whole matrix), so that no replacing of rows and columns is necessary. Thus, the next step is to add the second row to the third row, which cancels the 1 in the second column below the upper-left corner of the considered part of the matrix. We obtain

$$\partial_1 \sim \begin{bmatrix} 1&0&1&0&0 \\ 0&1&1&1&0 \\ 0&0&0&1&1 \\ 0&0&0&1&1 \end{bmatrix}.$$

Now the considered part of the matrix is obtained by removing again the first row and the first column. In other words, we consider the matrix above with the first two rows and the first two columns removed. In the remaining part of the matrix, there is 0 in the upper-left corner (which is the crossing of the third row and the third column). Hence, we must replace the third and the fourth columns of the matrix, and then add the third row to the fourth row to cancel the 1 below.

$$\partial_1 \sim \begin{bmatrix} 1&0&0&1&0 \\ 0&1&1&1&0 \\ 0&0&1&0&1 \\ 0&0&1&0&1 \end{bmatrix} \sim \begin{bmatrix} 1&0&0&1&0 \\ 0&1&1&1&0 \\ 0&0&1&0&1 \\ 0&0&0&0&0 \end{bmatrix}.$$

The last line has become zero, so we are done. The final matrix is in the normal form, and we can read off the rank and defect of ∂_1. The rank is 3, because there are 3 ones along the diagonal, and the defect is 2, because there are two columns remaining. Thus,

$$b_0 = 3,$$

$$z_1 = 2,$$

is the rank of the group B_0 of 0-boundaries, and the rank of the group Z_1 of 1-cycles, respectively.

Similarly, we perform the Gauss elimination method for the matrix ∂_2 of the 2-boundary operator in the running example. In the first step, the first row is added to the second and the third row to cancel the 1's below the upper-left corner. Then, the considered part of the matrix is obtained by removing the first row and the second column. The 1 at the crossing of the second row and the second column is used to cancel all the 1's below it by adding the second row to fourth and fifth row.

$$\partial_2 = \begin{bmatrix} 1 & 0 \\ 1 & 1 \\ 1 & 0 \\ 0 & 1 \\ 0 & 1 \end{bmatrix} \sim \begin{bmatrix} 1 & 0 \\ 0 & 1 \\ 0 & 0 \\ 0 & 1 \\ 0 & 1 \end{bmatrix} \sim \begin{bmatrix} 1 & 0 \\ 0 & 1 \\ 0 & 0 \\ 0 & 0 \\ 0 & 0 \end{bmatrix}$$

The last matrix is the normal form of ∂_2. Hence, the rank and defect of ∂_2 is read off that matrix. It gives

$$b_1 = 2,$$

$$z_2 = 0,$$

is the rank of the group B_1 of 1-boundaries, and the rank of the group Z_2 of 2-cycles, respectively.

5.6   Final Step - Betti Numbers in Homology

Finally, all the acquired information is combined to obtain the Betti numbers in homology of the simplicial complex assigned to a software graph, as in Sect. 4.9. In general,

$z_0 = n_0$ as always,
b_{p-1} and z_p are obtained from ∂_p for $1 \leq p \leq t$,
$b_t = 0$,

where t is the top dimension such that there are no $(t+1)$-simplices from which the t-boundaries could come from. From these values, as in Sect. 4.9, the Betti numbers are

$$\beta_p = z_p - b_p$$

for all $p = 0, 1, \ldots, t$.

In the case of the running example, we have calculated the following

$z_0 = n_0 = 4$ as always,

$b_0 = 3$ is obtained from ∂_1,

$z_1 = 2$ is obtained from ∂_1,

$b_1 = 2$ is obtained from ∂_2,

$z_2 = 0$ is obtained from ∂_2,

$b_2 = 0$ because 2 is the top dimension, as there are no 3-simplices.

Therefore, the Betti numbers are

$$\beta_0 = z_0 - b_0 = 4 - 3 = 1,$$
$$\beta_1 = z_1 - b_1 = 2 - 2 = 0,$$
$$\beta_2 = z_2 - b_2 = 0 - 0 = 0.$$

6 Exercise

The goal of the following exercise is to write a code performing the algorithm described in the previous sections. The idea is that writing the code will make the abstract topological notions more familiar to students, and testing the code on different types of graphs will give them the flavor of the difficulties in terms of computational power and sustainability of the algorithm. It is important that the solution is fast enough to handle software graphs, which do not contain too large complete subgraphs.

Exercise: Given a software graph in the form described below, write a code in the programming language of your choice, which computes the Betti numbers, i.e., the ranks of homology groups, for that software.

The input to your code is a software graph given as a list of edges. The vertices are labeled by positive integers up to n_0, which denotes the number of vertices. The edges are pairs of such integers. Let n_1 denote the number of edges. The list of edges is a matrix (an array) with three columns and n_1 rows, where each row represents an edge:

The first integer in a row is the starting vertex of an edge.

The second integer in a row is the ending vertex of an edge.

The third number in a row is the weight of an edge.

For our task the directions of the edges are ignored, multiple edges are ignored and the weights are ignored. Hence, the code should first clean up the list of edges by removing the weights, i.e., erasing the third column, and removing the duplicate edges in the list. Note that the order of vertices of an edge is irrelevant as the graph is unordered.

The steps of the algorithm described in Sect. 5 should be implemented in the code as the solution of the exercise. These steps lead to the solution.

Step 0. Write the code which transforms the cleaned up input list of edges into the adjacency matrix A of a given graph.

Step 1. Write the code which transforms the cleaned up input list of edges into the incidence matrix of a given graph, which is at the same time the matrix ∂_1 of the 1-boundary operator.

Step 2. Write the code which computes iteratively the matrix of higher-dimensional boundary operators ∂_p and at the same time the ordered list S_p of p-simplices.

Step 3. Write the code which implements the Gauss elimination algorithm and determines the normal form of a given matrix. Apply it to the matrices of boundary operators ∂_p.

Final Step. Write the code that reads off the rank and defect from a normal form of a matrix. Apply it to the normal forms of the boundary operators ∂_p and compute the Betti numbers.

The code that solves the exercise can be tested on any simple graph. However, for arbitrary graphs there is no solution that is fast enough to determine the Betti numbers for the simplicial complex assigned to a graph as in the previous sections. The problem occurs if the graph is highly connected in terms of complete subgraphs. The solution code should work fast enough to handle the software graphs, in which there are usually only smaller complete subgraphs.

At the summer school, the attendees were given a sample software graph as the test input. It was a software graph with 3833 vertices and 17602 edges. The top dimension of the graph turned out to be only $t = 3$, so that there are no complete subgraphs with more than 4 vertices. The Betti numbers are

$$\beta_0 = 7,$$
$$\beta_1 = 12591,$$
$$\beta_2 = 1995,$$
$$\beta_3 = 46.$$

The interested reader who solved the exercise is urged to either contact the authors for the sample graph input to test their solution, or test the solution on any other software graph available.

7   Conclusion

Software is the main technology driving digitalization in many domains as support in decision-making processes. It abstracts the physical world and manages such abstract resources and it implements machine learning and artificial intelligence algorithms to provide analysis of big data. The way how we engineer software solutions has a significant impact on addressing sustainability goals in the application domain where software is implemented but also within the software engineering industry. Engineering sustainable software is becoming crucial for future technology advancements. The main obstacle in engineering sustainable software is the human ability to engineer complex systems.

In this lecture, we define complex systems and specific challenges of modeling modern software systems and their consequences on sustainable software behavior. We explain the problem of modeling complex software behavior from the perspective of modeling local system properties that are measured on software parts and global system properties that are measured in the system operation. Furthermore, we introduce the software structure as one of the key instruments we model relations between local and global system properties and explain its graph representation.

Finally, we introduce students to Topological Data Analysis (TDA) as an analysis tool that may be very useful in modeling complex systems as a complementary tool to various existing graph algorithms. TDA is capable of describing the topological space of structure that may introduce an additional useful dimension to explain algorithmic decisions in various applications. Here we introduce the key concepts of TDA and guide students on how to implement these abstract topological notions.

In our running case, we showed how TDA may be used in the analysis of software structures with the help of TDA aiming to address concerns of sustainable software evolution. The results of our case of using TDA to model software structures in evolution are presented in [37]. By experimenting with different graphs the students may understand the difficulties in terms of computational power and sustainability of the algorithm by testing the code on different types of graphs, and how this implementation may impact sustainable software structures.

References

1. Aliance Next Generation Mobile Networks (NGMN): Green future networks: network energy efficiency (2021). https://www.ngmn.org/publications/green-future-networks-network-energy-efficiency.html
2. Allili, M., Kaczynski, T., Landi, C., Masoni, F.: Acyclic partial matchings for multidimensional persistence: algorithm and combinatorial interpretation. J. Math. Imaging Vis. **61**(2), 174–192 (2019)
3. Bendich, P., Edelsbrunner, H., Kerber, M.: Computing robustness and persistence for images. IEEE Trans. Vis. Comput. Graph. **16**(6), 1251–1260 (2010)
4. Callahan, D., Carle, A., Hall, M.W., Kennedy, K.: Constructing the procedure call multigraph. IEEE Trans. Softw. Eng. **16**(4), 483–487 (1990)
5. Cang, Z., Mu, L., Wu, K., Opron, K., Xia, K., Wei, G.W.: A topological approach for protein classification. Comput. Math. Biophys. **3**(1) (2015)
6. Carlsson, G.: Topology and data. Bull. Amer. Math. Soc. (N.S.) **46**(2), 255–308 (2009)
7. Carlsson, G., Vejdemo-Johansson, M.: Topological Data Analysis with Applications. Cambridge University Press, Cambridge (2022)
8. Choudhary, A., Kerber, M., Raghvendra, S.: Improved approximate rips filtrations with shifted integer lattices and cubical complexes. J. Appl. Comput. Topol. **5**(3), 425–458 (2021)
9. Dey, T.K., Mandal, S., Mukherjee, S.: Gene expression data classification using topology and machine learning models. BMC Bioinform. **22-S**(Suppl. 10), 627 (2021)

10. Edelsbrunner, H., Harer, J.: Computational Topology – An Introduction. American mathematical society (2010)
11. Edelsbrunner, H., Morozov, D.: Persistent homology: theory and practice. In: Proceedings of the European Congress of Mathematics 2012, pp. 31–50. EMS Press, Berlin (2014)
12. Fremerey, C., Mueller, M., Clausen, M.: Towards bridging the gap between sheet music and audio. In: Knowledge representation for intelligent music processing. Dagstuhl Seminar Proceedings (DagSemProc), vol. 9051, pp. 1–11. Schloss Dagstuhl – Leibniz-Zentrum für Informatik, Dagstuhl, Germany (2009)
13. Galinac Grbac, T.: The role of functional programming in management and orchestration of virtualized network resources. Part I. System structure for complex systems and design principles. In: Composability, Comprehensibility and Correctness of Working Software, 7th Winter School. Revised Selected Papers, LNCS, vol. 11916, Springer, Cham (2023). to appear. https://arxiv.org/abs/2107.12136
14. Galinac Grbac, T., Domazet, N.: The role of functional programming in management and orchestration of virtualized network resources. Part II. Network evolution and design principles. In: Composability, Comprehensibility and Correctness of Working Software, 8th Summer School. Revised Selected Papers, LNCS, vol. 11950, Springer, Cham (2023). to appear. https://arxiv.org/abs/2107.12227
15. Gasarch, W., Fasy, B.T., Wang, B.: Open problems in computational topology. SIGACT News **48**(3), 32–36 (2017)
16. Giray, G., Bennin, K.E., Köksal, Ö, Babur, Ö, Tekinerdogan, B.: On the use of deep learning in software defect prediction. J. Syst. Softw. **195**, 111537 (2023)
17. Hatcher, A.: Algebraic Topology. Cambridge University Press, Cambridge (2002)
18. Hilty, L.M., Aebischer, B.: ICT for sustainability: An emerging research field. In: Hilty, L.M., Aebischer, B. (eds.) ICT Innovations for Sustainability, pp. 3–36. Springer, Cham (2015). https://doi.org/10.1007/978-3-319-09228-7
19. Hilty, L.M., Arnfalk, P., Erdmann, L., Goodman, J., Lehmann, M., Wäger, P.A.: The relevance of information and communication technologies for environmental sustainability – a prospective simulation study. Environ. Modell. Softw. **21**(11), 1618–1629 (2006)
20. International Organization for Standardization (ISO), International Electrotechnical Commission (IEC): ISO/IEC 25002:2024 Systems and software engineering – Systems and software Quality Requirements and Evaluation (SQuaRE) – Quality model overview and usage. International standard, 1st edn. (2024)
21. Kang, L., Xu, B., Morozov, D.: Evaluating state space discovery by persistent cohomology in the spatial representation system. Front. Comput. Neurosci. **15**, 616748 (2021)
22. Kim, H.S., et al.: Topological data analysis can extract sub-groups with high incidence rates of type 2 diabetes. Int. J. Data Min. Bioinform. **22**(1), 44–60 (2019)
23. Lago, P., Koçak, S.A., Crnković, I., Penzenstadler, B.: Framing sustainability as a property of software quality. Commun. ACM **58**(10), 70–78 (2015)
24. Lago, P., Meyer, N., Morisio, M., Müller, H.A., Scanniello, G.: Leveraging "Energy Efficiency to Software Users": Summary of the Second Greens Workshop, at ICSE 2013. SIGSOFT Software Enginerring Notes **39**(1), 36–38 (2014)
25. Leykam, D., Angelakis, D.G.: Topological data analysis and machine learning. Adv. Phys. X **8**(1), 2202331, 24 p. (2023)
26. Malott, N.O., Chen, S., Wilsey, P.A.: A survey on the high-performance computation of persistent homology. IEEE Trans. Knowl. Data Eng. **35**(5), 4466–4484 (2023)

27. Mauša, G., Galinac Grbac, T.: Co-evolutionary multi-population genetic programming for classification in software defect prediction: an empirical case study. Appl. Soft Comput. **55**, 331–351 (2017)
28. Munch, E.: A user's guide to topological data analysis. J. Learn. Anal. **4**(2), 47–61 (2017)
29. Munkres, J.R.: Elements of Algebraic Topology. Addison-Wesley Publishing Company, Menlo Park, CA (1984)
30. Munkres, J.R.: Topology. Prentice Hall Inc., Upper Saddle River, NJ (2000)
31. Pandey, S.K., Mishra, R.B., Tripathi, A.K.: Machine learning based methods for software fault prediction: a survey. Expert Syst. Appl. **172**, 114595 (2021)
32. Penzenstadler, B., Raturi, A., Richardson, D., Tomlinson, B.: Safety, security, now sustainability: the nonfunctional requirement for the 21st century. IEEE Softw. **31**(3), 40–47 (2014)
33. Petrić, J., Galinac Grbac, T.: Software structure evolution and relation to system defectiveness. In: Proceedings of the 18th International Conference on Evaluation and Assessment in Software Engineering, EASE 2014, pp. 34:1–34:10. ACM (2014)
34. Petrić, J., Galinac Grbac, T., Dubravac, M.: Processing and data collection of program structures in open source repositories. In: Proceedings of the 3rd Workshop on Software Quality Analysis, Monitoring, Improvement and Applications (SQAMIA 2014). CEUR Workshop Proceedings, vol. 1266, pp. 57–66. CEUR (2014)
35. Pita Costa, J., Škraba, P.: A topological data analysis approach to the epidemiology of influenza. In: SIKDD15 Conference Proceedings (2015)
36. Pita Costa, J., Galinac Grbac, T.: The topological data analysis of time series failure data in software evolution. In: Proceedings of the 8th ACM/SPEC on International Conference on Performance Engineering Companion, pp. 25–30. ICPE 2017 Companion, ACM, New York, NY, USA (2017)
37. Puh, E., Grbac, T.G., Grbac, N., Budimac, Z., Vranic, V., Lang, J.: Preliminary study of higher dimensional software structures. In: Proceedings of the 10th Workshop on Software Quality Analysis, Monitoring, Improvement and Applications (SQAMIA 2023). CEUR Workshop Proceedings, vol. 3588, pp. 13–25. CEUR (2023)
38. Valverde, S., Solé, R.V.: Network motifs in computational graphs: a case study in software architecture. Phys. Rev. E Stat. Nonlinear Soft Matter Phys. **72 2 Pt 2**, 026107–1, 026107–8 (2005)
39. Vranković, A., Galinac Grbac, T., Car, Ž: Software structure evolution and relation to subgraph defectiveness. IET Softw. **13**(5), 355–367 (2019)

Energy Debt: Foundations, Techniques and Tools

Marco Couto, Rui Rua, and João Saraiva[(✉)]

Department of Informatics, University of Minho HASLab/INESC TEC,
Braga, Portugal
{rui.rua,saraiva}@di.uminho.pt

Abstract. This tutorial presents the foundations, techniques and tools on energy debt. Energy debt adapts the technical debt metaphor to the green software realm. Thus, we define energy debt as the additional estimated energy cost of executing a software system, due to energy consumption inefficiencies in the software's source code, when compared to the estimated energy cost of executing the energy-optimal version of that same software.

To express energy debt we compiled a catalog of Android-specific energy code smells which are presented in the current state-of-the-art literature on green software, together with the energy savings reported in the studies where such smells have been presented.

Finally, we present the *E-Debitum* tool, an energy debt estimation plugin for *SonarQube* that can be used by developers and researchers to estimate energy debt on Android applications.

Keywords: Technical Debt · Energy Debt · Energy Efficient Software

This work received financial support from national funds through the Portuguese funding agency FCT (*Fundação para a Ciência e a Tecnologia*) within the project **UID/EEA/50014/2013**.

The work was further supported by the COST Action 19135: *CERCIRAS—Connecting Education and Research Communities for an Innovative Resource Aware Society* and by a PhD scholarship with reference **SFRH/BD/146624/2019**.

This work received financial support through the Erasmus+ Strategic Partnership for Higher Education *SusTrainable—Promoting Sustainability as a Fundamental Driver in Software Development Training and Education* (project number 2020-1-PT01-KA203-078646), funded by the European Union and coordinated by the University of Coimbra, Portugal.

The information and views set out in this publication are those of the authors and do not necessarily reflect the official opinion of the European Union. Neither the European Union institutions and bodies nor any person acting on their behalf may be held responsible for the use which may be made of the information contained therein.

1 Introduction

The current widespread use of mobile devices often leads to heavy dependence and the pervasive low-battery anxiety among their users [1]. As a consequence energy consumption became a key concern both for hardware manufacturers, and for researchers and software developers [2]. As a result of recent research works, several energy-inefficient programming practices have been reported in literature [3–8] which have a significant impact on the energy consumption of software.

Energy-inefficient programming practices should be avoided/eliminated when developing/maintaining energy-efficient software solutions. When inefficient practices occur in a software system we usually say that it has Technical Debt (TD). Technical Debt is a metaphor describing the gap between the current state of a software system and the ideal state of that same software. The key idea of technical debt is that software systems may include artifacts that can be hard to understand/maintain/evolve, causing higher costs in future software development and maintenance activities. These extra costs can be seen as a type of debt that developers owe the software system. Although technical debt is still a recent area of research, it has gained significant attention over the past years: A recent systematic mapping study [9] identified ten different types of technical debt, namely *requirements, architectural, design, code, test, build, documentation, infrastructure, versioning,* and *defects* technical debt. Commonly used approaches to identify technical debt consist in the detection of source code smells [10] which is a quick indication of possible problems in the software system under consideration [11,12].

In the context of green software, research works show that developers fall into energy-greedy practices and tendencies due to the lack of knowledge and the lack of tools to help understand, locate, and optimize energy inefficiencies [13]. Additionally, other practitioners and decision-makers also lack the necessary tools to help interpret how energy inefficiencies can impact their product lifecycle, and what they should focus on tackling in order to reduce energy costs [14]. Other studies show that sustainability in software engineering is under-represented in education and advocates its introduction in higher education curricula [15].

Indeed, energy-greedy programming practices, also called energy smells, do often occur in software systems. This tutorial paper defines *energy debt* as the additional estimated energy cost of executing a software system, due to the occurrence of energy smells in the software's source code, when compared to the estimated energy cost of executing the non-energy smelly (*i.e.* energy optimal) version of that same software [16]. To express energy debt we consider a catalog of energy code smells presented in the current state-of-the-art literature on green software, together with the energy savings reported in the studies where such smells have been presented [8,17]. Thus, the energy debt of a program is computed after knowing the number of occurrences and their locations in the program's source code: energy smells inside loops/recursion, single statements, or inside dead code do have different debt weights.

In order to estimate the energy debt of a software system, we have implemented the automatic detection of the full catalog of energy smells in the *E-Debitum* [18] tool: a plugin to the *SonarQube* framework. Moreover, we have conducted a the preliminary experiment of using E-Debitum in detecting energy smells and computing the evolution of the energy debt on consecutive releases of an open-source (GitHub) software system.

This paper is structured as follows: Sect. 2 thoroughly describes the foundations of energy debt, and how it should be expressed/calculated. In Sect. 3 we present a catalog of energy smells that give an indication of energy consumption inefficiencies in the software. Section 4 presents the *E-Debitum* tool that is an energy debt estimation plugin for *SonarQube*. Finally, in Sect. 5 we present our related work, and the conclusions are presented in Sect. 6.

2   Energy Debt: Foundations

This section introduces the concept of energy debt: a new metric, reflecting the implied cost in terms of energy consumption over time, of choosing a flawed implementation of a software system rather than a more robust, yet possibly time-consuming, approach. A flawed implementation is considered to contain code smells, known to have a negative influence on energy consumption.

Similar to technical debt, if energy debt is not properly addressed, it can accumulate an energy "interest". This interest will keep increasing as new versions of the software are released, and eventually reach a point where the interest will be higher than the initial energy debt. Addressing the issues/smells at such a point can remove energy debt, at the cost of having already consumed a significant amount of energy which can translate into high costs.

We present all underlying concepts of energy debt, bridging the connection with the existing concept of technical debt. We describe our approach with a preliminary motivational example, showing how to compute the energy debt of a real-world application. A prototype is under development, which already includes the detection of several Android energy smells, and calculates the energy debt and energy interest of Android applications.

2.1   Introduction to Technical Debt

Technical Debt describes the gap between the current state of a software system and the ideal state of that same software. The key idea of technical debt is that software systems may include artifacts that can be hard to understand/maintain/evolve, causing higher costs in future software development and maintenance activities. These extra costs can be seen as a type of debt that developers owe the software system.

As software evolves, it is common to take on debt from several sources, such as technological obsolescence and changes in its environment, as well as due to more mundane reasons, like poor programming practices (code smells). In fact,

TD is a well-known and widely used method of measuring and managing code quality.

Thus, technical debt involves the following concepts:

- *Repayment Effort:* It takes into account the design and structure of the code to estimate the human cost to elevate the current software to its ideal state.
- *Maintenance Effort:* Allowing TD to build tends to increase the effort required to evolve the software due to rising human costs:
- *Interest:* Existing technical debt also tends to become more costly as the software is built around it. This extra cost is commonly known as *Interest.*

These concepts are clearly presented in Fig. 1.

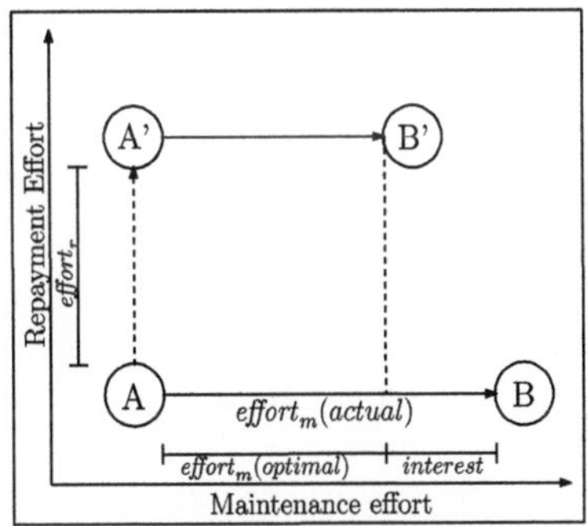

Fig. 1. Technical Debt: Repayment and Interest.

Existing technical debt is often estimated through the discovery of poor code design known as "code smells" [10–12]. For each type of code smell, there is a refactor, that eliminates the smell and this improves the software readability and modularity [10].

Although technical debt is still a recent area of research, it has gained significant attention over the past years: A recent systematic mapping study [9] identified ten different types of technical debt, namely *requirements, architectural, design, code, test, build, documentation, infrastructure, versioning,* and *defects* technical debt. In fact, TD is a concern both for researchers and software developers.

2.2 Introducing Energy Debt Concepts

The current widespread use of non-wired computing devices is also making energy consumption a key aspect not only for hardware manufacturers, but also for researchers and software developers [2]. Indeed, several *energy inefficient* programming practices have been reported in literature, namely, energy patterns for mobile applications [5,8], the energy impact of code smells [19–21], energy-greedy API usage patterns [3], energy (inefficient) data structures [4], programming languages [22,23], etc. which do have a significant impact on the energy consumption of software.

All these research works show that energy-greedy programming practices, also called energy smells, do often occur in software systems. These can be attributed to the current lack of knowledge software developers have in order to build energy-efficient software, and the lack of supporting tools [2].

This chapter defines *energy debt* as the additional estimated energy cost of executing a software system, due to the occurrence of energy smells in the software's source code when compared to the estimated energy cost of executing the non-energy smelly (*i.e.* energy ideal) version of that same software. To express energy debt we consider a set of energy code smells presented in the current state-of-the-art literature on green software, together with the energy savings reported in the studies where such smells have been presented. Thus, the energy debt of a program is computed after knowing the number of occurrences and their locations in the program's source code: energy smells inside loops/recursion, single statements, or inside dead code do have different debt weights.

Concept Overview. Before we present the definition of energy debt, let us recall the metaphor of technical debt. Technical debt reflects the cost arising from performing additional work on a software system, due to developers taking "shortcuts that fall short of best practices" [24]. Hence, this cost can be defined as the technical effort, in working hours, required for fixing all issues associated with bad programming practices, in a given release. The cost keeps increasing, as new versions (with new issues) keep getting released, and if the initial issues are not properly addressed, they accumulate *interest* [25].

Based on the underlying concept of technical debt, we define *Energy Debt* as *the amount of unnecessary energy that a software system used over time to maintaining energy code smells for sustained periods.*

A visual comparison of the two concepts is depicted in Fig. 2. The left-hand side of the figure illustrates the well-known representation of technical debt, including the concepts of refactoring and maintenance effort, along with the definition of interest. On the right-hand side, we present the definition of energy debt, where we assume that evolving the software (i.e., introducing new features on new releases) will eventually result in the addition of new (energy) code smells, hence the Energy Debt (ED) increases per version.

The main difference between technical and energy debt, at this point, is the fact that the former can be presented as a unique cost-value expressing how much effort would be necessary to address the issues, whereas the same approach cannot be applied to the latter. The cost of maintaining energy code smells in a software release is always directly proportional to the amount of time that the same release operates. As an example, if two software systems S_1 and S_2 have the exact same energy code smells, the amount of excessive energy consumed by S_1 might be higher than S_2 if it is intended to be used longer, during the same timespan.

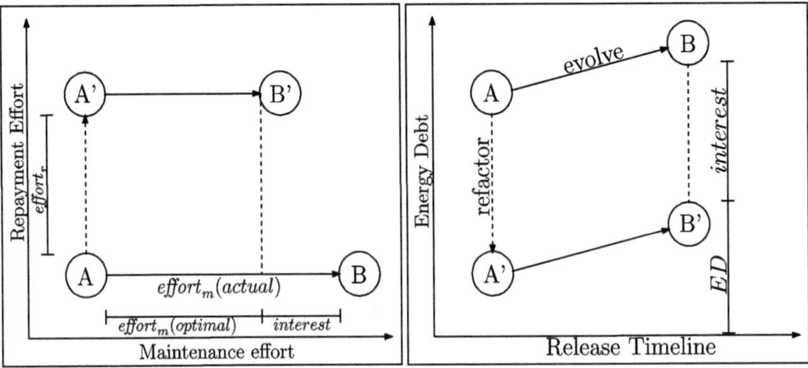

Fig. 2. Technical Debt vs Energy Debt (ED) Terminology.

Given the previous assumptions, we argue that the ED of a software release must be expressed not as a cost value, but as a cost function, which receives, as input, two variables: a software release r, and a usage time t. Equation 1 defines such a function, and it allows us to obtain, for a given release r, its ED after a given usage time of t:

$$ed(r, t) = cost(r) \times t \tag{1}$$

The $cost(r)$ function included in the equation represents the energy cost of release r, per unit of time. In other words, it relates to the existing number of energy code smells in that version, and the energy cost (per unit of time) of each one. The definition of that function is expressed as Eq. 2:

$$cost(r) = \sum_{i=1}^{N} w_i(r) \times E(i) \tag{2}$$

Here, N is the number of smells included in the considered catalog, while $w_i(r)$ returns a weight value for smell i, which is affected by the number of i smells found in release r and the context in which they were found (we will discuss this in greater detail in Sect. 3.3). $E(i)$ returns the expected energy debt per time unit of smell i, as defined in the smell catalog.

The formulas presented thus far assume that each considered energy code smell has an associated energy debt value, expressed in function of time units (for instance, per minute). Nevertheless, when studying the energy consumption impact of code smells, researchers often tend to present the potential gains/savings as an interval (i.e., highest and lowest observed energy saving).

The highest/lowest saving approach adds valuable information regarding potential energy savings. Fixing a certain smell can result in savings between, e.g., 150mJ and 3000 mJ per minute. When compared to another smell with savings between 300 mJ and 900 mJ per minute, we know that in a best-case scenario refactoring the first one would result in higher gains, but in a worst-case scenario, the second presents better savings. Hence, a developer can use this information to decide how to properly focus their attention when refactoring code smells, depending on the project goals [8].

In accordance with the previous assumption, we decided that our approach for energy debt should consider, for each code smell, two energy values: the highest (E_{max}) and lowest (E_{min}) observed energy savings. Since energy debt must be expressed as a function of the usage time, it is expected that E_{max} will be much higher with the increase in usage time, as depicted in Fig. 3.

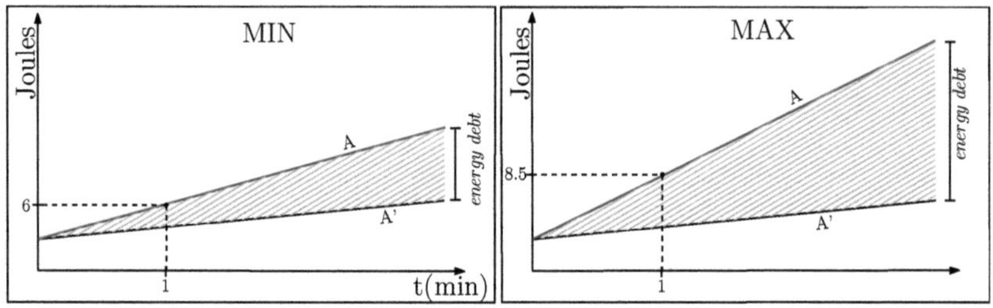

Fig. 3. Energy Debt Thresholds Increase Over Time.

There are two represented versions of a release in this figure: the optimal version, with all smells removed (A'), and the *energy smelly* version (A). The optimal version already has a constantly increasing energy consumption, as it would be expected. Energy debt can be summed up as the area between the line for A', and the (red) line for A, which becomes much larger when considering the maximum values. This will introduce changes to Eq. 1, which will consider two cost values, in the form of two functions:

$$ed(r,t) = \left(cost_{min}(r) \times t; cost_{max}(r) \times t \right) \tag{3}$$

The energy debt will therefore always be presented in the form of an interval. Consequently, each of the cost functions will need to refer to the proper energy debt per time unit. In other words, the $E(i)$ function in Eq. 2 will be $E_{min}(i)$ for the lowest savings, and $E_{max}(i)$ for the highest savings.

3   Energy Debt: Techniques

Very much like in the TD setting, in order to express the concept on energy debt for a programming language/software system, we need to define another key component: the energy smell catalog for that language/system. Because energy consumption especially relevant in the context of mobile software, in this document we will instantiate the concept of energy debt for this ecosystem, only.

Thus, in Sect. 3.1 we define an energy smell catalog for the Android ecosystem. This catalog contains all energy-related smells reported in the literature.

3.1   Energy Smell Catalog in the Android Ecosystem

As reported in several research works, developers fall into energy-greedy practices and tendencies due to the lack of knowledge, education and the lack of tools to help understand, locate, and optimize energy inefficiencies [13–15].

We have searched the literature for code patterns which have been tested for Android and that have proven to be energy inefficient; by an energy-inefficient pattern, we mean a pattern in which an alternative exists, one that preserves the application's functionality, while consuming less energy. We have found 9 independent works that identify 11 energy-inefficient patterns and that propose alternatives for them [5,6,19–21,26–29]. We shall refer to such a pattern as "**Energy Greedy Android Pattern**", or **EGAP** [8].

In this section, we include a brief description of every EGAP we found, where for each one we indicate which research work(s) detected the pattern, while explaining what makes it energy greedy, what the suggested alternative for it, and why such an alternative consumes less energy.

EGAP #1 - Draw Allocation This is the first of five EGAPs whose energy impact analysis was included in [5,29]. The authors aimed to understand how fixing code patterns detected by Android *lint*[1] can improve energy efficiency. *Lint*'s issues are divided into categories, such as *Performance* or *Security*. EGAP #1 (Draw Allocation), as well as EGAPs #3, #4, and #5 report *Performance* issues.

Draw Allocation occurs when new objects are allocated along with draw operations, which are very sensitive to performance. In other words, it is a bad practice to create objects inside the **onDraw** method of a class which extends a **View** Android component, as we see in the following snippet:

```
public class CloudMoonView extends View {

    @Override
    protected void onDraw(Canvas canvas) {
        RectF rectF1 = new RectF(); ✗

        . . .

        if(!clockwise) {
            rectF1.set(X2-r, Y2-r, X2+r, Y2+r);

            . . .
        }
    }

}
```

---

[1] *Lint* is a code analysis tool, provided by the Android SDK, which reports upon finding issues related to the code structural quality. Website: developer.android.com/studio/write/lint.

The recommended alternative for this EGAP, as of [5,29], is to move the allocation of independent objects outside the method, turning it into a static variable, as shown next:

```java
public class CloudMoonView extends View {

  RectF rectF1 = new RectF();  ✔

  @Override
  protected void onDraw(Canvas canvas) {
    ...
    if(!clockwise) {
      rectF1.set(X2-r, Y2-r, X2+r, Y2+r);
      ...
    }
  }

}
```

EGAP #2 - Wakelock This is the second Android *lint* performance issue [5, 21,26,29]. Basically, *lint* detects whenever a wake lock, a mechanism to control the power state of the device and prevent the screen from turning off, is not properly released, or is used when it is not necessary.

The following snippet shows an example of a wake lock being acquired, but not released when the activity pauses.

```java
public class DMFSetTempo extends Fragment {

  PowerManager.WakeLock wakeLock;

  public void onClickBtStart(View view) {
    wakelock.acquire();  ✔
  }

  @Override()
  public void onPause() {
    super.onPause();      ✗
  }
}
```

The alternative here would be to simply add a **release** instruction as shown next:

```
public class DMFSetTempo extends Fragment {
  PowerManager.WakeLock wakeLock;

  public void onClickBtStart(View view) {
    wakelock.acquire();       ✔
  }

  @Override()
  public void onPause() {
    super.onPause();
    if (wakeLock.isHeld()) wakeLock.release();   ✔
  }
}
```

EGAP #3 - Recycle This is another Android *lint* performance issue [5,29]. It detects when some collections or database-related objects, such as `TypedArrays` or `Cursors`, are not recycled nor closed after being used. When this happens, other objects of the same type cannot efficiently use the same resources.

The following snippet shows a `Cursor` instance being used without being recycled:

```
public Summoner getSummoner(int id) {
  SQLiteDatabase db = this.getReadableDatabase();

  Cursor c = db.query(TABLE_FAV, new String[] { ... };
  ...
  return summoner;  ✘
}
```

The alternative in this case would be to include a `close` method call before the method's return:

```
public Summoner getSummoner(int id) {
  SQLiteDatabase db = this.getReadableDatabase();

  Cursor c = db.query(TABLE_FAV, new String[] { ... };
  ...
  c.close();       ✔
  return summoner;
}
```

EGAP #4 - Obsolete Layout Parameter The fourth Android *lint* performance issue [5,29], `Obsolete Layout Parameter`, is the only one that is not Java-related. The view layouts in Android are specified using `XML`, and they tend to suffer several updates. As a consequence, some parameters that have no effect on the view may still remain in the code, which causes excessive processing at runtime. The alternative is to parse the XML syntax tree and remove these useless parameters.

The next snippet shows an example of a view component with parameters that can be removed:

```
<TextView android:id="@+id/centertext"
  android:layout_width="wrap_content"
  android:layout_height="wrap_content"
  android:text="remote files"
  layout_centerVertical="true"     ✘
  layout_alignParentRight="true">  ✘
</TextView>
```

EGAP #5 - View Holder The last Android *lint* performance issue [5,29], whose alternative intends to make a smoother scroll in *List Views*. The process of drawing all items in a *List View* is costly, since they need to be drawn separately. However, it is possible to make this more efficient by reusing data from already drawn items, which reduces the number of calls to `findViewById()`, known to be energy greedy [3].

In order to better describe this EGAP, we introduce the following snippet:

```
public View getView(int p, View cView, ViewGroup par) {
  LayoutInflater inflater = (LayoutInflater) context
     .getSystemService(Context.LAYOUT_INFLATER_SERVICE);

  cView = inflater.inflate(R.layout.apps, par, false);
  TextView txt =
        (TextView) cView.findViewById(R.id.label); ❶
  ImageView img =
        (ImageView) cView.findViewById(R.id.logo); ❷
  return row;
}
```

Every time `getView()` is called, the system searches on all the view components for both the `TextView` with the id "label" (❶) and the `ImageView` with the id "logo" (❷), using the energy greedy method `findViewById()`. The alternative version is to cache the desired view components, with the following approach:

```
static class ViewHolderItem {
  TextView txtView; ImageView imgView;
}

public View getView(int p, View cView, ViewGroup par) {
  ViewHolderItem hld; LayoutInflater inflater = ...

  if (cView == null) {                                    ❸
    cView = inflater.inflate(...);
    hld = new ViewHolderItem();
    hld.txtView =
           (TextView) cView.findViewById(...);            ❹
    hld.imgView =
           (ImageView) cView.findViewById(...);           ❺
    cView.setTag(hld);
  } else {
    hld = (ViewHolderItem) cView.getTag();                ❻
  }
  TextView txt = hld.txtView;
  ImageView img = hld.imgView;
  ...
}
```

Condition ❸ evaluates to true only once, which means instructions ❹ and ❺ execute once, i.e., `findViewById()` executes twice, and its results are stored in the `ViewHolderItem` instance. The following calls to `getView()` will use cached values for the view components `txt` and `img` (❻).

EGAP #6 - `HashMap` Usage This EGAP is related to the usage of the `HashMap` collection [6,19–21]. In fact, as stated in the Android documentation page, the usage of `HashMap` is discouraged, since the alternative `ArrayMap` is allegedly more energy-efficient, without decreasing the performance of `map` operations[2].

The alternative is to simply replace the type `HashMap`, whenever it is used, with `ArrayMap`.

EGAP #7 - Excessive Method Calls Unnecessarily calling a method can penalize performance, since a call usually involves pushing arguments to the call stack, storing the return value in the appropriate processor's register, and cleaning the stack afterward. This penalty was explored by [6,27], showing that the energy consumption in Android applications can be decreased by removing method calls inside loops that can be extracted from them. An example of an extractable method call would be one which receives no arguments, and is accessed by an object that is not altered in any way inside the loop.

---

[2] *ArrayMap* documentation: http://bit.ly/32hK0y9.

The alternative is to replace the method call with a variable that is declared outside the loop, and is initialized with the return value of the method call extracted.

EGAP #8 - Member Ignoring Method This EGAP addresses the issue of having a non-static method inside a class, and which could be static instead [6,21], i.e., it does not access any class fields, it does not directly invoke non-static methods, and it is not an overriding method. Static methods are stored in a memory block separated from where objects are stored, and no matter how many class instances are created throughout the program's execution, only an instance of such method will be created and used. This mechanism helps in reducing energy consumption.

EGAPs #9, #10 and #11 - Resource Leak Resource Leak [26,28] simultaneously refers to three EGAPs which are all particular cases of `Wakelock`, where a system resource is not properly released. Here, we consider the *Sensor, Camera*, and *Media* resources, which differ from `Wakelock` in the way resources are released. Since improving EGAPs on different resources can produce different energy gains, we decided to separate EGAPs #2 from #9, #10, and #11.

3.2 Energy Debt Smells Catalog

The energy smell catalog should map lower and upper-bounded energy estimations to each considered smell, These estimations need to be dependent on the elapsed time, i.e., they will be (as described above) functions that multiply the provided time value by a consumption constant.

Hence, by examining the reports of the smells presented in Sect. 3.1, we can extract this information, and associate to each pattern an approximation to their maximum and minimum cost (i.e. potential savings) per time unit. The results of this task (i.e., the energy debt supported catalog) are shown in Table 1.

Table 1. Energy Debt Smell Catalog

	Abbrev.	E_{min} (mJ per min)	E_{max} (mJ per min)
`DrawAllocation`	**DA**	36	158
`Wakelock`	**WL**	10	194
`Recycle`	**RC**	15	533
`ObsoleteLayoutParam`	**OLP**	94	561
`ViewHolder`	**VH**	892	2105
`HashMapUsage`	**HMU**	28	229
`ExcessiveMethodCalls`	**EMC**	557	9529
`MemberIgnoringMethod`	**MIM**	89	7844

3.3 Counting Expenses and Estimating Debt

The next step towards estimating energy debt is to define a strategy to analyze the occurrence of such smells in a given release. The starting point for this task will be to use a common source code analysis tool capable of detecting code smells. There are several ways to achieve this. For instance, *SonarQube*[3], which is a widely used tool for technical debt estimation, provides an API for defining detection rules for issues/smells of different languages.[4]

Detecting smell occurrences, however, is a necessary but not the sole requirement to properly analyze its impact on energy debt. A smell can be detected, for instance, inside a block of dead/unreachable code, or it can be placed inside a procedure that may only be executed once in a software lifecycle (e.g. an initial setup). On the other hand, a code smell can also be part of a mechanism designed to be re-utilized several times, such as a loop or a thread. These scenarios should be considered when estimating energy debt, and since our approach implies using statical analysis mechanisms, we can follow already defined strategies for static energy analysis.

A very common and well-established approach for these situations is to define weights for smells, depending on the context in which they occur. For instance, Jabbarvand et al. [30] defined a strategy for weighing instructions that might be repeated. First, it extracts the full method *call graph* of a program, and provide for each method an energy score; such a score depends on 3 things: *(i)* how many paths can be taken to reach that node from the root node, *(ii)* whether it is found inside a loop, and if so *(iii)* what is the expected loop's bound; this statically obtained information is then used to increase/decrease the energy score of the node.

Several strategies have been suggested for this task, all of which accepted by the community. This leads us to believe that, although it is important to weight code smells depending on the occurrence context, several factors can influence the decision on what approach to follow (e.g. how much information is extracted from the smell detection tool, or trading off information detail with the analysis time). Hence, we argue that the selected strategy is also context-dependent, and can be as simple or as detailed as desired. Nevertheless, whatever approach one follows, an update to Eq. 2 is necessary to consider it. As an example, we considered a simplification of the strategy from Jabbarvand et al. [30]:

$$w(i,r) = \sum_{j=1}^{C} paths(j) \times LB \qquad (4)$$

In this equation:

– C is the number of i smells found in the release r;

- $paths(j)$ represents the number of paths in the *call graph* through which the jth occurrence of smell i is reachable;
- LB will be 1 **if** the jth of the smell is outside a loop, or a constant indicating the loop bound; it can be inferred if possible, or pre-established.

In order to better explain how all these concepts connect with each other, when aiming at estimating the energy debt of different software releases, we have prepared a running example, depicted in Fig. 4. In this example, we have a catalog with 3 smells, each one with the energy gains thresholds defined (values are in milliJoules (mJ) per minute), and 3 releases with the analysis report for each. The report is a list of the detected smells, where for each one there is information regarding *(i)* the number of paths through which the smell is reachable *(paths)*, and *(ii)* whether it was found inside a loop $(LB > 1)$ or not $(LB = 1)$.

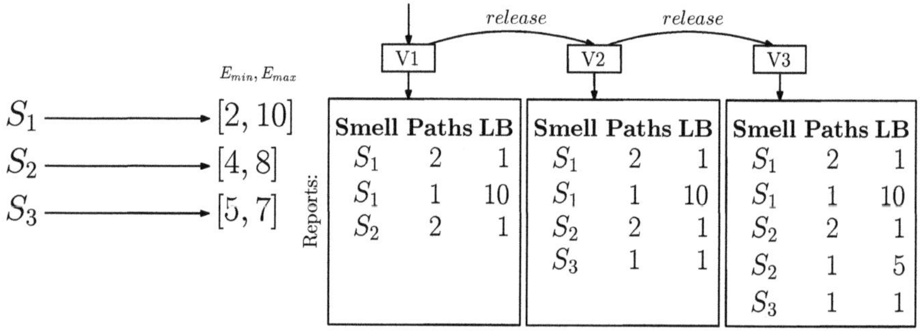

Fig. 4. Estimating Energy Debt per Release.

Using the formula from Eq. 4, we can determine the weight to be applied to each smell. For example, for release $v1$, the weights for smell $s1$ and $s2$ would be:

$$w(s1, v1) = (2 \times 1) + (1 \times 10) = 12$$
$$w(s2, v1) = (2 \times 1) = 2$$

We can apply the computed weights in the formula from Eq. 2, to obtain an estimated value for the energy debt of release $v1$. As previously mentioned, our energy debt definition considers two reference values: the lowest and highest estimated energy debt. This means that the *cost* function in Eq. 2 must be computed twice: the first using the lowest estimated gains per smell (E_{min}), and the second using the highest (E_{max}). Once again, for release $v1$, we would have the following $cost_{min}$ and $cost_{max}$ values:

$$cost_{min}(v1) = (w(s1, v1) \times E_{min}(s1)) + (w(s2, v1) \times E_{min}(s2))$$
$$\Leftrightarrow cost_{min}(v1) = (12 \times 2) + (2 \times 4) = 32$$
$$cost_{max}(v1) = (w(s1, v1) \times E_{max}(s1)) + (w(s2, v1) \times E_{max}(s2))$$
$$\Leftrightarrow cost_{max}(v1) = (12 \times 10) + (2 \times 8) = 136$$

These two reference values represent the energy debt for release $v1$. This means that energy debt can vary from a minimum of 32 to a maximum of 136 milliJoules per minute. As explained previously, energy debt is expressed as a function of usage time. Therefore, if one wants to know how much debt this release accumulates after being used for, e.g., one hour, this can be estimated as follows:

$$ed(v1, 60min) = \left(cost_{min}(v1) \times 60; cost_{max}(v1) \times 60 \right)$$

$$\Leftrightarrow ed(v1, 60min) = \left(1,920mJ; 8,160mJ \right)$$

The estimated values indicate an energy debt varying between 1.92 and 8.16 Joules per hour. This means that, for every hour that release $v1$ is being used, it could be consuming **at least** $1.92J$ less, and the savings could be **up to** $8.16J$.

Finally, it is important to point out that the accuracy of the estimated thresholds relies on the adequacy/robustness of the analysis components, namely the smells catalog, the code analysis tool, and the weighing function for repeated smell executions. It is possible to use our energy debt approach to compute reference values for the energy inefficiency of a release, rather than to produce extremely accurate estimates of the potential energy savings per usage time. It depends on how one wants to apply the concept.

3.4 Paying Interests

The concept of *interest* in technical debt has already been formulated [25], and its practical application has also been studied [31–33]. The concept is based on the fact that, as a software system evolves (i.e., new versions are released), the cost/effort of adding features to a new release (*maintenance effort*, expressed as working hours) keeps increasing if the task of addressing the technical debt keeps being postponed. Maintaining a release with technical debt requires more effort than to maintain the same release without it; the effort difference between the two is called the *technical debt interest*.

The left-hand side of Fig. 2 illustrates the interest concept. There is a software version, A, containing code smells, and therefore technical debt. At this point, a decision can be made on what to prioritize: *(i)* invest effort in fixing the smells (*repayment effort*) and release an optimal version of that release, A', without technical debt, or *(ii)* release the version with the smells. If the priority is *(ii)*, then the evolution effort to a new release B will be higher. This additional effort could be avoided, but the priority was releasing a new version, which can happen for a wide variety of reasons (e.g. client demands, faster market reach, etc.); this resembles the idea of accumulation of debt, and debt needs to be re-payed.

Chatzigeorgiou et al. [25] presented a technique to predict the technical debt *breaking point*, i.e., when the accumulated interest is higher than the initial effort to remove the technical debt (i.e., the *principal*). With this, it is possible to present developers with another choice: if technical debt keeps being "ignored", then they have approximately until release number N to properly deal with it; otherwise, from that moment on, the additional *maintenance effort* will always be higher than the effort to deal with the *principal*. In conclusion, at that point they are wasting development time.

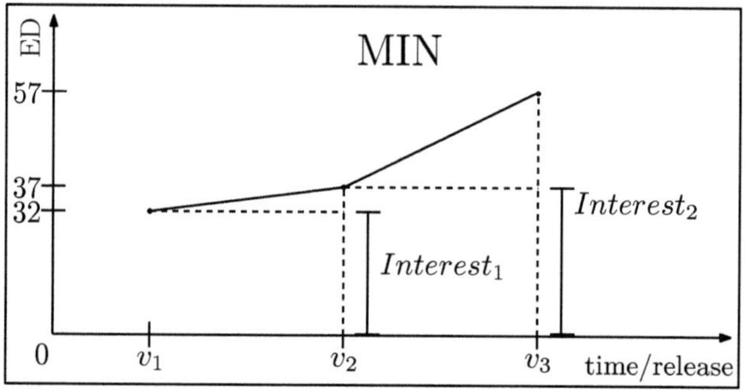

Fig. 5. Accumulation of Interest.

When considering energy instead of technical debt, the interest concept needs another definition. First, it will not indicate how much more maintenance effort is being applied, since energy debt does not measure effort, but the drainage of a resource. In that sense, *energy debt interest* is the amount of excessive energy consumed over time, that could be avoided if the issues were properly addressed earlier. In simple terms, it is the accumulated energy debt after n releases. This concept complements energy debt in the sense that it can be used to estimate the "real-world" cost of not fixing the smells, which can be monetary (as energy costs money) or uptime-related (if the analyzed software is targeted for IoT/mobile devices).

Figure 5 illustrates our perception of energy interest, using our example from Fig. 4. Again, 3 software releases are considered: $v1$, $v2$, and $v3$[5]. For this example, we are assuming that, as new versions are released, the issues from previous versions were not addressed. Hence, *energy debt* is always increasing. For $v1$, we consider that no interest was accumulated, due to the fact that energy debt depends on usage time. Hence, at the exact instant when the version was released, it was never used.

---

[5] The presented values refer to the minimum estimated debt.

For release $v2$, we know that it added a new smell, $s3$. If **all** smells from the previous version ($v1$) were fixed upon release, then this $v2$'s minimum energy debt would be 5. However, since $v1$ contained smells, from the time interval comprised between the two releases, the software was excessively consuming 32 mJ for each minute it was being used (i.e., the energy debt from $v1$). Therefore, for release $v2$, the accumulated debt (i.e., the interest) is 32 mJ per minute.

When considering release $v3$, however, the reasoning to infer the interest requires adjustments. For once, $v3$ has two predecessors, while $v2$ has only one. Between $v1$ and $v2$, the energy debt was 32, and between $v2$ and $v3$ it was 37. To estimate how much debt was accumulated, we should infer a value based on the two. One possible way to tackle this is to compute the average of **all** energy debts from previous releases. In this particular case, the minimum accumulated interest would be $\frac{(32+37)}{2} = 34.5$.

Finally, it is important to interpret interest similarly to how energy debt is interpreted: a minimum/maximum energy being excessively consumed **per usage time**. Hence, we argue that, when considering the interest for the nth release, the expected usage time should be higher than the one for any previous release. Therefore it is guaranteed that, even though energy debt is reduced from one release to another, the interest will be inflated for later releases.

3.5 Preliminary Experiment

At this point, we have already presented the concepts that define energy debt. In order to validate this concept, this section will describe an experiment performed over an existing software system, retrieved from GitHub, where we estimate the energy debt for all its releases. We start by presenting the software used for the experiment and the used approach for detecting the energy code smells; afterward, having detected the energy smells, we applied an energy debt analysis on each of the software's releases and presented the obtained results.

The Experimental Software System. In order to fully showcase the energy debt analysis described in the previous sections, we obtained a software system to use for a preliminary experiment, which followed two main requirements. First, it needed to be a system with several explicit releases. This way we could analyze the evolution of energy debt over time, and also present the energy interest values for each release. Second, it must be an Android application. This is due to the fact that 5 of the smells in our catalog are Android-specific, and we wanted to explore the catalog as much as possible.

We searched for applications on GitHub, as it is a well-known and widely used platform to host open-source projects. Using GitHub's search engine, we searched for repositories with two characteristics: *(i)* the repository's main language should be Java, as it is the development language for Android, and *(ii)* there should be specific references to the Android framework (i.e., Java imports to Android APIs, since we were filtering projects by code). GitHub's search engine only allows the retrieval of the first 1000 search results, hence we only selected the repositories referenced in those 1000 results containing several releases of Android applications.

Applications without smells had no interest to be analyzed, so defining an approach for detecting smells was necessary at this point. For this purpose, we used our tool called *E-debitum*[6], which is an extension to *lint*[7], the code inspection tool for Android. *Lint* has already built-in detection rules for 5 of the 8 smells in our catalog, thus we developed new rules for the remaining 3 smells. Our tool runs an optimized *lint* check on an Android application's code (i.e., only searching for the 8 energy smells, instead of performing a full analysis), and processes the output to indicate how many occurrences of each smell were found.

Table 2. Detected smells on *EscapeApp* releases

smell count	*versions*				
	v0.4	**v0.5**	**v0.6**	**v0.7**	**v0.8**
	OLP: 5	*OLP:* 9	*OLP:* 12	*OLP:* 17	*OLP:* 17
	EMC: 1	*EMC:* 2	*EMC:* 1	*EMC:* 1	*EMC:* 1
			HMU: 8	*HMU:* 8	*HMU:* 8
			MIM: 2	*MIM:* 3	*MIM:* 3

The *EscapeApp*[8], a client application used for a virtual reality escape game, which had 5 explicit releases. Table 2 presents the results of the *lint* analysis for this application. For each release, we indicate the number of occurrences of each detected smell, and how that number varied from the previous release.

Energy Debt Analysis. The smell detection performed on *EscapeApp* only provided the list of existing smells on each release. The *lint* tool does not offer a built-in mechanism to report the context of each detected smell, so information such as loop bounds could not be obtained automatically. As previously explained in Sect. 3.3, this information enhances the energy debt analysis. Thus we performed this analysis manually for each release and looked at each smell report to determine if it was being used within a loop. If so, a loop-bound value was inferred considering the loop's code. Although we argue that it is not critical to have this analysis in order to estimate energy debt, for the *SonarQube* extension that we are working on, we intend to include the smell detection with context information regarding reachable paths and loop bounds inference to further enhance the analysis.

Based on the report presented in Table 5 and the manual code inspection performed, we can estimate the energy debt interval for each release. For instance,

---

[6] *E-debitum* webpage: https://github.com/e-debitum/E-Debitum-tool.
[7] *Lint* webpage: http://tools.android.com/tips/lint.
[8] *EscapeApp* code: https://github.com/alanvanrossum/kroketapp.

after inspecting each smell occurrence in release $v0.4$, we realized that the **EMC** smell was the only one found within a loop. After carefully inspecting and understanding the code, we estimated a loop bound of 10 iterations. Hence, the $cost_{min}$ and $cost_{max}$ values for release $v0.4$ can be calculated as follows:

$$cost_{min}(v0.4) = (5 \times E_{min}(OLP)) + (w_{EMC} \times E_{min}(EMC))$$
$$\Leftrightarrow cost_{min}(v0.4) = 5 \times 94 + 10 \times 557$$
$$\Leftrightarrow cost_{min}(v0.4) = 5,027 \; mJ \; (per \; min)$$

$$cost_{max}(v0.4) = (5 \times E_{max}(OLP)) + E_{max}(EMC)$$
$$\Leftrightarrow cost_{max}(v0.4) = 5 \times 561 + 10 \times 9529$$
$$\Leftrightarrow cost_{max}(v0.4) = 98,095 \; mJ \; (per \; min)$$

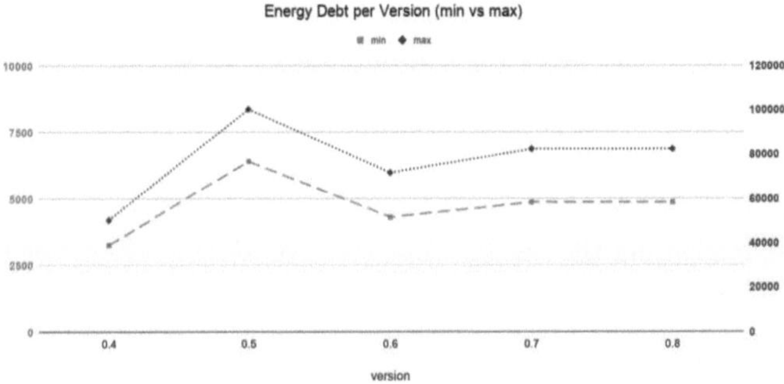

Fig. 6. Minimum and Maximum Energy Debt for *EscapeApp* releases.

The computed energy debt for each of the five versions is illustrated in Fig. 6. As expected, the energy debt increases from release $v0.4$ to $v0.5$, as the total number of smells also increases. An interesting aspect, however, is the fact that the energy debt decreases after the release $v0.5$. If we look at the detection report on Table 5, we see that it has 2 **EMC** smells. This smell has the highest value for maximum energy debt. It is also the smell with the highest weight, since it is always found inside loops. In that sense, addressing this smell before releasing $v0.6$ eventually paid off, even though other smells were introduced.

Nevertheless, it is important to also look at the concrete values of both the maximum and minimum energy debt. From release $v0.5$ to $v0.6$, energy debt might be reduced **at most** by $\approx 68 \; J$, which is a considerable improvement. It is, however, the most profitable scenario in potential, and if we consider a more cautious analysis to the data (i.e., looking at the minimum energy debt variations) we can see that energy debt might only reduced by $\approx 4 \; J$. Once again, it is up to the developers and/or product managers to decide how they look at the trade-off between refactoring effort and potential energy savings.

Estimating the accumulated interest for each release is also a straightforward task. If we consider, for example, releases $v0.5$ and 0.7, the minimum interest would be 6.04 J per minute for the former, and 8.67 J per minute for the latter, as the following calculations demonstrate:

$$interest(v0.5) = average(ed(v0.4)) = \left(6.04\ J\ ;\ 98.095\ J \right)$$

$$interest(v0.7) = average(ed(v0.4), ed(v0.5), ed(v0.6))$$

$$\Leftrightarrow interest(v0.7) = \left(8.67\ J\ ;\ 140.198\ J \right)$$

With these reference values, one could easily provide an estimate on how much energy was excessively consumed from the first release until release $v0.5/v.07$. Let us assume that the total amount of time an application is estimated to be used between these releases was, for instance, 5 full days (120 hours or $7,200$ minutes). This would mean that when $v0.5$ was released, the energy debt interest would vary between a minimum of $43,488$ J, to a maximum of $706,284$ J. When considering $v0.7$, with 3 prior releases, the expected usage time increases by: $3 \times 7,200 = 21,600$ minutes. Therefore, the interest values increase significantly as expected: a minimum of $187,272$ J and a maximum of $3,028,276$ J. Even though this is merely a motivational case study, it is safe to consider that neglecting the energy debt for several releases proves to be highly costly.

4   Energy Debt: Tools

Although energy smells at large may occur in any programming language, a number of language-specific smells exist. With this in mind, this proof of concept was written in Java.

To achieve our goals, the *E-Debitum* tool[9] was implemented through two separate SonarQube plugins, through the aid of templates which will be used to handle the necessary logic involved in integrating the built tool into the platform. The first plugin implements a set of rules corresponding to the aforementioned energy smells. As such, it is dubbed as the rules plugin. It is used during the analysis process to detect and document any and all energy smells present in the code, alongside any other issues actively being searched for by the platform.

The second plugin, dubbed the metrics plugin, is active immediately after the code scan is complete and is responsible for measuring the minimum/maximum estimated values of Energy Debt in the program. It tallies up the number of instances of each given code smell and, using the estimated joule per

---

[9] Github: https://github.com/e-debitum/E-Debitum-tool.

minute expenditure they cause, stores the total debt value in its best/worst-case scenarios.

These functionalities were separated into two to better improve the readability of the plugin's code, as well as to allow SonarLint users to make use of the rules as they write code, as opposed to exclusively discovering the energy smells upon analysis. It is worth noting that in this case, however, they would only see the rules but not fully know their impact as a whole.

4.1   Rules Plugin

To create the first SonarQube plugin, an official Java-specific rules template[10] was used to save development time. As such, five files will be created for each smell-detecting rule implemented.

For starters, a test file has to be written. This will contain code to be used to test the rule. It holds a set of methods in which instances of both compliant and non-compliant code are written to show SonarQube what patterns to search for when testing for that given smell and which to eke out. Afterward, a rule class will be developed, which handles the logic when SonarQube detects a possible instance of non-compliant code. It will check to confirm that it is not a false positive and, if it does, report the issue. Finally, a test class will be created, which simply houses the unit test for the rule. This will verify that instances where the test file and rule-class flag code as non-compliant match up when building the plugin.

Additionally, an HTML and JSON file will be made to display the details of the rule to the end user within SonarQube. These include a short description of the rule, along with an example case of non-compliant code and its respective fix, as well as the rule's related technical debt and tags.

Once implemented, each rule must be activated within the plugin by adding it to the *RulesList* class baked into the plugin template.

4.2   Metrics Plugin

For this plugin, another template was used, making use of the Sonar API[11]. To do this, three Java classes need to be developed. The first of these implements Sonar API's Plugin interface. This class is the entry point for all extensions made to SonarQube. In this case, two extensions will be added, which are the other two necessary classes.

The first extension is an implementation of the MeasureComputer interface. This class is responsible for describing its output to the system and reading which and how many energy smells were detected and summing up the associated maximum and minimum estimated debt. Lastly, the final class necessary and

---

[10] *SonarQube* Java rules template plugin: https://github.com/SonarSource/sonar-custom-rules-examples/tree/master/java-custom-rules.

[11] *SonarQube* custom plugin example plugin: https://github.com/SonarSource/sonar-custom-plugin-example.

the second extension added is an implementation of the Metrics interface. This class simply defines the metadata for minimum and maximum energy debt as variables in SonarQube.

4.3   Experimental Validation

To test the efficacy of the implemented plugins, a set of Android applications was acquired by searching GitHub's library of open-source projects. From these, a handful of highly downloaded Android apps were selected with *(i)* a fluctuating number of smells per release and *(ii)* a significant number of releases across its lifespan. As such, three applications that met this criteria were chosen.

PDF Viewer Plus[12] is an open source PDF viewer which allows the reading and sharing of PDF files, with the ability to customize the UI with a variety of themes, which had 17 separate releases. Table 3 presents the results of the tool analysis.

Table 3. Detected smells on Pdf Viewer Plus releases

	versions																
	1.0	1.1	1.2	1.3	1.4	1.5	1.6	1.7	1.8	1.9	2.0	2.1	2.2	2.3	2.4	2.5	2.6
smell count	MIM:1	MIM:0	MIM:0	MIM:0	MIM:0	MIM:0	MIM:0	MIM:0	MIM:0	MIM:0	MIM:0	MIM:0	MIM:0	MIM:0	MIM:0	MIM:0	MIM:0
			VH:1	VH:1	VH:1	VH:1	VH:1	VH:2	VH:2	VH:2	VH:2	VH:2	VH:2	VH:2	VH:1	VH:1	VH:1

Despite the high number of releases of the app, the number of smells detected remained relatively low, with only one to two smells being detected across nearly its entire lifespan. Likewise, its energy debt remained stable all throughout, keeping a low range between its best and worst-case scenarios at a maximum average of 4.09 J/min and a minimum average of 1.18 J/min. Using SonarQube's Activity tab, we can see the plot of the app's code smells - in this case, exclusively energy smells - along with their maximum and minimum energy debt, as seen in Fig. 7.

Malse Geluiden[13] is another open-source app available in the Google Play-Store. Through it, users can play a set of sounds and animations on their device, as well as share them through *WhatsApp*. This app was tested across 7 explicit releases, which were run through SonarQube. Table 4 presents the four energy smells and its occurrences found throughout the analysis of this app.

This app finds an increasing number of energy smells with each release and a rising overall level of debt, as well as a wider difference between its maximum and minimum values, achieving an average of 45.76 J/min and 6.66 J/min, respectively, as seen in Fig. 8.

---

[12] *PDF Viewer Plus* code: https://github.com/JavaCafe01/PdfViewer.
[13] *Malse   Geluiden*   code: https://apkpure.com/nl/malse-geluiden/nl.meganetjes. malsegeluiden.

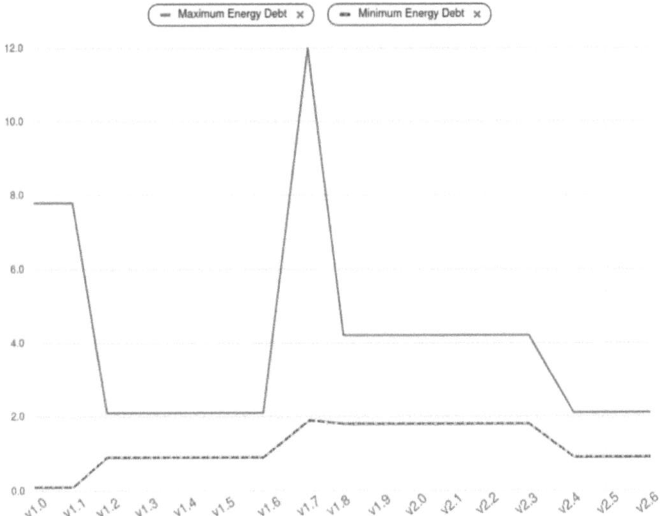

Fig. 7. Minimum and Maximum Energy Debt for *PDF Viewer Plus* releases over time.

Table 4. Detected smells on *MalseGeluiden* releases

	versions						
	1.0	**1.1**	**1.2**	**1.2.2**	**1.2.3**	**1.2.4**	**1.2.5**
smell count	MIM:1	MIM:2	MIM:2	MIM:2	MIM:2	MIM:2	MIM:2
	VH:6	VH:6	VH:6	VH:6	VH:6	VH:6	VH:6
	EMC:4	EMC:1	EMC:1	EMC:1	EMC:1	EMC:1	
		RC:6	RC:6	RC:6	RC:6	RC:6	

Lastly, *EscapeApp*[14] is a client application used for a virtual reality escape game, which contains three different smells occurrences spread over five releases, as seen in Table 5.

The smell detection performed on *EscapeApp* displays a wide gap of around 100 J/min between its best and worst case, with a maximum and minimum averages of 139.20 J/min and 8,64 J/min, respectively, as seen in Fig. 9.

5 Related Work

Technical debt refers to the pitfalls of creating sub-optimal software to fit a shorter interval, introduced by Cunningham [34]. This is a common practice employed to meet short-term development deadlines, with the intent of completing it at a later time. As software evolves, it is liable to take on debt from

---

[14] *EscapeApp* code: https://github.com/alanvanrossum/kroketapp.

Table 5. Detected smells on *EscapeApp* releases

smell count	versions				
	v0.4	**v0.5**	**v0.6**	**v0.7**	**v0.8**
smell count	EMC: 1	EMC: 2	EMC: 1	EMC: 1	EMC: 1
			HMU: 8	HMU: 8	HMU: 8
			MIM: 2	MIM: 3	MIM: 3

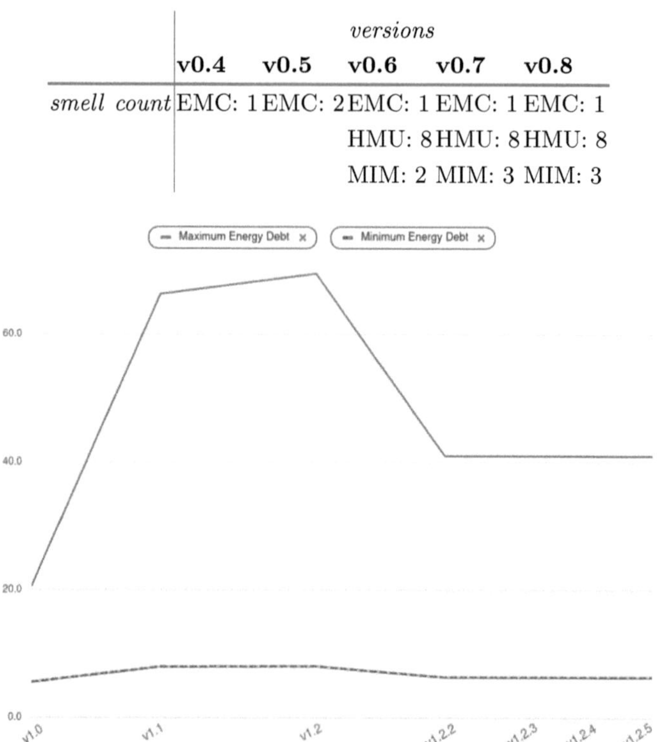

Fig. 8. Minimum and Maximum Energy Debt for *MalseGeluiden* releases over time.

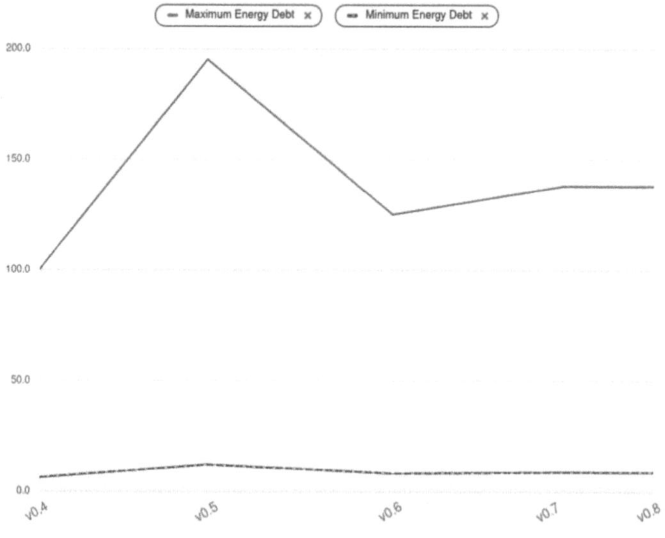

Fig. 9. Minimum and Maximum Energy Debt for *EscapeApp* releases over time.

several sources: "technological obsolescence, change of environment, rapid commercial success, the advent of new and better technologies, and so on – in other words, the invisible aspects of natural software aging and evolution." [35]. However, more common issues can also plague software development, born from poor coding practices or general ignorance, accentuating different aspects of technical debt [36].

As already known, allowing technical debt to continuously build up without a level of debt management raises the risk of producing unmanageable and inefficient code, which can hamper the addition of new or updating existing functionalities. This makes it so the longer such code goes unattended, the more resources will be needed to correct it and with diminishing returns [25].

One such inefficiency in software is high energy consumption. In fact, profiling, analyzing and improving software energy efficiency has become a very active research field [37]. Studies have shown that developers are aware of the energy consumption problem, and often times seek help in solving such issues [2,13,14]. Currently, there is a broad range of work done on understanding what aspects of programming languages can contribute to high energy costs such as different data structures [4,38–40], languages [22,41], memoization [42], design and coding patterns [43,44], code refactoring [45,46], and even the testing phase [47].

Specific to the Android ecosystem, there has been research in topics such as the classification of Android apps as being more/less energy efficient [30], identifying energy green APIs [3], estimating energy consumption in code fragments [48,49], etc. In fact, research in the reduction of energy consumption in the Android system is most likely the most explored environment in this research field over the past decade [5,6,19–21,26,28,29,50,51]. The results and patterns of most of these studies can be integrated into our energy smell catalog in order to be used in the calculation of energy debt.

Additionally, much research has been conducted in providing several approaches to the measurement of energy consumption. For example for Android energy analysis there is: *eCalc* [52], *vLens* [53], *eProf* [54], *Trepn* [30,42,49,55–57], *GreenOracle* [58], *GreenScaler* [59]. There is also work in automatic tools to help detect energy greedy code spots [60–63], automatically refactoring for the most energy-efficient data structure [64,65], or automatically refactoring energy greedy Android patterns [6,8,29].

The aforementioned works on the different approaches to reducing the energy consumption of software systems do not yet translate their potential gains across a period of time into the actual energy savings (or costs) a developer or business can have on his/her software by applying such transformations. It is our belief that, with this work, we have helped close this gap in not only knowing if an alternative solution is more energy efficient but by how much can we save (in energy consumption or even monetary savings) over time if and when we adopt the energy-efficient alternative.

6   Conclusions

This tutorial presented the concept of energy debt as the additional energy cost over time of a software system due to the occurrences of energy code smells in its source code. This new concept is analogous to the Technical debt, unifying software maintenance and software (energy) performance. It is expressed as a function considering *(i)* the number of smells, *(ii)* the context in which they were detected, and *(iii)* the expected usage time of the application. Energy debt interest is also expressed as the accumulation of energy debt per release, which could be avoided by eliminating energy smells in previous releases.

Currently, we are extending an existing catalog of reported state-of-the-art energy code smells, and their known energy costs per usage time, which can be considered when calculating energy debt. Additionally, an extension of the concept of energy debt is being developed within the *SonarQube* framework, where it will support the inference of the context and number of detected smells, based on our catalog.

References

1. Zhang, Y., Tang, G., Huang, Q., Wu, K., Wu, Y., Wang, Y.: Investigating low-battery anxiety of mobile users. In: 2022 IEEE International Conferences on Internet of Things (iThings) and IEEE Green Computing & Communications (GreenCom) and IEEE Cyber, Physical & Social Computing (CPSCom) and IEEE Smart Data (SmartData) and IEEE Congress on Cybermatics (Cybermatics), pp. 272–279 (2022). https://doi.org/10.1109/iThings-GreenCom-CPSCom-SmartData-Cybermatics55523.2022.00022
2. Pinto, G., Castor, F.: Energy efficiency: a new concern for application software developers. Commun. ACM **60**(12), 68–75 (2017). https://doi.org/10.1145/3154384
3. Linares-Vásquez, M., Bavota, G., Bernal-Cárdenas, C., Oliveto, R., Di Penta, M., Poshyvanyk, D.: Mining energy-greedy API usage patterns in android apps: an empirical study. In: Proceedings of 11th Working Conference on Mining Software Repositories, MSR 2014, ACM, pp. 2–11 (2014)
4. Pereira, R., Couto, M., Cunha, J., Fernandes, J.P., Saraiva, J.: The influence of the java collection framework on overall energy consumption. In: Proceedings of 5th International Workshop on Green and Sustainable Software, GREENS '16, ACM, pp. 15–21 (2016)
5. Cruz, L., Abreu, R.: Performance-based guidelines for energy efficient mobile applications. In: Proceedings of the 4th International Conference on Mobile Software Engineering and Systems, MOBILESoft '17, IEEE Press, pp. 46–57 (2017)
6. Carette, A., Younes, M.A.A., Hecht, G., Moha, N., Rouvoy, R.: Investigating the energy impact of Android smells. In: 2017 IEEE 24th International Conference on Software Analysis, Evolution and Reengineering (SANER), pp. 115–126 (2017)
7. Oliveira, W., Oliveira, R., Castor, F., Fernandes, B., Pinto, G.: Recommending energy-efficient java collections. In: 2019 IEEE/ACM 16th International Conference on Mining Software Repositories (MSR), pp. 160–170 (2019). https://doi.org/10.1109/MSR.2019.00033

8. Couto, M., Fernandes, J.P., Saraiva, J.: Energy refactorings for android in the large and in the wild. In: 2020 IEEE 27th International Conference on Software Analysis, Evolution and Reengineering (SANER), London, Ontario, Canada (2020)

9. Li, Z., Avgeriou, P., Liang, P.: A systematic mapping study on technical debt and its management. J. Syst. Softw. **101**(C), 193–220 (2015). https://doi.org/10.1016/j.jss.2014.12.027

10. Fowler, M., Beck, K., Brant, J., Opdyke, W., Roberts, D.: Refactoring: Improving the design of existing code (2002)

11. Fontana, F.A., Ferme, V., Spinelli, S.: Investigating the impact of code smells debt on quality code evaluation. In: Third International Workshop on Managing Technical Debt (MTD), vol. 2012, pp. 15–22 (2012). https://doi.org/10.1109/MTD.2012.6225993

12. Lahti, J.R., Tuovinen, A.-P., Mikkonen, T.: Experiences on managing technical debt with code smells and antipatterns. In: 2021 IEEE/ACM International Conference on Technical Debt (TechDebt), pp. 36–44 (2021). https://doi.org/10.1109/TechDebt52882.2021.00013

13. Pinto, G., Castor, F., Liu, Y.D.: Mining questions about software energy consumption. In: Proceedings of the 11th Working Conference on Mining Software Repositories, MSR 2014, Association for Computing Machinery, pp. 22–31 (2014). https://doi.org/10.1145/2597073.2597110

14. Manotas, I., et al.: An empirical study of practitioners' perspectives on green software engineering. In: 2016 IEEE ACM 38th International Conference on Software Engineering (ICSE), pp. 237–248. IEEE (2016)

15. Torre, D., Procaccianti, G., Fucci, D., Lutovac, S., Scanniello, G.: On the presence of green and sustainable software engineering in higher education curricula. In: Proceedings of the 1st International Workshop on Software Engineering Curricula for Millennials, SECM '17, IEEE Press, pp. 54–60 (2017). https://doi.org/10.1109/SECM.2017.4

16. Couto, M., Maia, D., Saraiva, J., Pereira, R.: On energy debt: managing consumption on evolving software. In: Proceedings of the 3rd International Conference on Technical Debt, TechDebt '20, Association for Computing Machinery, New York, NY, USA, pp. 62–66 (2020). https://doi.org/10.1145/3387906.3388628

17. Feitosa, D., Cruz, L., Abreu, R., Fernandes, J.P., Couto, M., Saraiva, J.: Patterns and energy consumption: design, implementation, studies, and stories. In: Calero, C., Moraga, M.Á., Piattini, M. (eds.) Software Sustainability, Springer, pp. 89–121 (2021). https://doi.org/10.1007/978-3-030-69970-3_5

18. Maia, D., Couto, M., Saraiva, J., Pereira, R.: E-debitum: managing software energy debt. In: Proceedings of the 35th IEEE/ACM International Conference on Automated Software Engineering, ASE '20, Association for Computing Machinery, New York, NY, USA, pp. 170–177 (2021). https://doi.org/10.1145/3417113.3422999

19. Morales, R., Saborido, R., Khomh, F., Chicano, F., Antoniol, G.: EARMO: an energy-aware refactoring approach for mobile apps. IEEE Trans. Software Eng. **44**(12), 1176–1206 (2018)

20. Saborido, R., Morales, R., Khomh, F., Guéhéneuc, Y.-G., Antoniol, G.: Getting the most from map data structures in Android. Empir. Softw. Eng. **23**(5), 2829–2864 (2018)

21. Palomba, F., Di Nucci, D., Panichella, A., Zaidman, A., De Lucia, A.: On the impact of code smells on the energy consumption of mobile applications. Inf. Softw. Technol. **105**, 43–55 (2019)

22. Pereira, R., et al.: Energy efficiency across programming languages: how do energy, time, and memory relate?. In: Proceedings of the 10th ACM SIGPLAN International Conference on Software Language Engineering, SLE 2017, ACM, pp. 256–267 (2017)
23. Pereira, R., et al.: Ranking programming languages by energy efficiency, science of computer programming 205, cited by: 46; all open access. Green Open Access (2021). https://doi.org/10.1016/j.scico.2021.102609
24. Allman, E.: Managing technical debt. Commun. ACM **55**(5), 50–55 (2012). https://doi.org/10.1145/2160718.2160733
25. Chatzigeorgiou, A., Ampatzoglou, A., Ampatzoglou, A., Amanatidis, T.: Estimating the breaking point for technical debt. In: 2015 IEEE 7th International Workshop on Managing Technical Debt (MTD), pp. 53–56 (2015). https://doi.org/10.1109/MTD.2015.7332625
26. Vekris, P., Jhala, R., Lerner, S., Agarwal, Y.: Towards verifying android apps for the absence of no-sleep energy bugs. In: Proceedings of the 2012 USENIX Conference on Power-Aware Computing and Systems, HotPower'12, USENIX Association (2012)
27. D. Li, W. G. J. Halfond, An Investigation into Energy-saving Programming Practices for Android Smartphone App Development, in: Proc. of 3rd Int. Workshop on Green and Sustainable Software, GREENS 2014, ACM, 2014, pp. 46–53
28. Jiang, H., Yang, H., Qin, S., Su, Z., Zhang, J., Yan, J.: Detecting energy bugs in android apps using static analysis. In: Duan, Z., Ong, L. (eds.) Formal Methods and Software Engineering, pp. 192–208. Springer International Publishing (2017)
29. Cruz, L., Abreu, R.: Using Automatic Refactoring to Improve Energy Efficiency of Android Apps, CoRR abs/1803.05889 (2018)
30. Jabbarvand, R., Sadeghi, A., Garcia, J., Malek, S., Ammann, P.: EcoDroid: an approach for energy-based ranking of android apps. In: Proceedings of 4th International Workshop on Green and Sustainable Software, GREENS '15, IEEE Press, pp. 8–14 (2015)
31. Ampatzoglou, A., Michailidis, A., Sarikyriakidis, C., Ampatzoglou, A., Chatzigeorgiou, A., Avgeriou, P.: A framework for managing interest in technical debt: an industrial validation. In: Proceedings of the 2018 International Conference on Technical Debt, TechDebt '18, Association for Computing Machinery, pp. 115–124 (2018). https://doi.org/10.1145/3194164.3194175
32. Tsintzira, A., Ampatzoglou, A., Matei, O., Ampatzoglou, A., Chatzigeorgiou, A., Heb, R.: Technical debt quantification through metrics: an industrial validation. In: 15th China-Europe International Symposium on Software Engineering Education (CEISEE' 19) (2019)
33. Ampatzoglou, A., Ampatzoglou, A., Avgeriou, P., Chatzigeorgiou, A.: Establishing a framework for managing interest in technical debt. In: Proceedings of the Fifth International Symposium on Business Modeling and Software Design - Volume 1: BMSD,, INSTICC, SciTePress, pp. 75–85 (2015). https://doi.org/10.5220/0005885700750085
34. Cunningham, W.: The WyCash portfolio management system. SIGPLAN OOPS Mess. **4**(2), 29–30 (1992). https://doi.org/10.1145/157710.157715
35. Kruchten, P., Nord, R.L., Ozkaya, I.: Technical debt: from metaphor to theory and practice. IEEE Softw. **29**(6), 18–21 (2012). https://doi.org/10.1109/MS.2012.167
36. Tom, E., Aurum, A., Vidgen, R.: An exploration of technical debt. J. Syst. Softw. **86**(6), 1498–1516 (2013). https://doi.org/10.1016/j.jss.2012.12.052
37. Rua, R.: Green software in the large: Energy-driven techniques, tools and repositories, Ph.D. thesis, Universidade do Minho (2024)

38. Pinto, G., Castor, F.: Characterizing the energy efficiency of java's thread-safe collections in a multi-core environment. In: Proceedings of SPLASH'2014 workshop on Software Engineering for Parallel Systems (SEPS), SEPS, Vol. 14 (2014)
39. Lima, L.G., Soares-Neto, F., Lieuthier, P., Castor, F., Melfe, G., Fernandes, J.P.: Haskell in green land: analyzing the energy behavior of a purely functional language. In: 2016 IEEE 23rd International Conference on Software Analysis, Evolution, and Reengineering (SANER), Vol. 1, pp. 517–528 (2016)
40. Hasan, S., King, Z., Hafiz, M., Sayagh, M., Adams, B., Hindle, A.: Energy profiles of java collections classes. In: Proceedings of the 38th International Conference on Software Engineering, ACM, pp. 225–236 (2016)
41. Couto, M., Pereira, R., Ribeiro, F., Rua, R., Saraiva, J.: Towards a green ranking for programming languages. In: Proceedings of the 21st Brazilian Symposium on Programming Languages, SBLP 2017, ACM, pp. 7:1–7:8 (2017)
42. Rua, R., Couto, M., Pinto, A., Cunha, J., Saraiva, J.: Towards using memoization for saving energy in android. In: Proceedings of the XXII Iberoamerican Conference on Software Engineering, CIbSE, pp. 279–292 (2019)
43. Sahin, C., et al.: Initial explorations on design pattern energy usage. In: 2012 First International Workshop on Green and Sustainable Software (GREENS), pp. 55–61. IEEE (2012)
44. Rua, R., Saraiva, J.: A large-scale empirical study on mobile performance: energy, run-time and memory. Empirical Softw. Eng. **29**(1) (2023). https://doi.org/10.1007/s10664-023-10391-y
45. Sahin, C., Pollock, L., Clause, J.: How do code refactorings affect energy usage?. In: Proceedings of 8th ACM/IEEE International Symposium on Empirical Software Engineering and Measurement, ESEM '14, ACM, pp. 36:1–36:10 (2014)
46. Verdecchia, R., Saez, R.A., Procaccianti, G., Lago, P.: Empirical evaluation of the energy impact of refactoring code smells. In: Penzenstadler, B., Easterbrook, S., Venters, C., Ahmed, S.I. (eds.) ICT4S2018. 5th International Conference on Information and Communication Technology for Sustainability, Vol. 52 of EPiC Series in Computing, pp. 365–383 (2018). https://doi.org/10.29007/dz83
47. Li, D., Jin, Y., Sahin, C., Clause, J., Halfond, W.G.J.: Integrated energy-directed test suite optimization. In: Proceedings of 2014 International Symposium on Software Testing and Analysis, ISSTA 2014, ACM, pp. 339–350 (2014)
48. Couto, M., et al.: Detecting anomalous energy consumption in android applications. In: Quintão Pereira, F.M. (ed.), Programming Languages, Vol. 8771 of LNCS, Springer Int. Publishing, pp. 77–91 (2014)
49. Hu, Y., Yan, J., Yan, D., Lu, Q., Yan, J.: Lightweight energy consumption analysis and prediction for Android applications, Science of Computer Programming (2018) 132–147. Special Issue on TASE 2016
50. Cruz, L., Abreu, R.: Catalog of energy patterns for mobile applications. Empir. Softw. Eng. **24**(4), 2209–2235 (2019)
51. Abreu, R., Zoeteweij, P., Van Gemund, A.J.C.: On the accuracy of spectrum-based fault localization. In: Testing: Academic and Industrial Conf. Practice and Research Techniques - MUTATION, 2007. TAICPART-MUTATION 2007, pp. 89–98 (2007)
52. Hao, S., Li, D., Halfond, W., Govindan, R.: Estimating android applications' CPU energy usage via bytecode profiling. In: 2012 First International Workshop on Green and Sustainable Software (GREENS), pp. 1–7 (2012)
53. Li, D., Hao, S., Halfond, W.G.J., Govindan, R.: Calculating source line level energy information for android applications. In: Proceedings of 2013 International Symposium on Software Testing and Analysis, ISSTA 2013, ACM, pp. 78–89 (2013)

54. Pathak, A., Hu, Y.C., Zhang, M.: Where is the energy spent inside my app?: fine grained energy accounting on smartphones with eprof. In: Proceedings of 7th ACM European Conference on Computer Systems, EuroSys '12, ACM, pp. 29–42 (2012)
55. Hoque, M.A., Siekkinen, M., Khan, K.N., Xiao, Y., Tarkoma, S.: Modeling, profiling, and debugging the energy consumption of mobile devices. ACM Comput. Surv. **48**(3), 39:1–39:40 (2015)
56. Rua, R., Couto, M., Saraiva, J.: GreenSource: a large-scale collection of android code, tests and energy metrics. In: 2019 IEEE/ACM 16th International Conference on Mining Software Repositories (MSR), pp. 176–180 (2019). https://doi.org/10.1109/MSR.2019.00035
57. Rua, R., Fraga, T., Couto, M., Saraiva, J.: Greenspecting android virtual keyboards. In: Proceedings of the IEEE/ACM 7th International Conference on Mobile Software Engineering and Systems, MOBILESoft '20, Association for Computing Machinery, pp. 98–108 (2020). https://doi.org/10.1145/3387905.3388600
58. Chowdhury, S., Hindle, A.: GreenOracle: estimating software energy consumption with energy measurement corpora, pp. 49–60 (2016). https://doi.org/10.1145/2901739.2901763
59. Chowdhury, S., Borle, S., Romansky, S., Hindle, A.: GreenScaler: training software energy models with automatic test generation. Empirical Softw. Eng. **24**(4), 1649–1692 (2019). https://doi.org/10.1007/s10664-018-9640-7
60. Pereira, R., Carção, T., Couto, M., Cunha, J., Fernandes, J.P., Saraiva, J.: Helping programmers improve the energy efficiency of source code. In: Proceedings of the 39th International Conference on Software Engineering Companion, ICSE-C 2017, ACM, pp. 238–240 (2017)
61. Pereira, R., et al.: SPELLing out energy leaks: aiding developers locate energy inefficient code. J. Syst. Softw. **161**, 110463 (2020). https://doi.org/10.1016/j.jss.2019.110463
62. Rua, R., Saraiva, J.: PyAnaDroid: a fully-customizable execution pipeline for benchmarking android applications. In: 2023 IEEE International Conference on Software Maintenance and Evolution (ICSME), pp. 586–591 (2023). https://doi.org/10.1109/ICSME58846.2023.00077
63. Rua, R., Saraiva, J.: E-MANAFA: energy monitoring and analysis tool for android. In: Proceedings of the 37th IEEE/ACM International Conference on Automated Software Engineering, ASE '22, Association for Computing Machinery, New York, NY, USA (2023). https://doi.org/10.1145/3551349.3561342
64. Pereira, R., Simão, P., Cunha, J., Saraiva, J.: jStanley: placing a green thumb on java collections. In: Proceedings of the 33rd ACM/IEEE International Conference on Automated Software Engineering, ACM, New York, NY, USA, pp. 856–859 (2018). https://doi.org/10.1145/3238147.3240473
65. Manotas, I., Pollock, L., Clause, J.: SEEDS: a software engineer's energy-optimization decision support framework. In: Proceedings of the 36th International Conference on Software Engineering, ACM, pp. 503–514 (2014)

Applying Soft Computing for Sustainability: The Effects of Feature Reduction on Machine Learning Model Energy Consumption

Marko Njirjak[1,2], Erik Otović[1,2], and Goran Mauša[1,2]([envelope])

[1] Faculty of Engineering, University of Rijeka, Vukovarska 58, 51000 Rijeka, Croatia
{marko.njirjak,erik.otovic}@uniri.hr
[2] Center for Artificial Intelligence and Cybersecurity, University of Rijeka, 51000 Rijeka, Croatia
goran.mausa@uniri.hr

Abstract. Soft computing methods have proved to be a viable way of tackling some of the most difficult scientific challenges and practical problems we face nowadays. Recent advances in machine learning have led to an increased usage of artificial intelligence in various domains, such as medicine, mechanical engineering, robotics, and transportation. The quality of machine learning models is usually assessed only through their capability of solving a specific task; however, we argue that energy consumption ought to be taken into consideration as well, as the current trend of creating larger and more complex models has had a big influence on the energy footprint such machine learning models exhibit. Depending on the problem being tackled, the amount of data used, the complexity of the model, and the utilized hardware, the energy consumption can rise to 3.6×10^9 J. One way to reduce such high energy consumption is to pre-process the training data prior to model training and retain only the most important features describing each data instance. Here, we conduct three case studies examining the energy efficiency of machine learning models for multivariate regression, multiclass classification, and binary classification. Using our previous work on therapeutic peptide prediction as an experiment baseline, we conclude that by reducing the number of features characterizing each peptide from 94 to 45, 119 J of energy are

This work received financial support through the Erasmus+ Strategic Partnership for Higher Education *SusTrainable—Promoting Sustainability as a Fundamental Driver in Software Development Training and Education* (project number 2020-1-PT01-KA203-078646), funded by the European Union and coordinated by the University of Coimbra, Portugal.
The information and views set out in this publication are those of the authors and do not necessarily reflect the official opinion of the European Union. Neither the European Union institutions and bodies nor any person acting on their behalf may be held responsible for the use which may be made of the information contained therein.

C. Grelck et al. (eds.), *Promoting Sustainability as a Fundamental Driver in Software Development Training and Education*, LNCS 15670, pp. 153–174, 2026.
https://doi.org/10.1007/978-3-032-22278-7_6

saved per model training cycle while retaining a similar degree of accuracy. Considering the iterative nature of finetuning the machine learning models in numerous training cycles and retraining when additional data is gathered, we believe such an approach is of crucial importance for the sustainability of developing machine learning-based systems.

Keywords: machine learning · data pre-processing · feature selection · efficiency

1   Introduction

Soft computing is a set of techniques based on probabilistic modelling and uncertainty which represent the means to tackle analytically intractable problems. Artificial intelligence (AI) related techniques like evolutionary computing, swarm intelligence, neural networks and machine learning (ML) algorithms fall under the umbrella of soft computing [11,17]. This is a powerful set of tools that are capable of solving difficult problems, but such techniques often come with the high cost of energy consumption and execution time, hence the sustainability of AI is becoming a very important topic [43].

As the demand for computing power is growing, it is important to ensure its sustainability. One factor to achieving this is optimizing software meant to run in large data centers that operate continuously and consume significant amounts of energy [13]. By optimizing the energy efficiency of computer programs, we can achieve savings in energy consumption for both running the hardware and the cooling systems. In the context of machine learning, large quantities of data typically need to be stored and transferred within and between data centers. This data is then used to train deep learning algorithms, a process that can take several days or even weeks to complete. By reducing data dimensionality and identifying relevant features, we can transfer and use only a subset of the data for training the algorithm. This can result in faster data transfers, shorter training times and lower overall energy consumption. Another application where software energy efficiency plays an important role are battery-powered devices [14,18]. The battery life of portable and IoT devices running AI-based algorithms can be increased by minimizing the software energy consumption, which in turn allows for the use of smaller batteries. Given the vast number of wearable and IoT devices being produced, such optimizations could lead to more efficient use of battery resources, potentially reducing the demand for their production and indirectly mitigating the environmental impact of battery-powered devices [13].

Contrary to the common belief that energy consumption of a computer program is linearly dependent on execution time, the actual relationship between these two factors is largely determined by the type of load [21]. For instance, CPU-intensive loads are linearly proportional to the execution time, whereas other types of loads that frequently interact with memory, other programs, or devices may exhibit a different relationship. As such, the energy efficiency of an algorithm cannot always be estimated from its execution time and the energy efficiency of different approaches to solving the same problem has to be experimentally assessed.

Several studies suggested that adequate coding practices manage to reduce the energy consumption of basic programming patterns [23,25,28,31,33]. Similarly, the green principles of data mining, which are recognized as the main drivers of sustainability in the area, focus on data collection and warehousing, data pre-processing and smart model tuning [37]. When preparing the dataset for training machine learning models for new and unknown problems, we tend to collect as much data as possible. However, this approach comes at the cost of including some redundant, contradictory and/or missing data points which deteriorate the model performance and need to be filtered out. Overly increased feature space of the dataset leads to the curse of dimensionality, but it also slows down the training phase because the model has to process irrelevant characteristics that are used to represent each sample in the dataset. The removal of such features directly increases the memory efficiency, but it may also improve the predictive performance [41].

In this tutorial, we will focus more in-depth on the phase of data pre-processing, which is a crucial step in creating high-quality data and more accurate prediction models. Section 2 addresses various aspects of green data mining with careful consideration of the model's accuracy, training time and power consumption in several case studies. It contains examples of sustainability science studies that are more interested in the benefits of using prediction models than in the models themselves. Inspired by these examples, we replicated one of the studies that used a publicly available dataset and added the measurement of energy that is spent on training the models. Section 2.1 presents three of our case studies that demonstrate the benefits of data cleaning, feature extraction and feature selection. Although general conclusions are difficult to obtain, these case studies did use datasets that come from very different application areas, from software defect prediction [26] to therapeutic peptide discovery [12] and other publicly available datasets of the UCI repository [10], thus showing that data pre-processing techniques, such as feature reduction, are widely applicable and can bring obvious benefits.

Following these motivating results, Sect. 3 gives the practical example of this tutorial that presents a feature selection based on the wrapper method that was proven useful in a recently published study that used the Recurrent Neural Network (RNN) model for the *in silico* screening of therapeutic peptides [30]. The original study used Bura supercomputer (Bull DLC B720 servers, 24 cores, 62GiB RAM) to train and fine-tune a binary classifier for determining the presence of antimicrobial and antiviral activities in chemical compounds comprised of short chains of amino acids. The execution time of feature selection with this powerful configuration was approximately 4 h. After selecting the subset of the 10 most informative features in the sequential properties model, the performance of RNN was improved by 3% in terms of area under the receiver operating characteristic curve (AUC) and the training time was reduced by 4 s. Although spending 4 h to save 4 s does not seem like a price worth paying for, it must taken into account that 4 s are saved for a single model training. The study used 2 times repeated 10-fold cross-validation to obtain repeated measurements and each time it performed a grid search for 5 RNN parameters which required a total of 243 possible configurations to be evaluated using nested 5-fold cross-validation. Therefore, there is a potential to save approximately 81 min ($4\,s \times 5\,folds \times 243\,RNN configurations$) per each iteration of outer cross-validation, which is 27 h ($81\,min \times 20\,folds$) saved for the entire run of the experiment. Due to a very large number of repeated training cycles, the total amount of time saved by the feature selection surpassed the time spent on feature selection by a factor of 6.75. Moreover, regular retraining of ML models using techniques such as online learning [24] is crucial for preventing model drift which can result in performance degradations [36,42]. Therefore, the benefits of feature selection can be observed throughout the whole lifecycle of the ML model.

2   Green Data Mining

The applications of soft computing can nowadays be found in all research areas and one of them is the field of sustainability science, which seeks to understand the fundamental character of interactions between Nature and Society [9]. This field faces the challenge of solving problems burdened with high complexity, uncertainty, and little information, and it relies on new methodological approaches, tools, and techniques.

An example of the application of soft computing for sustainability science is the case study aiming to reduce vessel fuel consumption by optimizing its operational conditions [8]. A set of 41 features, which consisted of navigational parameters like position, direction and acceleration, ship's characteristics like generator power, and atmospheric parameters, was used to predict shaft power, torque, and main engine consumption. Three ML models Regularised Least Squares (RLS), Lasso Regression (LAR) and Random Forrest (RF) were employed for the task and the dimensionality of the dataset was reduced based on three feature selection techniques: Brute Force, Regularisation and a RF Based Method. The results of this case study confirmed the superiority of statistical methods over mechanistic models in their ability to accurately predict the performance of the vessel. Furthermore, the application of a naive-gray box model to the problem of trim optimisation identified the possibility of decreasing the fuel consumption by 2.3% without the need to install further equipment on board.

Another example is the recently published study of seawater quality prediction based on ML that is supported by feature interpretation and spatiotemporal analysis [15]. The need to combine knowledge and data collected from two of Croatia's public Institutes (the "Institute of Oceanography and Fisheries" and the "Croatian Meteorological and Hydrological Institute") and Solcast (a privately-held Australian company) with the microbiological measurements of local public health institute increased the complexity of this study and its implications. The concentration of bacteria E. Coli and Enterococcus was predicted based on 14 environmental features such as sea and air temperature, salinity, solar irradiance, cumulative precipitation data, wind speed and direction, measured at the location of certain popular beaches in the city of Rijeka. The use of spatial models based on standard ML models like RF, Support Vector Machine (SVM) and state-of-the-art algorithms such as CatBoost and XGBoost, together with the SHapley Additive exPlanations (SHAP) feature interpretation technique allowed the identification of key features that contribute to the pollution prediction opening ways for enhancing future measurements and potentially improving public health through timely and accurate predictions of hazardous bacteria levels.

Table 1. The performance of ML models in classifying energy efficiency dataset in terms of accuracy (original VS replicated) and energy consumed for cross-validation. The green cell colour depicts the best performance, red colour indicates the worst performance and yellow color is used to point out interesting trade-off between consumed energy and obtained accuracy.

ML Model	Accuracy [%] Replicated	Accuracy [%] Original [39]	Energy [J] New
J48	**84.9**	**70.6**	**292.49**
Naive Bayes	81.7	36.7	307.04
Bagging	79.9	65.8	227.12
JRip	72.0	58.2	206.26
Bayes Net	70.1	53.7	245.26
Kstar	65.1	63.4	353.15

Although the sustainability of computational techniques was not the focus of these two studies, they did approach the problem of data dimensionality, i.e. they applied feature selection [8] and feature interpretation [15]. However, the choice to use exhaustive search [8] stands out as a concept that is not in accordance with the principles of green data mining [37]. Needless to say, none of these studies analyzed the energy consumption of their prediction algorithms nor made any effort to make the algorithms more energy efficient. To take into account this aspect, in our first case study we replicated another sustainability science-related experiment that compared the classification techniques using an energy efficiency dataset [39]. The mentioned dataset contained the information about energy efficiency of buildings, where the dependent variables were Heating load (y1) and Cooling load (y2), and the analyzed ML models were: decision tree J48, Naive Bayes, Bagging, JRip, Bayes Net and Kstar. The original performance of these six classifiers along with the performance obtained in our replicated study is presented in the second and third columns of Table 1. Although the reason for such poor performance in the original study remained unclear, the decision tree J48 was the best-performing classifier. Given the absence of any reference to hyperparameters for each classifier, it is likely that the authors of the initial study retained them at their default settings. In contrast, our replicated study used the trial-and-error approach aligned with standard ML guidelines to select appropriate values. The last column in Table 1 contains the average energy consumption for a CPU that was measured using the Intel Power Gadget during the training and testing phase in 5 times 10-fold cross-validation repeated measurements. The new measurements reveal that the best-performing algorithm, indicated in bold, is not the one that consumes the highest amount of energy, and surprisingly, the Kstar algorithm was the worst one both in terms of accuracy and energy consumption. In this particular case, J48 would probably be the best choice, whereas the second best place might belong to Bagging because it consumed a much lower amount of energy while having a similar performance to Naive Bayes, which is less than 2% more accurate.

2.1   Data Pre-processing

Motivated by the interesting results of our replicated study, we wanted to analyze more in-depth the impact of data pre-processing and data reduction on the energy consumption of various prediction models. In our second case study, we prioritized data cleaning, a foundational step in data pre-processing. Specifically, we focused on removing instances with missing values to enhance the quality of our analysis. We used the *Online retail* dataset from the UCI repository [10] that consists of 8 attributes and dependent variable with 38 categories, making this a multi-class classification problem. Using the Weka implementation of Bayes Net, Naive Bayes, J48 and ZeroR classifiers in a 5 times 10-fold cross-validation setting and measuring the energy consumption by the Intel Power Gadget, we obtained the results given in Table 2. Three performance indicators (accuracy, energy consumption and training time) were measured before and after the removal of missing values, where the right-hand side subcolumns represents the usage of clean data. The best performance for each indicator is given in bold.

Table 2. The performance of classifiers before([1]) and after([2]) data cleaning in terms of accuracy (Acc), energy consumption (E) and time (t) spent for cross-validation. The green cell colour indicates cases when data cleaning improved the performance.

ML model	Acc^1 [%]	Acc^2 [%]	E^1 [J]	E^2 [J]	t^1 [s]	t^2 [s]
Bayes Net	99.4	**99.9**	1052.8	624.4	67	57
Naive Bayes	99.9	95.8	875.4	447.0	41	49
J48	94.6	96.2	594.7	340.2	36	36
ZeroR	91.4	88.9	308.8	**166.1**	21	**16**

With more than 25% of instances that contained missing values, we managed to reduce the number of instances from 535,875 to 398,197 and this process cost 426.3 J and 77 s on a standard PC configuration. The most accurate classifier for this task was Bayes Net and this one proved to be faster, more accurate and less energy-demanding after the data cleaning. Interestingly, the energy spent for cleaning the data (426.3 J) is almost equal to the amount of energy saved for its training ($1,052.8\ J - 624.4\ J = 428.4\ J$). Although all classifiers managed to reduce energy consumption after the dataset was cleaned, the remaining three classifiers saved less than the cost of cleaning the data, and only ZeroR reduced the training time. However, it has to be taken into account that this case study did not perform other iterative procedures like hyper-parameter fine-tuning, feature selection and others, which would increase the number of training cycles and, consequently, energy consumption. Energy consumption and execution time tend to be proportional for CPU-intensive tasks, while the relationship between the two may be different for other types of load. The generalization of the prediction rules from the training data is usually the most expensive part of a learning algorithm and the reduction in the training set will reflect in shorter and/or lower energy consumption. As the examined algorithms fall into a category of *eager learning* algorithms which generalize from the training data before receiving queries for inference, the savings in time and/or energy will be perceived in the training phase. The exact savings of each algorithm depend on its implementation and the type of workload it incurs on the system. The impact of data cleaning may not always bring positive change in performance of the classifier and this, too, depends on the model of choice.

Another aspect of data pre-processing is to assess the importance of features used to describe each data instance and remove redundant descriptors from the dataset. Moreover, in one of our recent studies, we have shown that one-hot encoding may be replaced with binary encoding when applied to the prediction of therapeutic peptides, thus reducing the size of the dataset by a factor of 4 [12]. Although one-hot encoding is considered the best way to represent categorical data, the switch to binary encoding did not lower the predictive performance of the Support Vector Machine (SVM) classifier significantly and it maintained a high level of sensitivity. Moreover, coupling the binary encoding-based model with the physico-chemical features-based encoding model allowed for the lowest level of fall-out [12]. The physico-chemical features used in that study were the result of Principal Component Analysis (PCA)-based feature extraction that allowed a significant reduction in dataset size. For example, the Structural Topology category of 827 structural properties was represented with only 8 main components, the Topological Scale category of 67 topological descriptors was reduced to 5 components and the 58 physico-chemical properties of The Protein Fingerprint category was expressed by 8 components [29]. PCA transforms the set of variables into a new coordinate system, whose components are orthogonal and the first ones explain the most variance within the dataset. The small number of components thus retained more than 80% of variance in each of the mentioned feature categories.

The concept of feature extraction is based on transformation, whereas a more commonly used strategy is feature selection, which is based on eliminating the features that are not important to the classifier [20]. In addition to reducing training time and reducing the overfitting effect of learning models, this set of techniques yields a better understanding of the underlying effects that govern the phenomenon we are analyzing. Our last case study is based on a dataset systematically collected from the Eclipse JDT open source project that contains 48 attributes of software metrics and 3,420 instances of public Java classes with 61 : 39 ratio between fault-prone and non-fault-prone [26]. Using the statistical filter methods Chi-squared, ANOVA, Mutual information and Kendall's correlation we filtered the features and after a 10-fold cross-validation procedure obtained the results presented in Table 3. Filter methods offer a computationally efficient solution to issues that arrise due to sparse and high dimensional data by reducing the dimensionality of the datasets without significantly affecting classification performance [2]. This example reveals that reducing the number of features by more than 70% (from 48 to 13 ± 2) allows the logistic regression model to predict software defects with equal success. The best performance, indicated in bold, was achived by ANOVA filter with prior application of normalization and in that case the model without feature selection was slightly outperformed, proving that less can indeed be more for machine learning. In the following chapter, the tutorial will present how to use the wrapper method for feature selection.

Table 3. The performance of Logistic regression classifier with and without feature filtering in terms of accuracy (Acc) and area under the receiver operating characteristic curve (AUC). The green cell colour indicates the case when feature selection improved the performance.

Filtering technique	Train Acc [%]	Test Acc [%]	Test AUC [%]	
No filter	73.90 (1.3)	70.82 (3.8)	59	73
Chi-square	72.56 (1.5)	68.25 (5.3)	58	75
ANOVA	72.82 (1.3)	70.15 (5.4)	59	71
ANOVA & normalization	72.98 (1.7)	**71.05 (6.1)**	**60**	**75**
Mutual information	69.91 (1.9)	66.58 (5.9)	58	67
Kendall	71.77 (1.7)	68.92 (5.9)	59	69

3   Practical Example for Feature Selection

The following practical example is based on the experiment from the aforementioned research paper that presented a novel representation scheme for *in silico* screening of novel peptides. The study indicated that feature selection led to a higher predictive performance, but did not consider the energy consumption aspect of it. Here, we inspected the dependency of energy consumption and predictive performance on the number of selected features. This enabled us to use feature selection to preserve the predictive performance and reduce energy consumption as much as possible.

The original study used Bura supercomputer to perform feature selection and grid search to find optimal hyperparameters. On top of all that, the experiment was performed using 2 times repeated 10-fold cross-validation which required that everything was rerun 20 times. Such an experimental setup is infeasible for regular desktop computers so it had to be simplified for this practical example. With the measurement of energy consumption as our main focus and to speed up the training, we simplified the model and removed the tedious processes of cross-validation and grid search hyperparameter fine-tuning. Therefore, the predictive model was trained only once and the additional energy cost of cross-validation and grid search was estimated by multiplying the measurement from this example with the number of repeated training cycles. Moreover, the original study used four datasets, but this example focused only on the antimicrobial dataset for simplicity.

3.1 Machine Learning

Machine learning is a branch of computer science striving to create computer algorithms and models capable of mimicking human intelligence. ML is a diverse area of research and incorporates different algorithms and models, such as SVM [16], random forest [32], artificial neural network [3], decision tree [38], etc.

An artificial neural network (ANN) can be viewed as a non-linear function transforming input data into a decision space. When performing binary classification, the goal is to assign a class label, 1 or 0, to the input data. However, before an ANN can be used for classification, it has to be trained. A dataset containing positive and negative instances is required. During the training phase, the network strives to learn the intrinsic differences between the two classes. This can be done using a variety of algorithms, the most utilized one being backpropagation [34]. Backpropagation minimizes the error between the true class labels and the ones predicted by the ANN, iteratively nudging the network parameters towards an optimum in the parameter space. The following code demonstrates how to create a simple image classification ANN using Tensorflow [1] and Keras libraries [5].

```
# Define a model as a sequence of stacked layers.
model = keras.Sequential(
        [
            # Define an input layer where the input shape
            # is (height, width, channels).
            keras.Input(shape=input_shape),

            # First block of Conv2D and MaxPooling2D.
            # 2D convolutional layer detects specific
            # features/patterns in the input data, while
```

```
11        # MaxPooling2D reduces the spatial dimensions.
12        layers.Conv2D(32, kernel_size=(3, 3), activation=
   "relu"),
13        layers.MaxPooling2D(pool_size=(2, 2)),
14
15        # Second block of Conv2D and MaxPooling2D.
16        # Multiple blocks are usually stacked to increase
17        # the receptive field of the neural network.
18        layers.Conv2D(64, kernel_size=(3, 3), activation=
   "relu"),
19        layers.MaxPooling2D(pool_size=(2, 2)),
20
21        # 2D data is flattened to a long 1D vector which
22        # is then classified by a fully connected layer.
23        # Dropout is added to prevent overfitting.
24        layers.Flatten(),
25        layers.Dropout(0.5),
26        layers.Dense(num_classes, activation="softmax"),
27    ]
28 )
```

To evaluate the accuracy of our ML model, we use the Matthews correlation coefficient (MCC). MCC is calculated as shown in Eq. (1) and represents a correlation between two binary variables, in this case between the observed peptide labels and the true peptide label values. Additionally, MCC takes into account the possible imbalance between the classes, therefore it is suitable for imbalanced datasets.

$$MCC = \frac{TP \times TN - FP \times FN}{\sqrt{(TP + FP)(TP + FN)(TN + FP)(TN + FN)}} \tag{1}$$

where TP is the number of true positive instances, TN is the number of true negative instances, FP is the number of false positive instances and FN is the number of false negative instances.

3.2 Peptide Dataset

Peptides are short chains of amino acids and typically consist of up to 50 residues. Some peptides exhibit useful activities like antimicrobial activity. Because of their low toxicity and high specificity, they represent a promising alternative to the existing compounds in the drug design process. The chemical space of possible compounds is large and its research is mostly based on trial and error methods that require a lot of time-consuming laboratory work, which is why much fewer positive instances are known compared to negative instances. Recently, machine learning became a popular choice for the high throughput screening of peptides in order to quickly and efficiently identify potential chemical compounds that possess the desired activity.

The dataset we used consists of positive, antimicrobial peptides from DRAMP 2.0 data repository [19], and negative instances from UniProt [7]. In total, it contains 9409 peptides, the distribution of which is shown in Fig. 1.

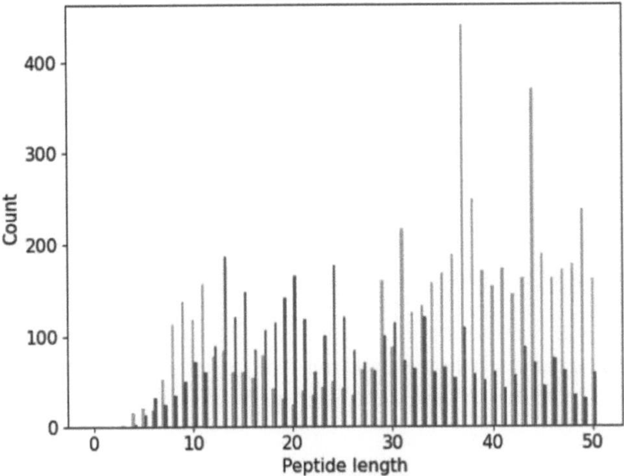

Fig. 1. Distribution of peptide lengths and class labels in the dataset. Positive instances are shown in blue and the negative instances are shown in red color.

To successfully conduct *in silico* experiments with peptides, an appropriate encoding scheme is required to represent chemical compounds in a computer. There are numerous ways to achieve the former, here we state a few:

- Amino acid composition (AAC) [6]
  A peptide is encoded as a vector that consists of N decimal values, where each vector member represents a relative frequency of a specific amino acid with respect to the sequence length. For example, if we consider the 20 proteinogenic amino acids, and a peptide AAQR, the resulting AAC vector is [0.5, 0, 0, 0, 0, 0, 0, 0, 0, 0, 0, 0, 0, 0.25, 0.25, 0, 0, 0, 0, 0]. The first position in a vector represents the alanine amino acid. Since there are two alanines in a sequence, and the sequence length is four, it yields a relative frequency of 0.5. The same principle applies to the other two amino acids in the sequence.
- Sparse [40]
  A peptide is represented as an array of one-hot vectors of length 20, where

each position, except one, is set to 0. The sparse representation of the AAQR peptide is a matrix:

1	0	0	0	0	0	0	0	0	0	0	0	0	0	0	0	0	0	0	0
1	0	0	0	0	0	0	0	0	0	0	0	0	0	0	0	0	0	0	0
0	0	0	0	0	0	0	0	0	0	0	0	0	1	0	0	0	0	0	0
0	0	0	0	0	0	0	0	0	0	0	0	0	0	1	0	0	0	0	0

The first two rows encode alanine amino acid, while the third and fourth rows encode glutamine (Q) and arginine (R) respectively.

- Simplified molecular-input line-entry system (SMILES) [27]
  Describes the structure of chemical species using short ASCII strings. SMILES representation of nicotine, shown in Fig. 2, is CN1CCC[C@H]1c2cccnc2.

Here, we will use a representation scheme we have proposed in our recent paper, which maximizes the accuracy of ML models [30]. Each peptide is represented as an array of 94-D vectors, where each vector encodes one amino acid in a peptide sequence.

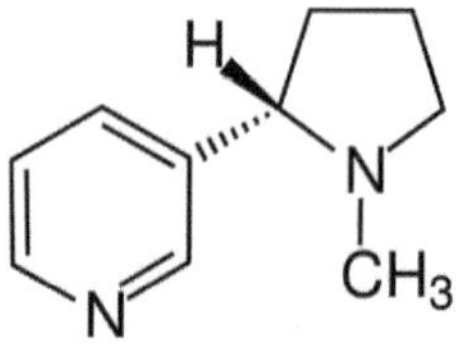

Fig. 2. Chemical structure of nicotine ($C_{10}H_{14}N_2$) by n.d., 21 August 2017. Public Domain.

3.3   Feature Selection

The number of trainable parameters in a neural network depends on the number of features which then leads to a longer training time and higher power consumption. Using too many features can also have a negative impact on the predictive performance of neural networks if not enough training data is provided. Feature selection is generally a process which aims to reduce the number of features by removing uninformative, redundant or highly correlated features.

One such method is Permutation Feature Importance (PFI) [4] which determines the importance of each feature by making it informative and observing the difference in the predictive performance - more important features will cause a bigger drop in predictive performance. The feature can be turned into an informative feature by shuffling the values of that feature in the dataset. In this way, the distribution of values for that feature remains the same, but the connection between the feature and the predicted variable is lost. However, if there is a correlation between features, then the model can still access the removed feature through other features and the computed feature importance may not reflect the true importance of the features. Therefore, this method assumes that features are not correlated.

Antimicrobial?

	Feature 1	Feature 2	Feature 3
Peptide 1	0.5	0.1	0.4
Peptide 2	0.3	0.9	0.1
Peptide 3	0.3	0.2	0.9
Peptide 4	0.8	0.7	0.9
...	...	...	...

Yes/No

Fig. 3. Demonstration of Permutation Feature Importance (PFI) for Feature 1.

The algorithm starts by training a model with all available features, evaluates the importance of each feature, sorts them by their importance and chooses the top N most important features. For example, consider Fig. 3. A model pre-trained on a dataset with all features present is depicted with a brain icon. The output from the model is the probability that a given peptide has antimicrobial activity. The predictive performance of a pre-trained model is evaluated on a validation set. To determine the importance of the first feature, the algorithm shuffles its values and evaluates the model using the modified dataset. The values for the first feature are then returned to their original values, and the same procedure is repeated for every next feature. The obtained scores can be subtracted from the baseline score or can be used directly as indicators of feature importance. Furthermore, it can be seen that the importance of each feature has to be assessed before they can be sorted and the most important features can be selected. Hence the PFI computation is not influenced by the number of features that need to be selected.

3.4   Energy Consumption Measurement

RAPL (Running Average Power Limit) interface is a feature implemented on most processors that provides the average power consumption for a system. Power consumption estimates are available for each CPU package and each core in the package. Additionally, it can estimate the power DRAM power consumption if the system supports this feature. In most cases, the consumption is estimated by a pre-calibrated software model which depends on hardware performance counters and I/O models, but there are examples of systems that implement real power measuring methods [35]. RAPL was originally introduced by Intel in the Sandy Bridge generation and recently has been implemented on the newest AMD processors.

Linux kernel provides RAPL information in terms of read-only system files that are automatically updated when new estimates are available. These files can be located in /sys/class/powercap/intel-rapl/ directory in which each CPU package will be given its own directory (e.g. estimates for the first package are inside intel-rapl:0 directory). Estimates are stored in text files which can be read only by users with root privileges. For example, the user can issue the following command to obtain the total energy consumed by the single package system since boot:

```
1  sudo cat /sys/class/powercap/intel-rapl/intel-rapl:0/
       energy_uj
```

The returned value represents energy expressed in microjoules.

PyRAPL package for Python implements functions for reading RAPL estimates and provides a relatively easy interface for code profiling. It can be installed from the PyPI repository with

```
1  pip install pyrapl
```

In order to use it, it is necessary to import and set up the module, instantiate the PyRAPL Measurement object and call begin and end methods to start and end the measurement. This is demonstrated in the following code snippet.

```
1  import pyRAPL
2
3  pyRAPL.setup()
4  meter = pyRAPL.Measurement('bar')
5  meter.begin()
6  # ...
7  # Block of code whose power consumption
8  # is to be evaluated.
9  # ...
10 meter.end()
```

Even though the package provides the functions to measure the power consumption for a piece of code, it relies on the total energy consumed by the system during the execution of the selected piece of code. Therefore, to obtain estimates that are as good as possible, users are advised to stop or close all unnecessary programs and not use the computer system during the measurement.

3.5   Exercise Execution

The goal of the practical exercise is to experimentally compare the energy consumption during ML model training with all the available peptide descriptors, and the energy consumption observed when training the model with only a subset of the most informative descriptors.

A practical exercise is shaped as a Jupyter notebook [22], which offers a convenient way to execute code and embed additional explanations and visualizations. To measure the systems's energy consumption, we use RAPL (see Sect. 3.4). Firstly, to demonstrate how energy consumption varies with the amount of workload, we compare the consumption of a simple busy wait loop and a sleep kernel call.

```python
1  # Busy wait
2  busy_wait_meter = pyRAPL.Measurement('busy_wait')
3  busy_wait_meter.begin()
4
5  start_time = time.time()
6  while time.time() - start_time < 2:
7      pass
8
9  busy_wait_meter.end()
```

```python
1  # Sleep
2  sleep_meter = pyRAPL.Measurement('sleep')
3  sleep_meter.begin()
4
5  time.sleep(2)
6
7  sleep_meter.end()
```

Busy wait loop consumes 24.1 J of energy, while the sleep call uses 4.7 J. Therefore, by utilizing sleep kernel call, we can cut down the energy consumption by 80.3%.

The first step in an ML training process is to load and curate the dataset. We use the antimicrobial dataset detailed in Sect. 3.2, and the 70:15:15 train-validation-test distribution.

```python
1  # Load sequences from a CSV file
2  import pandas as pd
3  data = pd.read_csv("datasets/antimicrobial.csv")
4
5  peptides = data["peptide"].to_numpy()
6  y = data["label"].to_numpy()
7
8  from sklearn.model_selection import train_test_split
9  peps_train, peps_test, y_train, y_test = train_test_split
       (peptides, y, test_size=0.3, random_state=42)
10 peps_test, peps_val, y_test, y_val = train_test_split(
       peps_test, y_test, test_size=0.5, random_state=42)
```

To represent chemical compounds in a computer, an appropriate representation scheme is necessary. To maximize the resulting accuracy, we used the sequential properties representation scheme described in Sect. 3.2.

```python
from sklearn.preprocessing import MinMaxScaler
from seqprops import SequentialPropertiesEncoder
encoder = SequentialPropertiesEncoder(scaler=MinMaxScaler
    (feature_range=(-1, 1)), max_seq_len=50, stop_signal=
    False)

X_train = encoder.encode(peps_train)
X_test = encoder.encode(peps_test)
X_val = encoder.encode(peps_val)

features_count = len(encoder.get_selected_properties())
print("Number of features: {}".format(features_count))
print("Training instances: {}".format(len(X_train)))
print("Validation instances: {}".format(len(X_val)))
print("Testing instances: {}".format(len(X_test)))
```

The model is trained on 6586 instances, validated on 1412 instances, and tested on 1411 instances. Firstly, we measure the energy consumption while training the model with all the available, 94 features.

```python
from utils import train_model
meter = pyRAPL.Measurement('meter')
meter.begin()

model = train_model(X_train, y_train, verbose=1)

meter.end()
no_fs_energy = meter.result.pkg[0] / 1000000
print("Training energy: {:.0f}J".format(no_fs_energy))
```

The training takes approximately one minute and consumes 652 J of energy. However, energy consumption is not the only important parameter, we also evaluate model accuracy using MCC detailed in Sect. 3.1. The model achieves a score of 0.75.

```python
from sklearn.metrics import matthews_corrcoef
y_pred = model.predict(X_test)
y_pred = [1 if probability > 0.5 else 0 for probability
    in y_pred]
no_fs_score = matthews_corrcoef(y_test, y_pred)
print("MCC score: {:.2f}".format(no_fs_score))
```

The next step is to calculate feature importance using PFI. PFI works by permuting the values of a single feature and observing how the MCC score changes after model training. A large change in the score indicates that the feature is important, while the opposite indicates the model finds the feature less informative. The 94 features are permuted sequentially, and each feature is given an importance score. The feature importance calculation consumes approximately 437 J of energy and is not dependent on the desired number of features.

```python
# Instantiate PyRAPL meter
meter = pyRAPL.Measurement('meter')
meter.begin()

# Obtain a list of feature names
feature_importance = []
feature_names = encoder.get_selected_properties()

# Compute feature importance for each feature
for idx, feature in enumerate(feature_names):
    # Shuffle current feature
    X_val_permuted = np.array(X_val)
    X_val_permuted[:, :, idx] = np.random.permutation(
    X_val[:, :, idx])

    # Make a prediction on a modified dataset
    y_pred = model.predict(X_val_permuted)
    y_pred = [1 if probability > 0.5 else 0 for
    probability in y_pred]

    # Compute MCC score
    score = matthews_corrcoef(y_val, y_pred)
    feature_importance.append(score)

# Print energy consumed during feature selection
meter.end()
pfi_energy = meter.result.pkg[0] / 1000000
print("PFI energy: {:.0f}J".format(pfi_energy))
```

The computed importance scores are stored in the array *feature_importance* and are visualized in Fig. 4. Feature names were omitted for clarity. The most informative features caused the highest drop in the MCC score. Therefore, the features at the bottom are the most informative ones, since their shuffling had the highest negative impact on the model's performance. We ran the experiment multiple times, each time with a different number of the most informative features to assess their impact on the predictive performance and energy required for model training. As the computation of PFI is not dependent on the number of features that need to be selected afterwards, its energy consumption can be seen as a constant for a specific experimental setup (dataset, model, hyperparameters, etc.). PFI scores from the previous step are used in this step to select the desired number of the most relevant features meaning that additional energy has to be spent only for the model training. The influence of the number of features on the predictive performance and energy required for training is presented in Fig. 5. It is clear that the energy consumption (green line) rises linearly with respect to the number of features used for training, which is expected. A larger

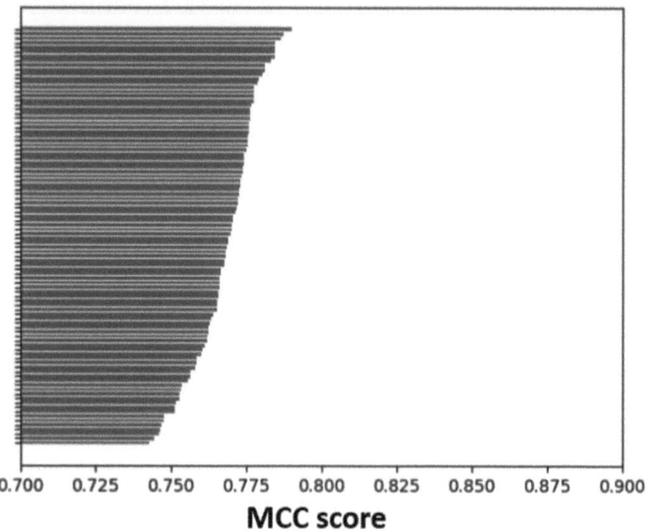

Fig. 4. Features are sorted by their importance. The vertical axis represents features while the horizontal axis represents the MCC score that was achieved when a corresponding feature was shuffled. Feature names on a vertical axis are omitted for simplicity.

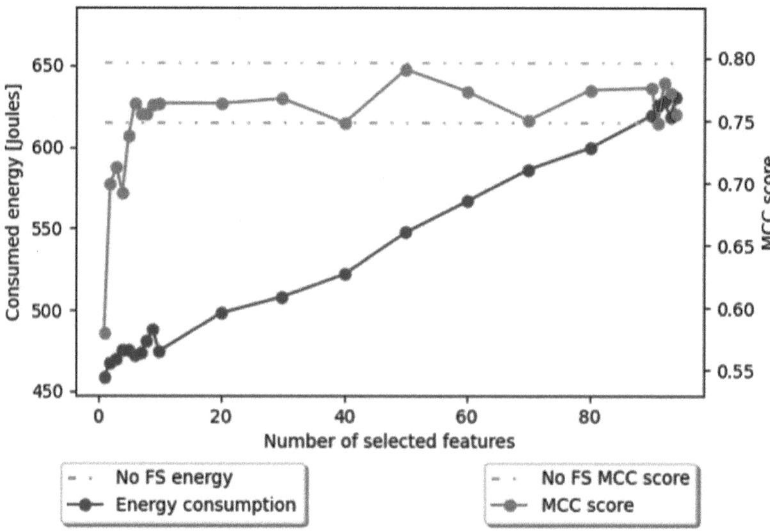

Fig. 5. Dependency of energy consumed during neural network training and achieved MCC score on the number of selected features. Energy required for PFI computation is not included in the measured energy as it is not dependent on the number of features. The baseline is shown with dashed lines and represents consumed energy and achieved MCC score when feature selection is not performed. (Color figure online)

number of features requires more computational energy to be invested during model training. The model accuracy represented through an MCC score (red line) rises from 1 to about 50 features and then stagnates. The optimal number of features to choose from depends on the requirements and potential application. Here, we consider the optimal number of features to be the one at which the highest MCC score is achieved, which is 45.

By choosing the 45 most important features, the total energy consumption excluding the energy required to calculate feature importance is 533 J, while the MCC score is 0.78. This leads to the conclusion that feature selection can decrease energy consumption while maintaining, or even enhancing, ML model accuracy.

4 Conclusion

The task of data scientists is to find new ways to improve the performance of existing models, but also to take into account the sustainability of such approaches. Before starting the ML training phase, it is important to remove irrelevant and redundant instances and features to improve the efficiency of prediction models. This tutorial demonstrated the benefits of using techniques for data cleaning, feature selection and feature extraction in terms of prediction performance, execution time and energy consumption. Although the statistical filter methods tend to be very fast for the task of feature selection, the practical exercise demonstrated the usage of a more complex wrapper method, which iteratively trains the prediction model with different subsets of features and searches for the best-performing combination.

The main results from the practical example are the observed dependency of predictive performance and energy consumption on the number of selected features. The predictive performance increases with the number of features until 45 features are selected. After that point, the predictive performance stagnates or increases insignificantly. On the other hand, energy consumption exhibits a linear trend with an increase in the number of features. Therefore, by using the 45 most informative features we managed to achieve predictive performance that slightly increased the MCC score by 0.03 when compared to the case when no feature selection was performed and at the same time reduced the energy consumption by 119 J. The saving in energy is proportional to the number of removed features.

We can conclude that there are multiple benefits of feature selection: it can increase the predictive performance of the model, reduce the training time and energy consumption, decrease the possibility of over-fitting, but it also makes a model simpler which contributes to its explainability.

References

1. Abadi, M., et al.: TensorFlow: large-scale machine learning on heterogeneous systems (2015). https://www.tensorflow.org/, software available from tensorflow.org
2. Abellana, D.P.M., Lao, D.M.: A new univariate feature selection algorithm based on the best-worst multi-attribute decision-making method. Decis. Anal. J. **7**, 100240 (2023)
3. Basheer, I.A., Hajmeer, M.: Artificial neural networks: fundamentals, computing, design, and application. J. Microbiol. Methods **43**(1), 3–31 (2000)
4. Breiman, L.: Random forests. Mach. Learn. **45**(1), 5–32 (2001)
5. Chollet, F., et al.: Keras (2015). https://keras.io
6. Chou, K.C.: Pseudo amino acid composition and its applications in bioinformatics, proteomics and system biology. Curr. Proteomics **6**(4), 262–274 (2009)
7. Consortium, U.: Uniprot: a hub for protein information. Nucleic Acids Res. **43**(D1), D204–D212 (2015)
8. Coraddu, A., Oneto, L., Baldi, F., Anguita, D.: Vessels fuel consumption: a data analytics perspective to sustainability. In: Soft Computing for Sustainability Science, pp. 11–48. Springer (2018)
9. Corona, C.C.: Soft Computing for Sustainability Science. Springer (2018)
10. Dua, D., Graff, C.: UCI machine learning repository (2017). http://archive.ics.uci.edu/ml
11. El-Alfy, E.S.M., Awad, W.: Computational intelligence paradigms: an overview. In: Improving Information Security Practices through Computational Intelligence, pp. 1–27 (2016)
12. Erjavac, I., Kalafatovic, D., Mauša, G.: Coupled encoding methods for antimicrobial peptide prediction: how sensitive is a highly accurate model? Artif. Intell. Life Sci. **2**, 100034 (2022)
13. Freitag, C., Berners-Lee, M., Widdicks, K., Knowles, B., Blair, G., Friday, A.: The climate impact of ICT: a review of estimates, trends and regulations (2021)
14. Georgiou, K., Xavier-de Souza, S., Eder, K.: The IoT energy challenge: a software perspective. IEEE Embed. Syst. Lett. **10**(3), 53–56 (2018)
15. Grbčić, L., et al.: Coastal water quality prediction based on machine learning with feature interpretation and spatio-temporal analysis. Environ. Modell. Softw. **155**, 105458 (2022)
16. Huang, S., Cai, N., Pacheco, P.P., Narrandes, S., Wang, Y., Xu, W.: Applications of support vector machine (SVM) learning in cancer genomics. Cancer Genomics Proteomics **15**(1), 41–51 (2018)
17. Ibrahim, D.: An overview of soft computing. Procedia Comput. Sci. **102**, 34–38 (2016), 12th International Conference on Application of Fuzzy Systems and Soft Computing, ICAFS 2016, 29-30 August 2016, Vienna, Austria
18. Jayaseelan, R., Mitra, T., Li, X.: Estimating the worst-case energy consumption of embedded software. In: 12th IEEE Real-Time and Embedded Technology and Applications Symposium (RTAS 2006), pp. 81–90. IEEE (2006)
19. Kang, X., et al.: Dramp 2.0, an updated data repository of antimicrobial peptides. Sci. Data **6**(1), 1–10 (2019)
20. Khaire, U.M., Dhanalakshmi, R.: Stability of feature selection algorithm: a review. J. King Saud Univ.-Comput. Inf. Sci. (2019)

21. v. Kistowski, J., Block, H., Beckett, J., Lange, K.D., Arnold, J.A., Kounev, S.: Analysis of the influences on server power consumption and energy efficiency for CPU-intensive workloads. In: Proceedings of the 6th ACM/SPEC International Conference on Performance Engineering, pp. 223–234. ICPE 2015. Association for Computing Machinery, New York, NY, USA (2015)
22. Kluyver, T., et al.: Jupyter notebooks – a publishing format for reproducible computational workflows. In: Loizides, F., Schmidt, B. (eds.) Positioning and Power in Academic Publishing: Players, Agents and Agendas, pp. 87–90. IOS Press (2016)
23. Lima, L.G., Soares-Neto, F., Lieuthier, P., Castor, F., Melfe, G., Fernandes, J.P.: Haskell in green land: analyzing the energy behavior of a purely functional language. In: 2016 IEEE 23rd International Conference on Software Analysis, Evolution, and Reengineering (SANER), vol. 1, pp. 517–528. IEEE (2016)
24. Lobo, J.L., Del Ser, J., Bifet, A., Kasabov, N.: Spiking neural networks and online learning: an overview and perspectives. Neural Netw. **121**, 88–100 (2020)
25. Longo, M., Rodriguez, A., Mateos, C., Zunino, A.: Reducing energy usage in resource-intensive java-based scientific applications via micro-benchmark based code refactorings. Comput. Sci. Inf. Syst. **16**(2), 541–564 (2019)
26. Mauša, G., Galinac Grbac, T., Dalbelo Bašić, B.: A systematic data collection procedure for software defect prediction. Comput. Sci. Inf. Syst. **13**(1), 173–197 (2016)
27. Minkiewicz, P., Iwaniak, A., Darewicz, M.: Annotation of peptide structures using smiles and other chemical codes-practical solutions. Molecules **22**(12), 2075 (2017)
28. Oliveira, W., Oliveira, R., Castor, F., Pinto, G., Fernandes, J.P.: Improving energy-efficiency by recommending java collections. Empir. Softw. Eng. **26**(3), 1–45 (2021)
29. Osorio, D., Rondón-Villarreal, P., Torres, R.: Peptides: a package for data mining of antimicrobial peptides. Small **12**, 44–444 (2015)
30. Otović, E., Njirjak, M., Kalafatovic, D., Mauša, G.: Sequential properties representation scheme for recurrent neural network-based prediction of therapeutic peptides. J. Chem. Inf. Model. **62**(12), 2961–2972 (2022)
31. Pinto, G., Liu, K., Castor, F., Liu, Y.D.: A comprehensive study on the energy efficiency of java's thread-safe collections. In: Proceeding of IEEE International Conference on Software Maintenance and Evolution (ICSME), pp. 20–31 (2016)
32. Qi, Y.: Random forest for bioinformatics. In: Ensemble Machine Learning, pp. 307–323. Springer (2012)
33. Rocheteau, J., Gaillard, V., Belhaj, L.: How green are java best coding practices? In: Proceedings of the 3rd International Conference on Smart Grids and Green IT Systems (SMARTGREENS), pp. 235–246 (2014)
34. Rojas, R.: The backpropagation algorithm. In: Neural Networks, pp. 149–182. Springer (1996)
35. Rotem, E., Naveh, A., Ananthakrishnan, A., Weissmann, E., Rajwan, D.: Power-management architecture of the intel microarchitecture code-named sandy bridge. IEEE Micro **32**(2), 20–27 (2012)
36. Schelter, S., Biessmann, F., Januschowski, T., Salinas, D., Seufert, S., Szarvas, G.: On challenges in machine learning model management. IEEE Data Eng. Bull. (2015)
37. Schneider, J., Basalla, M., Seidel, S.: Principles of green data mining. In: Proceedings of the 52nd Hawaii International Conference on System Sciences (2019)
38. Song, Y.Y., Ying, L.: Decision tree methods: applications for classification and prediction. Shanghai Arch. Psychiatry **27**(2), 130 (2015)
39. Toprak, A., Koklu, N., Toprak, A., Ozcan, R.: Comparison of classification techniques on energy efficiency dataset. Int. J. Intell. Syst. Appl. Eng. **5**, 81–85 (2017)

40. Usmani, S.S., Bhalla, S., Raghava, G.P.: Prediction of antitubercular peptides from sequence information using ensemble classifier and hybrid features. Front. Pharmacol. **9**, 954 (2018)
41. Wang, S., Tang, J., Liu, H.: Feature selection (2017)
42. Wu, Y., Dobriban, E., Davidson, S.: Deltagrad: rapid retraining of machine learning models. In: International Conference on Machine Learning, pp. 10355–10366. PMLR (2020)
43. van Wynsberghe, A.: Sustainable AI: AI for sustainability and the sustainability of AI. AI Ethics **1**(3), 213–218 (2021)

Multicriteria Decision Methods
for Sustainable Software Development

Luís Paquete[(✉)] [iD]

CISUC, Department of Informatics Engineering, University of Coimbra,
Coimbra, Portugal
`paquete@dei.uc.pt`

Abstract. Multicriteria decision methods aim at supporting decisions
when several conflicting points of view need to be considered. This
is often the case in the evaluation of alternative software engineering
projects, for which not only cost/profit matters, but also other less quantifiable criteria, such as team well-being and project *sustainability*. This
chapter provides an short overview of the main multicriteria decision
methods, discuss some of their weaknesses and strengths, and present a
didactic case study on ranking programming languages using outranking
methods.

Keywords: Multiple-Criteria Decision Analysis · Sustainability

1 Introduction

Decision problems arise very often in our daily life. For instance, when choosing
a path from work to home, one may take into account not only distance but also
sightseeing quality, road traffic, or some environmental aspect. One has to deal
with several possible choices that are, to some extent, incomparable because of
the multiple-criteria nature of the underlying problem.

This work was financed by national funds through the FCT (Foundation for Science and Technology, I.P.) within the scope of the project CISUC (project number UID/CEC/00326/2020).

This work received financial support through the Erasmus+ Strategic Partnership for
Higher Education *SusTrainable—Promoting Sustainability as a Fundamental Driver in
Software Development Training and Education* (project number 2020-1-PT01-KA203-
078646), funded by the European Union and coordinated by the University of Coimbra,
Portugal.

The information and views set out in this publication are those of the author and do
not necessarily reflect the official opinion of the European Union. Neither the European Union institutions and bodies nor any person acting on their behalf may be held
responsible for the use which may be made of the information contained therein.

C. Grelck et al. (eds.), *Promoting Sustainability as a Fundamental Driver in Software Development
Training and Education*, LNCS 15670, pp. 175–188, 2026.
https://doi.org/10.1007/978-3-032-22278-7_7

Software industry also deals with several criteria at different stages of the software engineering process. For example, the choice for the most appropriate software engineering practice must take into account whether, or not, it allows for agile development, customer interaction, early deployment, risk reduction, continuous testing, sustainability policy pursuance, and many other criteria.

Multicriteria Decision Aiding (MCDA), a field of knowledge that arose more 50 years ago, provides formal methods that allow to take informed decisions. However, in despite of its wide use in many industrial applications, it is not yet a current practice at software industry and it is not even taught at graduate IT courses. Application of these methods can only be found in specialized literature, see [1, 2, 4, 9, 12, 16–18]. For instance, Pereira et al. [13] propose to rank programming languages with respect to several criteria, such as energy and memory. Also, some of the problems that arise in Search-Based Software Engineering are shown to be naturally multicriteria; see survey of Yao [19]. Given the increasing relevance of sustainability as a criterion in the several steps of both software development and maintenance processes, it is more important than ever that software engineers know how to apply multicriteria decision methods.

This chapter aims to introduce non-experts with background on software engineering to multicriteria decision methods. It provides an introductory overview of these methods, discuss their weaknesses and strengths, and presents a didactic case study on ranking programming languages, based on the work of Pereira et al. [13], using one the most well-known methods.

2   Preliminaries

It is assumed that a Decision Maker has to take a choice from a finite set of alternatives, each of which is evaluated on the basis of a finite family of evaluation criteria. These criteria represent different evaluation aspects of the several alternatives. Furthermore, each criterion is associated with a monotone direction of preference. For example, the cost of a software project has a decreasing direction of preference, whereas sustainability has an increasing direction of preference.

Typically, decision aiding problems can be recast as choice, ranking or sorting problems. In *choice problems*, the goal is to select the best alternative or the best subset of alternatives by rejecting those that are worse [10]. In *ranking problems*, all alternatives are ranked from the best to the worst, allowing for ties and incomparabilities. In *sorting problems* [11], each alternative is assigned to one or more classes that have been pre-defined and ordered from the best to the worst according to some Decision Maker's model.

In order to deal with any of the three problems above, a performance table is constructed, which gives objective information about each alternative according to the several criteria at hand. From this information it is possible to extract a set of interesting alternatives based on the concept of *dominance relation*: an alternative a dominates b if and only if a is at least as good as b for all criteria and better for at least one of them. This is the most basic assertion of preference with respect to the set of alternatives considered in decision problem. Note that there might not exist a single *dominating* alternative, but a set of them that are noncomparable.

Table 1. A performance table for software projects in terms of cost and sustainability

Project	Cost (M. €)	Sustainability
1	1.1	3
2	2.0	1
3	3.2	4
4	5.3	7
5	7.1	6
6	10.2	9

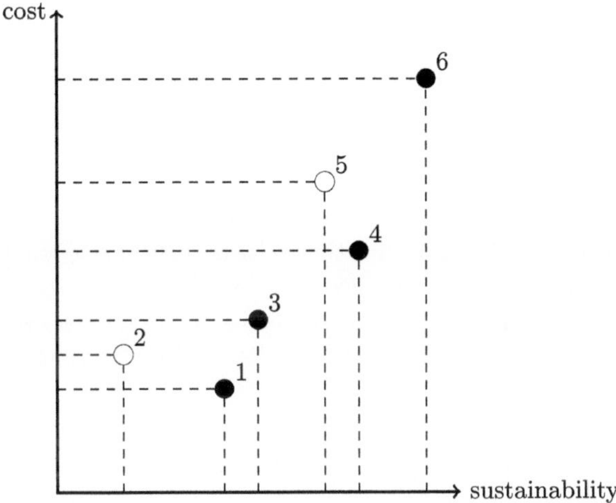

Fig. 1. Evaluation of software projects in terms of cost and sustainability.

Table 1 is a performance table for a hypothetical example of six software projects that are evaluated in terms of two criteria: cost (in millions of euros) and sustainability (the larger the value, the more sustainable the project is). Figure 1 plots the projects with respect to the two criteria. The black points, corresponding to projects 1, 3, 4, and 6, correspond to projects that are dominating whereas the white points, projects 2 and 5, correspond to dominated projects. A Decision Maker would never choose any project corresponding to a white point since it is dominated by at least one other project corresponding to a black point. For instance, the most costly of the dominated projects, point 5, is dominated by the second most sustainable project, point 4.

Although assertive, dominance relation has little discriminant power. There might exist too many dominating projects for the Decision Maker to choose from. In order to reduce the number of alternatives considerably, there is the need to aggregate the several criteria according to some *preference model.* Three aggregation methods can been considered in the MCDA literature:

- *Value function,* which consists of assigning to each alternative a real number that is representative of its assessment in the problem [8]. This is the case of the weighted-sum model and goal programming.
- *Outranking relation,* which is a binary relation that takes into account the number of criteria in favor, and not in favor, of an alternative over another [14]. This is the basis of PROMETHEE [3] and ELECTRE methods [6].
- *Decision rules,* which are logical statements that define some threshold requirements on the chosen criteria and provide a recommendation for an alternative that satisfies those requirements [7].

In the following sections, we illustrate the application of the first two models. For the third method, we refer to [7]. We assume a set of alternatives $A = \{a, b, c, \ldots\}$ that are evaluated on the basis of a set of n evaluation criteria $G = \{g_1, g_2, \ldots, g_n\}$. We denote by $g_j(a)$ the quantitative or qualitative performance of an alternative $a \in A$ on criterion $g_j \in G$. For simplification purpose, we assume that each criterion g_j is associated with an increasing direction of preference, that is, the larger is the value, the better it is.

We will say that alternative a *dominates* alternative b, $a \geq b$, if and only if $g_j(a) \geq g_j(b)$ for all $j = 1, \ldots, n$ and there exists at least a g_i such $g_i(a) > g_i(b)$. Depending of the context, we will also say that b is *dominated* by a. Note that this defines a binary relation that is transitive, anti-reflexive, and asymmetric, that is, a *strict partial order.* Moreover, an alternative a is *nondominated* if there exists no other alternative in A that dominates it. Moreover, alternatives a and b are (mutually) *incomparable* if and only if they do not dominate each other.

3   Value Function Methods

A value function method is expressed by a mapping $U : A \mapsto \mathbb{R}$ that assigns to each alternative in A a real number $U(a)$ that is representative of the assessment of a. A typical value function method is the *weighted-sum* function, which is defined as follows

$$U(a) = \sum_{i=1}^{n} \omega_j g_j(a) \tag{1}$$

where ω_j, $j = 1, \ldots, n$, the weight of criterion g_j, is a non-negative real number, and it expresses the relative importance of this criterion for the Decision Maker.

This function is scale-dependent and, usually, the following normalization is considered

$$g_j'(a) = \frac{g_j(a) - \min_j}{\max_j - \min_j} \tag{2}$$

where $\min_j$ and $\max_j$ are the minimum and the maximum value of the alternatives at criterion g_j, respectively.

The larger the weighted-sum value, the better the alternative is to the Decision Maker. However, in general, not all nondominated alternatives can attain the largest weighted-sum value for any possible choice of weights, which is a limitation of this method.

An alternative formulation is based on the *Tchebycheff* distance of an alternative with respect to a reference point $z = (z_1, \ldots, z_n)$ provided by the Decision Maker. The Tchebycheff method is formalized as follows

$$U(a) = \max_{i=1,\ldots,n} \omega_j \left(|z_j - g_j(a)| \right) \tag{3}$$

4 Outranking Methods - ELECTRE

In this section, we describe the ELECTRE, the first and one of the better known among the outranking methods. Other methods, such as Promethee, follow similar principles. We refer to [5] for more details about these methods and variants. Different from weighted-sum and Tchebycheff methods, ELECTRE has the advantage of being scale independent.

In ELECTRE, an alternative a *outranks* an alternative b, or a is at least as good as b, if

1. a majority of criteria supports this proposition, known as the *concordance principle*, and
2. the opposition of the minority is not too strong, known as the *non-discordance principle*.

ELECTRE uses the notion of *indifference* and *veto* at each criterion to take into account imprecision in defining the relative performance of alternatives.

An *indifference threshold* q_j is defined for each $j = 1, \ldots, n$. We shall say that an alternative a is indifferent to b in criterion g_j if the following holds

$$|g_j(a) - g_j(b)| < q_j \tag{4}$$

From the indifference threshold at criterion g_j, a *concordance index* can be constructed that measures the strength of the assertion of outranking alternative a over b on that criterion as follows

$$c_j(a,b) = \begin{cases} 1 & g_j(b) \leq g_j(a) + q_j \\ 0 & g_j(b) > g_j(a) + q_j \end{cases} \tag{5}$$

If $c_j(a,b) = 1$, then criterion g_j is in favor of the assertion that a outranks b, otherwise, it is not. Note that an interpolated value of $c_j(a,b)$ between 0 and 1 can also be constructed depending of how large is the difference between $g_j(a)$ and $g_j(b)$.

Given the concordance indices for two alternatives in A for all criteria in G, a *comprehensive concordance index* $CCI(a, b)$ aggregates this information over all criteria as a weighted sum as follows

$$CCI(a, b) = \sum_{i=1}^{n} w_j c_j(a, b) \tag{6}$$

where $w_j > 0$, $j = 1.\ldots, n$ and $w_1 + w_2 + \cdots + w_n = 1$. Note that $CCI(a, b) \in [0, 1]$ and is scale independent. Finally, a *concordance test* is satisfied if $CCI(a, b) \geq \lambda$, where $\lambda \in [0.5, 1]$ represents the allowed portion of criteria that should be in favor of a outranking b.

A *veto* can be introduced to handle the cases in which some few criteria are against the assertion that a outranks b. A *discordance index* $d(a, b)$ is constructed for each alternative a and b in A and for each criterion g_j as follows

$$d_j(a, b) = \begin{cases} 1 & g_j(b) \geq g_j(a) + v_j \\ 0 & g_j(b) < g_j(a) + v_j \end{cases} \tag{7}$$

where v_j is the *veto threshold*. If $d_j(a, b) = 1$, then criterion g_j is against the assertion that a outranks b, otherwise, it is not. Similarly to the concordance index, an interpolated value of $d_j(a, b)$ can be considered.

The comprehensive concordance index for two alternatives can now be combined with the corresponding discordance indices for all criteria in G, leading to a *credibility index* $CrI(a, b)$. Several formulation of this index are known. Here, we consider the following

$$CrI(a, b) = CCI(a, b) \times (1 - \max(d_1(a, b), d_2(a, b), \ldots, d_n(a, b))) \tag{8}$$

In this case, it is enough to have a criterion g_j for which $d_j(a, b) = 1$ to obtain $CrI(a, b) = 0$. This formulation gives full power to a discordant criterion. Finally, if $CrI(a, b) \leq \lambda$, we conclude that a outranks b. Note that this binary relation is not transitive.

The credibility indices for all alternatives can be recast as a directed graph where the alternatives are vertices, and an alternative b is connected to alternative a by an arc if a outranks b. From this graph, it is possible to better understand the relation between the different alternatives, find those that are preferable, that is, those with null indegree, and establish an hierarchy of alternatives.

5   Case Study: Selecting Programming Languages

In this section, we describe a didactic case study based on the work in Pereira et al. [13]. This work uses time, memory and energy spent by 10 programs implemented in 27 different programming languages. For our case study, we have chosen 6 well-known languages: C, C++, C#, Python, Go, and Java. We

Table 2. Energy, time and score for programming languages

	Energy	Time	Score
C	1.00	1.00	94.70
C++	1.34	1.56	92.40
Java	1.98	1.89	95.40
Go	2.23	2.83	77.70
C#	3.14	3.14	82.40
Python	75.88	71.90	100.00

Table 3. Scenario 1

	Energy	Time	Score
Indifference threshold q_j	1.00	1.00	5.00
Weight w_j	1/3	1/3	1/3

consider criterion *time* as the average amount of seconds and criterion *energy* as the average amount of Joules, as given in [13], each of which normalized with respect to the fastest and the most energy-efficient programming language.

In addition to these two criteria, we have considered the score of programming languages in 2021 according to IEEE [15] as a third criterion. According to the authors, the score of a programming language should be correlated to its popularity among professional programmers and employers. This score is based on the information collected from multiple sources, such as Stack Overflow, Google, IEEE's Xplore article database, IEEE Jobs Site, CareerBuilder, and among others. The values range from 1 (least popular) to 100 (most popular). The authors did not provide details on how the scores were computed.

Table 2 shows the *performance table* with respect to the chosen programming languages, sorted with respect to the energy criterion. Noteworthly, C is both the fastest and most energy-efficient programming language. Python is, by far, the least time consuming and the least energy-efficient programming language, but it is the most popular among the six. Moreover, we see that energy and time are strongly correlated, although no correlation seems to exist between each of these two criteria and score.

In the following we describe the application of ELECTRE method under six possible preference scenarios and discuss the corresponding findings. Graphs obtained from comprehensive concordance indices and credibility indices are in Appendix.

Scenario 1. Table 3 shows the values of the indifference threshold q_j and the weight w_j for each criterion for the computation of the comprehensive concordance index for each pair of alternatives. We assume that the weights are the same to the three criteria. We consider $\lambda = 1$, which is the most restrictive sce-

182   L. Paquete

Table 4. CI table for energy

	C	C++	C#	Python	Go	Java
C	1	1	1	1	1	1
C++	1	1	1	1	1	1
C#	0	0	1	1	1	0
Python	0	0	0	1	0	0
Go	0	0	1	1	1	0
Java	1	1	1	1	1	1

Table 5. CI table for time

	C	C++	C#	Python	Go	Java
C	1	1	1	1	1	1
C++	1	1	1	1	1	1
C#	0	0	1	1	1	0
Python	0	0	0	1	0	0
Go	0	0	1	1	1	0
Java	1	1	1	1	1	1

Table 6. CI table for score

	C	C++	C#	Python	Go	Java
C	1	1	1	0	1	1
C++	1	1	1	0	1	1
C#	0	0	1	0	1	0
Python	1	1	1	1	1	1
Go	0	0	1	0	1	0
Java	1	1	1	1	1	1

Table 7. CCI table for Scenario 1

	C	C++	C#	Python	Go	Java
C	–	1.00	1.00	0.67	1.00	1.00
C++	1.00	–	1.00	0.67	1.00	1.00
C#	0.00	0.00	–	0.67	1.00	0.00
Python	0.33	0.33	0.33	–	0.33	0.33
Go	0.00	0.00	1.00	0.67	–	0.33
Java	1.00	1.00	1.00	1.00	1.00	–

nario, that is, only total concordance among the three criteria is considered for outranking. Also, we assume that no veto information is available.

The concordance indices (CI) for energy, time, and score are presented in Tables 4, 5 and 6, respectively. The comprehensive concordance index (CCI) table under this scenario is presented in Table 7. The computation of these tables are left as exercises to the reader. Figure 2 shows the graph that is obtained from Table 7 by setting $\lambda = 1$. This means that only the 1s in the latter table will be considered for the graph construction. The graph suggests that C, C++, and Java are mutually indifferent, and are preferable to C# and Go under this preference scenario. Moreover, Java is preferable to Python, but the latter is incomparable with respect to the remaining programming languages. For this reason, one would be tempted to choose Java. From Table 1, we see that the latter falls slightly behind C and C++ in terms of energy and time, but it is more popular.

Scenario 2. In this scenario, we consider the information provided in Table 3, except that $\lambda = 2/3$, that is, at least 2 out 3 criteria can be used for outranking, which is a relaxation of the previous scenario. No veto information is considered. The resulting graph is shown in Fig. 3. As expected, it is denser and it shows that Python is clearly the least preferred among the six programming languages. For this reason, under this preference scenario, there is no longer a particular reason to prefer Java over C++ or C.

Scenario 3. Table 8 shows the values of the indifference threshold q_j and the weight w_j for the computation of the comprehensive concordance indices. Differently from the two previous scenarios, the score has a very large weight as compared with the two other criteria. We consider the restrictive scenario of $\lambda = 1$, and that no veto information is available. Only the comprehensive concordance

Table 8. Scenario 3

	Energy	Time	Score
Indifference threshold q_j	1.00	1.00	5.00
Weight w_j	0.01	0.01	0.98

Table 9. CCI table for Scenario 3

	C	C++	C#	Python	Go	Java
C	–	1.00	1.00	0.02	1.00	1.00
C++	1.00	–	1.00	0.02	1.00	1.00
C#	0.00	0.00	–	0.02	1.00	0.00
Python	0.98	0.98	0.98	–	0.98	0.98
Go	0.00	0.00	1.00	0.02	–	0.01
Java	1.00	1.00	1.00	1.00	1.00	–

index needs to be calculated (see Table 9). The graph in Fig. 4 is obtained from this table by setting $\lambda = 1$. It is very similar to the graph of Fig. 2, except that Python and Java are now indifferent (see the bidirectional arc between the two).

Scenario 4. Similarly to Scenario 2, we consider the same information of Table 8, except that we allow $\lambda = 2/3$. No veto information is considered. Figure 5 shows the graph obtained from Table 9 by setting $\lambda = 2/3$. Under this scenario, there are many incomparable cases, but Python is the most preferred programming language. We recall that Python is the most popular among the six programming languages.

Table 10. Scenario 5

	Energy	Time	Score
Indifference threshold q	1.00	1.00	5.00
Veto threshold v	3.00	3.00	6.00
Weight w	1/3	1/3	1/3

Scenario 5. In this scenario, we consider Scenario 2 plus veto information for each criterion (see Table 10). We also set $\lambda = 2/3$. The discordance index (DI) tables for energy, time, and score are presented in Tables 11, 12 and 13, respectively. The comprehensive concordance index table under this scenario is presented in Table 14. The graph in Fig. 6 is obtained from the credibility index extracted from the discordance index tables and the comprehensive concordance table by setting $\lambda = 2/3$. The conclusions are similar to that of Scenario 1, with C, C++ and Java being the most preferred programming languages and mutually indifferent.

Table 11. DI table for Energy

	C	C++	C#	Python	Go	Java
C	0	0	0	0	0	0
C++	0	0	0	0	0	0
C#	0	0	0	0	0	0
Python	1	1	1	0	0	1
Go	0	0	0	0	0	0
Java	0	0	0	0	0	0

Table 12. DI table for Time

	C	C++	C#	Python	Go	Java
C	0	0	0	0	0	0
C++	0	0	0	0	0	0
C#	0	0	0	0	0	0
Python	1	1	1	0	1	1
Go	0	0	0	0	0	0
Java	0	0	0	0	0	0

Table 13. DI table for Score

	C	C++	C#	Python	Go	Java
C	0	0	0	0	0	0
C++	0	0	0	1	0	0
C#	1	1	0	1	0	1
Python	0	0	0	0	0	0
Go	1	1	0	1	0	1
Java	0	0	0	0	0	0

Table 14. CCI table for Scenario 5

	C	C++	C#	Python	Go	Java
C	–	1.00	1.00	0.67	1.00	1.00
C++	1.00	–	1.00	0.00	1.00	1.00
C#	0.00	0.00	–	0.00	1.00	0.00
Python	0.00	0.00	0.00	–	0.00	0.00
Go	0.00	0.00	1.00	0.00	–	0.00
Java	1.00	1.00	1.00	1.00	1.00	–

Table 15. DI table for Score

	C	C++	C#	Python	Go	Java
C	0	0	0	0	0	0
C++	0	0	0	0	0	0
C#	0	0	0	0	0	0
Python	0	0	0	0	0	0
Go	0	0	0	1	0	0
Java	0	0	0	0	0	0

Table 16. CCI table for Scenario 6

	C	C++	C#	Python	Go	Java
C	–	1.00	1.00	0.67	1.00	1.00
C++	1.00	–	1.00	0.00	1.00	1.00
C#	0.00	0.00	–	0.00	1.00	0.00
Python	0.00	0.00	0.00	–	0.00	0.00
Go	0.00	0.00	1.00	0.00	–	0.00
Java	1.00	1.00	1.00	1.00	1.00	–

Scenario 6. In this scenario, we consider the same information of Table 10, except that the veto threshold of criterion score is now 20.00, which relax the veto power in this criterion with respect to Scenario 5. Only the discordance index for score needs to be recomputed, see Table 15, which shows that veto has less effect as compared with Table 13. The comprehensive concordance index under this scenario is presented in Table 16. The graph in Fig. 7 is obtained from the discordance index tables and the comprehensive concordance index table by setting $\lambda = 2/3$. The conclusions are similar to that of Scenario 2, with Python being the least preferred option.

6   Conclusions

Multicriteria decision methods allow software engineers to make more informed decisions when faced with multiple criteria. Although these methods have been known for over 50 years, they are rarely used in software development practice. The strength of these methods lies not in providing a single solution but in offering a way to better understand the trade-offs between different criteria. They help decision-makers gain a deeper understanding of the problem at hand.

This overview, along with the didactic example, only covers the basics of these methods. For more detailed and fundamental results, readers are encouraged to refer to the book by Ehrgott et al. [5]. To learn about the ELECTRE method and its variants in more detail, see [6].

Acknowledgements. L. Paquete wishes to thank J.R. Figueira from the University of Lisbon for a fruitful discussion on the application of ELECTRE.

A   Graphs Representations

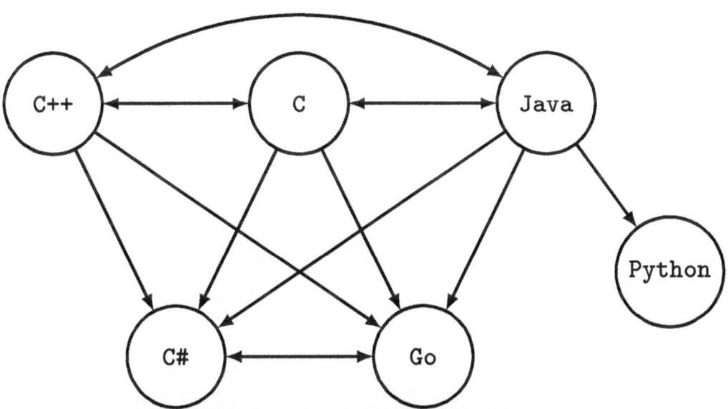

Fig. 2. Graph representation under Scenario 1.

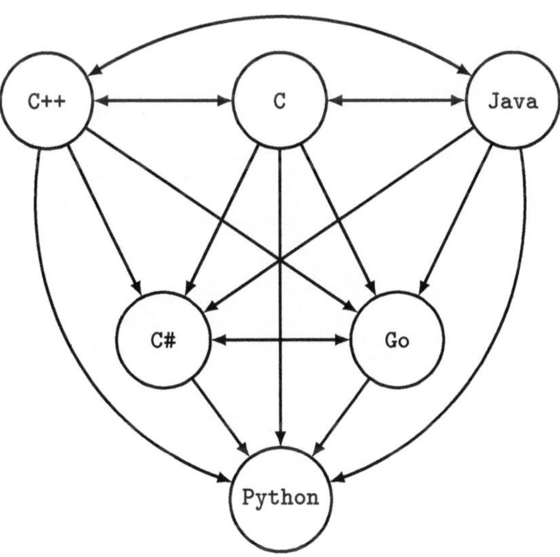

Fig. 3. Graph representation under Scenario 2.

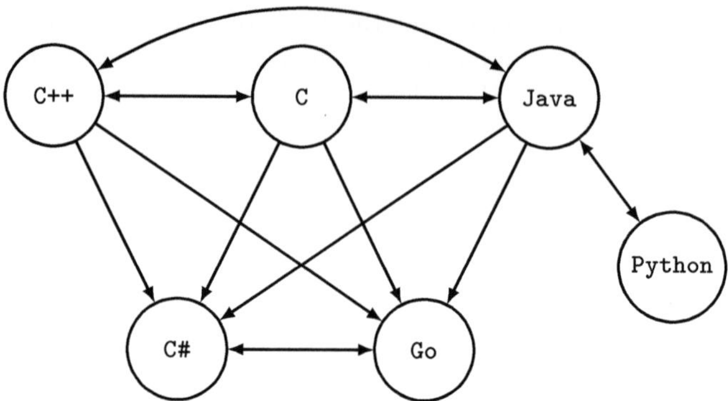

Fig. 4. Graph representation under Scenario 3.

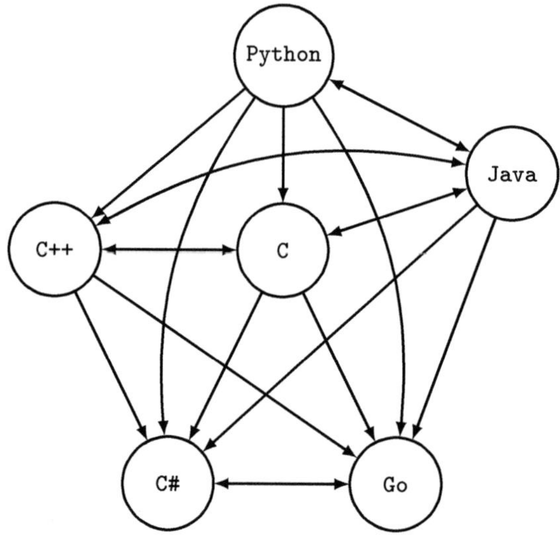

Fig. 5. Graph representation under Scenario 4.

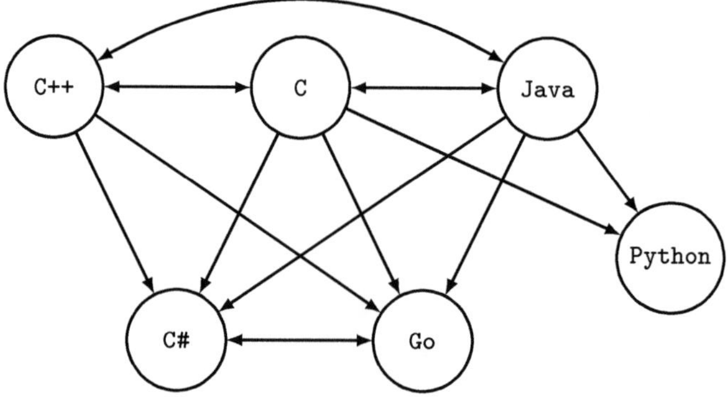

Fig. 6. Graph representation under Scenario 5.

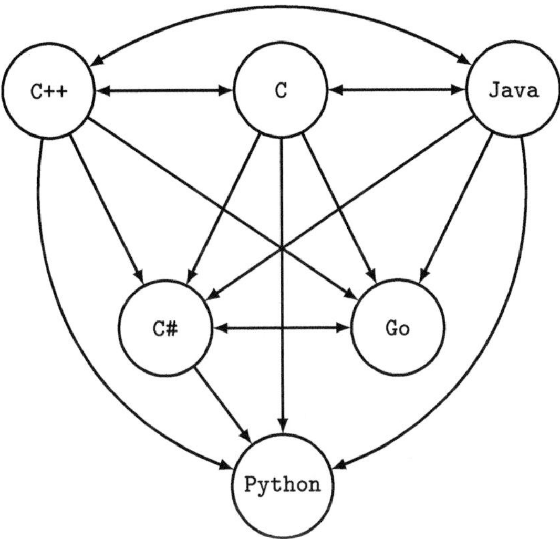

Fig. 7. Graph representation under Scenario 6.

References

1. Abdulwareth, A.J., Al-Shargabi, A.A.: Toward a multi-criteria framework for selecting software testing tools. IEEE Access **9**, 158872–158891 (2021)
2. Barcus, A., Montibeller, G.: Supporting the allocation of software development work in distributed teams with multi-criteria decision analysis. Omega **36**(3), 464–475 (2008). Special Issue on Multiple Criteria Decision Making for Engineering
3. Brans, J., Vincke, P.: A preference ranking organisation method: the PROMETHEE method for MDCM. Manage. Sci. **31**(6), 1647–656 (1985)
4. Büyüközkan, G., Ruan, D.: Evaluation of software development projects using a fuzzy multi-criteria decision approach. Math. Comput. Simul. **77**(5–6), 464–475 (2008)
5. Ehrgott, M., Figueira, J.R., Greco, S.: (eds.): Trends in Multiple Criteria Decision Analysis. Springer (2010)
6. Figueira, J., Greco, S., Roy, B., Słowińskiv, R.: An overview of ELECTRE methods and their recent extensions. J. Multicrit. Decis. Anal. **20**, 61–85 (2013)
7. Greco, S., Mousseau, V., Słowińskiv, R.: Rough sets theory for multicriteria decision analysis. Eur. J. Oper. Res. **175**(1), 247–290 (2001)
8. Keeney, R., Raiffa, H.: Decisions with Multiple Objectives: Preferences and Value Trade-Offs. Wiley, New York (1976)
9. Lai, V.S., Wong, B.K., Cheung, W.: Group decision making in a multiple criteria environment: a case using the AHP in software selection. Eur. J. Oper. Res. **137**(1), 134–144 (2002)
10. Malekmohammadi, B., Zahraie, B., Kerachian, R.: Ranking solutions of multi-objective reservior operation optimization models using multi-criteria decision analysis. Expert Syst. Appl. **38**(6), 7851–7863 (2011)
11. Morais, D., de Almeida, A., Figueira, J.: A sorting model for group decision making: a case study of water losses in brazil. Group Decis. Negot. **23**(5), 937–960 (2014)

12. Otero, C.E., Otero, L.D., Weissberger, I., Qureshi, A.A.: A multi-criteria decision making approach for resource allocation in software engineering. In: Al-Dabass, D., Orsoni, A., Cant, R.J., Abraham, A. (eds.) Proceedings of the 12th UKSim, International Conference on Computer Modelling and Simulation, Cambridge, UK, 24–26 March 2010, pp. 137–141. IEEE Computer Society (2010)
13. Pereira, R., et al.: Ranking programming languages by energy efficiency. Sci. Comput. Program. **205**, 102609 (2021)
14. Roy, B.: Multicriteria Methodology for Decision Aiding. Nonconvex Optimization and Its Applications. Kluwer Academic, Dordrecht (1996)
15. Cass, S., Kulkami, P.E.G.: Top programming languages 2021. IEEE Spectrum (2021). https://spectrum.ieee.org/top-programming-languages
16. Santhanam, R., Kyparisis, J.: A multiple criteria decision model for information system project selection. Comput. Oper. Res. **22**(8), 807–818 (1995)
17. Trendowicz, A., Kopczynska, S.: Adapting multi-criteria decision analysis for assessing the quality of software products. current approaches and future perspectives. Adv. Comput. **93**, 153–226 (2014)
18. Wang, J., Lin, Y.: A fuzzy multicriteria group decision making approach to select configuration items for software development. Fuzzy Sets Syst. **134**(3), 343–363 (2003)
19. Yao, X.: Some recent work on multi-objective approaches to search-based software engineering. In: Ruhe, G., Zhang, Y. (eds.) Search Based Software Engineering - 5th International Symposium, SSBSE 2013, St. Petersburg, Russia, 24–26 August 2013. Proceedings. LNCS, vol. 8084, pp. 4–15. Springer (2013)

Using Strong Types in C++
for Long-Term Code Management

Richárd Szalay and Zoltán Porkoláb[(✉)]

Department of Programming Languages and Compilers, Institute of Computer
Science, Faculty of Informatics, Budapest, Hungary
{szalayrichard,gsd}@inf.elte.hu

Abstract. When using programming languages, type systems are crucial tools in the hands of developers to guarantee an elevated level of safety of their programs. However, type systems are not used to their full extent in practice, and trade-offs are made. This mainly manifests in developers overusing built-in and well-known library types, such as `int` or `string`, instead of types that express the domain of the represented values with a finer granularity. The results of this range from a hindrance to code comprehension to subtle security vulnerabilities going undetected for potentially years. Education is a crucial element in combating the trend of quality deterioration in software projects. By teaching the new generation of developers the importance of stronger and safer types, we hope to ensure more sustainable development processes in the future.

In this paper, we detail how the education of strong typing was introduced in our *Advanced C++* M.Sc. subject. While certain ideas of strong typing are actionable from the very early subjects in education, the hands-on experience with improving an existing, hard-to-understand code ensures students internalise the ideas better. We discuss the lifting of an example software project to strong types step-by-step.

Keywords: Sustainable development · Strong typing · Programming languages · C++ · Practical education · Tutorial guide

This chapter was prepared with the professional support of the Doctoral Student Scholarship Programme of the Co-operative Doctoral Programme of the Ministry of Innovation and Technology, financed from the National Research, Development, and Innovation Fund of Hungary.

This work received financial support through the Erasmus+ Strategic Partnership for Higher Education *SusTrainable—Promoting Sustainability as a Fundamental Driver in Software Development Training and Education* (project number 2020-1-PT01-KA203-078646), funded by the European Union and coordinated by the University of Coimbra, Portugal.

The information and views set out in this publication are those of the authors and do not necessarily reflect the official opinion of the European Union. Neither the European Union institutions and bodies nor any person acting on their behalf may be held responsible for the use which may be made of the information contained therein.

1   Introduction

Several large and mainstream programming languages, including C++, exhibit a statically checked and strongly typed type system. Under these conditions, the compiler checks whether an expression or instruction is allowed to be evaluated, and this check if performed during compilation. Developers and architects can thus use the type system to guarantee a degree of safety of their program. C++ comes with a small set of *fundamental types* defined in the Standard [12] to be supplied built into the compiler – as opposed to being written as code in a library. These fundamental types include, most notably, integer (`int`) and floating-point (`float`) numerals, the `char` (character) type, and trivial type constructs such as pointers and arrays. Moreover, the *standard library* – whose content is determined and required by the Standard, but the implementation exists as C++ code – offers some additional generic types, such as `std:: string`.

Unfortunately, it has been observed that developers tend to overuse these fundamental and widely available types, e.g., by setting both the "size" of a machine part and the "orientation" of it to be a `double` or conflating the "name" of a person with their "address" in `string` variables, data fields. With the types of program elements kept at a coarse level, developers resort to embedding semantic information in the *names* of variables. While this is visually indicative to a knowledgeable developer, and there are ways to use human-assigned names meaningfully in an automated context [2], the mainstream languages and compilers ascertain no value to the identifier. This had resulted in researchers observing dormant security vulnerabilities in large, industrial projects [20], with potentially many more mistakes only waiting to be made and discovered [25]. The lack of elaborate, descriptive and enforced type granularity has led to serious accidents, like in the case of the *Mars Climate Orbiter* [22] where program modules related to physics calculations were interfacing by only passing "floating-point numbers" – i.e., `doubles` – but one part was using SI values, while another was written with Imperial measurements in mind.

The solution to the problem is using *strong typing* [14,15], which is a design principle that allows using the capabilities of the language and the type system to the full extent possible. Employing strong typing ensures that the compiler can better detect and thus prevent programmer errors. Strong typing, however, is not a discrete binary state of the system but a comparable measure: some software solutions to the same problem can be *"more strongly typed"* than another. At one of the extreme ends of the spectrum, a solution might not use "types" any more than required by the execution environment's hardware; while the other extreme would be a separate, distinct type for each value calculated in the program. To discretise this, we will discuss the problem on 4 levels of type strength. The exact details of strong typing techniques is present in Sect. 3.

Two key insights justify the education of type safety and strong types.

First, it is known from previous literature that, without expending additional effort by the developers to combat the trends, the quality of a software system generally deteriorates over time [16,33]. Usually, and unfortunately, only a small subset of developers are inclined – professionally or financially – to such expenditures. Often, the application of refactoring, which improves quality metrics, can generally be traced back to a few developers' personal motivations for doing so [13]. While reworking an existing software system to stronger types is a tedious task to do manually, creating automated tools as a generic solution is also non-trivial [27]. Existing solutions were created as specific algebra for particular domains, limiting their applicability [32].

Thus, the second reason is that by ensuring that new graduates are educated about strong typing, they can start their practice, whether industrial or research, by writing new code already satisfying the need for type safety.

2 Education of Strong Typing in a Large-Scale Academic Curriculum

First, we will detail how the education of strong typing was integrated into the larger framework of the Master's Degree (M.Sc.) curriculum at our University.

2.1 Laws and Curricular Framework

Higher education in Hungary is regulated by an associated legal framework. The general requirements of *Computer Science* (CS) education is described in the law, and universities have only some autonomy in defining the specifics of their education locally. In accordance with the Bologna framework [5], education is done in three tiers: Bachelor (B.Sc.), Master (M.Sc.) and Doctorate (Ph.D.). CS education in Hungary is mandated by law to be 6 semesters (180 ECTS credits) of B.Sc. followed by 4 semesters (120 ECTS credits) of M.Sc., which is in stark contrast when compared to engineering training which instead use a 7:4 semester (210:120 ECTS) division. The Faculty of Informatics's old CS curriculum was accredited in 2008 in which B.Sc. students were taught beginner C++ on a obligatory subject, and one of the M.Sc. specialisations contained an elective *"Programming languages"* block in which modern C++ topics were taught on *"Multi-paradigm programming"*. This subject was allocated 3 ECTS credits for a weekly 2-hour lecture and individual work, followed by an exam, which was a test of various forms of questions but contained no dedicated coding exercise other than questions that can be answered with just a few lines of code.

The curriculum for CS education was refactored, refreshed, and reaccredited since, and the first year of students under the new system started their studies in 2018. The B.Sc. study plan includes a single *"Programming Languages"* subject, on which students are taught the fundamentals of programming language elements and design, currently through the examples of Java. Beginner C++ education was resurrected as an elective subject. The Multi-paradigm programming subject had been renamed *"Advanced C++"* and given an additional 2 ECTS credits – now totalling 5 – for a weekly 2-hour practice lesson in addition to the lecture. While the system of multi-subject *blocks* had been removed from the framework, Advanced C++ remains an elective subject on the M.Sc. curriculum.

Approximately 600 students enrol CS each year on the Bachelor level and 100–120 on the Master level [4]. Bachelor students are enrolled based on their secondary school final exam (*Matura*) scores without a dedicated admission exam by the University. Knowledge of foreign languages (including English) is not a requirement for enrollment. To be given the Bachelor degree, the students must obtain a state-accredited certificate of language knowledge at the CEFRL B2 level from one foreign language. Most CS students will do an English exam or already have done it. Master students are enrolled based on an admission exam, and a relevant Bachelor degree is a requirement. As the Advanced C++ subject is only announced for local, Hungarian students, the previous figures include only the local enrolments. While the B.Sc. curriculum has a direct equivalent for international students, taught in English, the post-2018 M.Sc. curricula are different: several specialisations are announced only in English, and both international and local students may enrol on those. Those students are out of scope for this education report, as their education is usually done in Python or Java. They do not have a dedicated subject specifically for teaching a programming language.

Typically, the Faculty of Informatics offers CS education both full-time/day-time (lessons each day may start as early as 08:00) and part-time/evening classes (lessons each day start after 16:00), but not in correspondence (lessons only one or two days per month) form. Most Master courses are held in accordance with the night-time schedule, together with, and even for those officially enrolled as day-time students. Also, the COVID-19 situation [30] has forced the University to do fully remote or hybrid attendance education, depending on the circumstance.

As refactoring and type migration necessarily involve the understanding of an existing software system, **we expect students to have an intermediate command of English**, so they understand the documentation and comments of existing code, **and have absolved beginner C++ and Object-Oriented Programming skills**: they understand the fundamental data structures and operations, they know what is and how to write a class, how encapsulation and visibility works, and what C++ templates are. Unfortunately, we are not allowed to require that the M.Sc. students have studied the corresponding B.Sc. subject, "Basic C++", and direct subject-to-subject dependencies crossing study plan boundaries are not possible.

2.2 The *Advanced C++ Subject*

As discussed previously in Sect. 2.1, the Advanced C++ subject in the new curriculum was made into a 5 ECTS credit subject, now including effort for a weekly 2-hour long practice lesson. The lecture's material starts from the tricky parts of "beginner C++" – such as the pitfalls of special member functions – and encompasses template metaprogramming [31], lambdas, exception safety, concurrency, and `std::visit`, remained fundamentally unchanged, along with the shape of the exam. To ensure that the final grade reflects both the "theory" and the "practice" part, both examinations were calculated in the final grade with a 50% weight.

In Hungary, a 5-scale grading system is used, where grade 1 is *Fail* and grade 5 is *Excellent*. Assuming the exam (lecture) result and the practice is 100% total, the boundaries between grades were set as follows: 40% minimum for *Pass*, then 55%, 70%, and 85% and onwards for *Excellent*. Earning 20%p. (out of the 50%)[1] each on both the exam and the assignment was also mandatory – students doing a perfect exam but no assignment, still resulting in 50% total, would not pass.

We have opted **not to do** *synchronous (online)* lessons for the practice – i.e. students need not be available and attend a lesson each week during practice hours. (This is not true for the lecture, which was given synchronously, and a recording was made available for the students to watch later.) The teachers kept the officially scheduled time for the practice available as on-demand, and due to the COVID-19 situation, online videoconferencing-based consultation. The students were given, on 4 occasions, a presentation-like or discussion lesson during practice hours, and then the work of the semester was the fulfilment of an assignment project out of two choices. As both the fact that Advanced C++ has a practice lesson and the teaching of strong typing was a new addition to our material, we offered not just the strong typing project – discussed in Sect. 2.3 – but a *library-writing* exercise: students were instructed to create a program library, in the spirit of the *Standard Template Library (STL)*, using the tools and techniques presented on the lecture, that implements a container or data structure, for this semester, *Trie* (or *prefix tree*) [8].

Table 1. Breakdown on the results of the semester "**2020-21/II**"

(a) Student participation

All students	69	100.00%
Non-participant	31	44.93%
Abandoned subject	27	87.10%
No submission	4	12.90%
Old curriculum	1	3.23%
Participant	38	55.07%
Trie (library)	30	78.95%
Strong Typing	7	18.42%
Both	1	2.63%

(b) Grading outcome

All students	69	100.00%
0 (*Miss/Abandon*)	27	39.13%
1 (*Fail*)	5	7.25%
2 (*Pass*)	4	5.80%
3 (*Average*)	11	15.94%
4 (*Good*)	16	23.19%
5 (*Excellent*)	6	8.70%
0 and 1 (incomplete)	32	46.38%
2 to 5 (success)	37	53.62%

The Spring semester begins in early February, and the term time ends in the middle of May, after which the examination period lasts until the end of June. The assignments were published in late February and were due in early May, giving the students between 2 and 2.5 months to complete the assignment. The lecture exam was held in the first two weeks of the examination period

---

[1] 20% points, i.e., marks corresponding to 20% worth of the total, as opposed to earning "20% out of the 50%", which would mean earning a total of 10% only.

(late May), and students were allowed to take both chances, with the teachers considering the better of the two attempts. Students were also allowed to either hand the assignment in late at the very end of May if they missed the first deadline, or perform a resubmission based on the comments received on the initial submission.

Table 1 details the statistics of the semester. Unfortunately, the first two takeaways are that attrition of students is considerable (45% of the registered students did not partake in the assignment or even the lecture) [29] – especially for elective subjects – and even students who perform on the subject seem to take the perceived "easier" assignment. 1 student had registered on the subject while being on the older 2008 curriculum where the subject did not include a practice part, and thus this student was exempted from the assignments. 4 students took the exam but did not submit an assignment and thus failed the subject. There was only 1 student (2.63%) out of the 38 who submitted everything, but failed to acquire the bare minimum points to be given a pass. Everyone else succeeded in obtaining a passing grade.

Table 2. Breakdown on the results of the semester "**2021-22/II**"

(a) Student participation

All students	50	100%
Non-participant	14	28%
Abandoned subject	10	71.43%
No submission	4	28.57%
Old curriculum	—	—
Participant *(may overlap)*	40	80%
ID-BiMap (library)	30	75%
Comprehension exercise №1	28	70%
Comprehension exercise №2	20	50%

(b) Grading outcome

All students	50	100%
0 (*Miss/Abandon*)	10	20%
1 (*Fail*)	6	12%
2 (*Pass*)	7	14%
3 (*Average*)	8	16%
4 (*Good*)	14	28%
5 (*Excellent*)	5	10%
0 and 1 (incomplete)	16	32%
2 to 5 (success)	34	64%

In the following year, the COVID-19 situation turned out to be more manageable, but with *hybrid remote education* measurements still in place. In this semester, which statistics are detailed in Table 2, we did not offer Sect. 2.3's strong typing assignment due to the unfortunate feedback that students found it prohibitively challenging to immerse themselves into completely unfamiliar projects. A colleague of our research team designed tasks related specifically related to code comprehension, which we let the students partake in. (The details of those tasks are out of scope for this paper, and can instead be found in the publications of the aforementioned colleague, [6,7].) However, we kept the *library-writing* exercise, with students having to implement a bidirectional associative container where the keys are numeric IDs, complete with various methods from the usual *STL* `std::map`, and some additional custom requirements. The assignment's evaluation scheme focused on adherence to the conventions of the language, run-time efficiency, and the maintainability of the implementation, in which strong typing was also considered. The rest of the semester's structure, the material presented on the lectures, the offered deadlines, and the overall grading policy remained the same.

2.3   Practical Strong Typing in *Advanced C++*

During the first three weeks of the semester, students were given talks and held an oriented discussion in the time slot of the practice lesson. This involved showing them an overview presentation about the problem space of and the motivation for strong typing problems, from which the keystones of the main example is depicted in Listing 1. While the semester-long subject had ample time to fit that full-length presentation, a day-long or half-day-long training may only provide a **shortened version of such a presentation** to the students. An example of the code snippets the students were shown is depicted in Listing 1. As alluded to in Sect. 1, "strong typing" is better imagined as if it was laid out on a spectrum of various comparable "levels" of solutions.

For the strong typing assignment in semester "2020-21/II", students were instructed to find an open-source and freely modifiable C++ project on the Internet that is at least "sufficiently complex". In the project, they had to identify type system usage which was instinctually "weak", and fix it, carrying the fix through as much of the project as possible. "Carrying the fix through" meant that once the new, stronger type was introduced, it should also be given a useful encapsulated interface and existing program elements' types changed to use the new type, too. The requirements were not strict for this, as we left it up to the students' judgement to select a project where they feel their improvements would meaningfully benefit the community as a whole. The initial suggestion from the teachers was *Doxygen* [9], but we were up to accepting different projects. All 8 students who chose to participate in the assignment chose Doxygen, however.

Students were also instructed to keep a sort of logbook of their actions and questions, as we were interested in not just the end result of the changes but the thought process explored during the refactoring of the project. As most active community-driven software projects nowadays use the version control system Git, creating a fork of the official repository and committing the student's changes with meaningful messages was often a trivial way of keeping a logbook. The insights from their commit history were enhanced with the asynchronous discussion between the student and teachers during the evaluation of the assignment. Students were also suggested to use the *CodeCompass* [17,18] code comprehension tool to navigate the project. The University hosted the server with the parse of the latest Doxygen tagged release at that time available for exploration. Unfortunately, **these requirements**, especially the need of understanding a larger project, result in an increased workload that is **not feasible for a shorter teaching session.** Instead, students need to be given practical **knowledge through proctored examples.**

```
struct student {
    std::string name;
    std::string teacher;
};

int main() {
    std::vector<student> students = load_students();
    std::string name = read_name("Student name? ");
}
```

(a) Trivial example only using widely available *Standard Template Library (STL)* types that carry no domain-specific meaning, but only prescribe program behaviour.

```
struct student {
    person_name name;
    person_name teacher;
};

int main() {
    std::vector<student> students = load_students();
    person_name name = read_name("Student name? ");
}
```

(b) Introduction of a problem-specific entity domain, "names", over the low-level `string` type. Such a type might add potential new invariants into the design.

```
struct student {
    student_name name;
    teacher_name teacher;
};

int main() {
    std::vector<student> students = load_students();
    student_name name = read_student_name("Student name? ");
}
```

(c) Preventing argument selection defects [20], one of the problem classes that result from weak type granularity, by creating disjoint subdomains for `person_name`.

```
struct student {
    student_student_name_t name;
    student_teacher_name_t teacher;
};

int main() {
    load_students_return_collection_t students = load_students();
    read_student_name_return_t name = read_student_name("Student name? ");
}
```

(d) Contrived example that falls on the "too much" extreme of the spectrum, where each individual data member and variable is given an excessively descriptive, unique type.

Listing 1. Different levels of type distinction granularity for the same observable behaviour and program semantics in a small example.

In total, 7 students chose and succeeded with the assignment. One student's submitted Git repository fork was impossible to understand as it contained formatting commits and trash interleaved with the supposed changes. Coincidentally, this student was the same who chose *both assignments*, and as such, they obtained their passing grade from the "easier" library writing assignment.

2.4   Assignments Handed in by Students

Although we, the teachers, presented and discussed the generalised idea for what program elements would be good candidates for the refactoring exercise, we, on purpose, did not provide any specific locations or "class names" to the students. Instead, the identification of some pain points were also part of the exercise.

There are two main representation domains, each mapping to several problem domains, in Doxygen, that are good candidates, and the students were able to accurately identify examples. Doxygen is a self-contained tool that deals with creating HTML documentation with graphs (in Dot or SVG formats) from source code. To achieve this purpose, it deals primarily with **strings**, with the occasional **numeric values** for orientation, size, etc. For numeric values, several strong typing and dimensionality libraries exist [19]. 4 out of the 7 students chose a strong wrapper type over a numeric value and created associated interfaces, e.g., "width * height = area". **Examples in the context of numeric types are good for the introduction of strong typing, even in a hand-engineered, sterile context**.

The rest of the students, however, modified strings to stronger types. While these examples may be harder at first glance, they are more rewarding as it involves changing how the *values* of the type are handled in the program. These examples will be detailed in the following.

Hard-coded Strings to `enum`. Doxygen's documentation rendering uses CSS classes to highlight tokens of the source code differently. This is handled by the `FontClass` value, but in upstream Doxygen, it was only passed as a C string (`const char*`) only. As all potential values reaching the eventual printer logic was from a set of hard-coded values, the student introduced an `enum` with these values and changed the usage points accordingly. In several places, they also added `std::optional` to represent the lack of information which was previously represented by the "null pointer" value [23]. This allowed the decoupling of *"we cannot determine the font-class"* from *"the font class is nothing special"*. Moreover, there are several advantages to `enum`s: using an undefined value or not handling a defined value is detected by the compiler (e.g., via `-Wswitch`); and it is easier to compare as the values are well-defined integer constants, not pointers to some potentially changing memory address – which is how string literals are represented at run-time (Fig. 2).

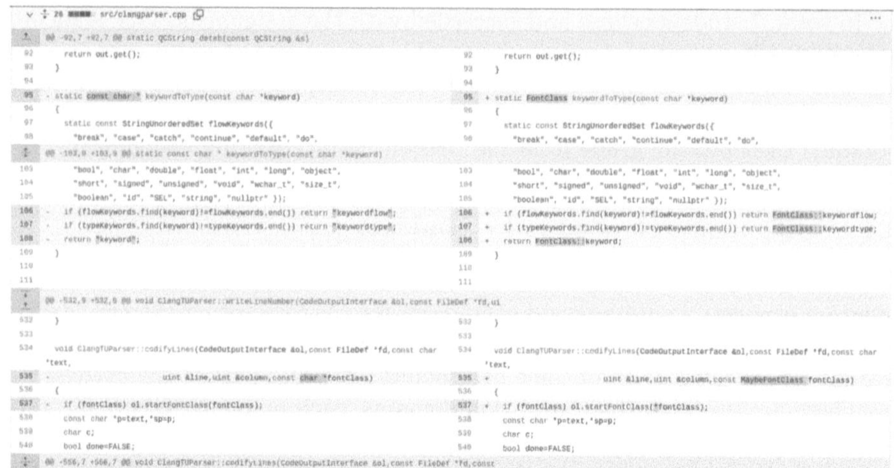

Fig. 1. Code changes resulting from propagating the introduction of a **enum** for classes in lieu of comparatively weaker typed "strings" on two program elements dealing with the affected problem domain. (Screenshot of the *GitHub* code difference rendering created from the student's repository.)

Figure 1 shows an example, rendered as a code difference, from the student's submission. **This is a beginner-level example that highlights the benefits of an otherwise seldom-used language element.**

URLs. As the output of Doxygen is a website, the system naturally also has to deal with a special subdomain of strings, URLs. URLs are structured data represented in textual format in a standardised way [1]. String types in almost all programming languages allow operations that are not meaningful for URLs, e.g., consider a call that, potentially by mistake, boils down to (`"http://example.com"`).`substr(3, 7)`. The result of this operation is `"p://e"`, something that is not useful when observing and invalid in the problem domain of URLs when manipulating. The student implemented a strong type, `URLString`, which can be constructed from a string and performs parsing of the constituents – e.g., the host name, the path, or the protocol – of the URL. Afterwards, client code might manipulate the individual constituents. When needed as a plain string, the instance can convert back to a single string value via calling the `operator()`. **This is a complex example, as the developers need to actively deeply investigate the usage of the value, and add problem-specific logic into the implementation;** however, this is the example in which "true" strong typing was created.

3   Techniques for Strong Typing

To succeed in strengthening the interface of our example project, the general techniques facilitating such approaches must be discussed first. The primary focus of this Tutorial will be the *interfaces of functions*, as this is the place where

```
177   void URLString::ParseProtocol(const std::string& loc_url)
178   {
179     size_t comma_location = loc_url.find(":");
180
181     _protocol = comma_location != std::string::npos
182       ? loc_url.substr(0, comma_location)
183       : "";
184   }
185
186   void URLString::ParseAuthority(const std::string& loc_url)
187   {
188     size_t authority_slashes_begin = GetProtocolSize();
189
190     if (authority_slashes_begin < loc_url.size() && loc_url.substr(authority_slashes_begin, 2) == "//")
191     {
192       size_t authority_begin_location = authority_slashes_begin + 2;
193
194       size_t authority_end_location = loc_url
195         .substr(authority_begin_location, loc_url.size() - authority_begin_location).find("/");
196
197       authority_end_location = authority_end_location != std::string::npos
198         ? authority_begin_location + authority_end_location
199         : loc_url.size();
200
201       std::string authority = loc_url
202         .substr(authority_begin_location, authority_end_location - authority_begin_location);
203
204       if (!CheckAuthorityValidCharacters(authority))
205         throw std::invalid_argument("Invalid URL: Invalid characters in authority.");
206
207       ParseUserInfo(authority);
208       ParseHost(authority);
209       ParsePort(authority);
210     }
211     else
212       _userinfo = _host = _port = "";
```

Fig. 2. Excerpt from the implementation code of a **new**, stronger **URLString** class created by the student: a domain- and problem-specific "string" type builds on top of the low-level Standard type to encode invariants and proper semantics. (Screenshot of the *GitHub* code viewer's rendering created from the student's repository.)

```
void submit_homework(double forGPA,
                     int year, int month, int day,
                     int hour, int minute);

submit_homework(8.0, 2022, 7, 5, 12, 0); // Good call.
submit_homework(5, 7, 2022, 8.0, 24, 0); // Bad call, that's accepted!
```

Listing 2. An ambiguous function interface that allows passing arguments in a bad order. All 6 parameters are back-and-forth swappable with one another.

```
typedef int      Hour;
using    Minute = int;  // Analogous syntax available since C++11.
void submit_homework(double forGPA,
                     int year, int month, int day,
                     Hour h,   Minute m); // ← Looks like different types.

// But raw 'int' literal values accepted without issue:
submit_homework(8.0, 2022, 7, 5, 12, 0);
```

Listing 3. Conventional use of **typedef**s to create aliasing names over the same type.

the largest enhancement of system safety can be achieved from a type system point of view [20]. The program in Listing 2 contains **several** opportunities for dangerous misuse, even including swapping the parameters [25].

Combating weakness in typing granularity requires effort from the developers involved. A system that had been designed without strong types in mind might not be trivial to improve in a later development cycle. For the purposes of type granularity, we distinguish **four – increasing – levels of type safety and strength:**

– Lack of type safety ("level zero")
– Strong type aliases
– Strong interfaces
– Strong types

3.1 "Weak" typedef

C and C++ students learn about the **typedef** or **using** declaration early in their studies which allows creating an *alias* for an existing type. However, the created symbol is a **weak alias**, which will allow bidirectional conversion, because there is in fact no distinct *"new type"* created by the code.   This technique, called *"weak" typedefs* to distinguish it from the *strong* version discussed below in Sect. 3.2, is commonly used in large projects for two primary purposes.

First, it enhances the code's readability **for human developers**, as the concept of *hours* and *minutes* are intuitive to developers. We still need to emphasise at this point that these aliases are completely transparent to the language and the compiler!

Second, it allows for the code to be more platform-independent, in case the type aliases are created based on platform-specific conditions. For example, system errors are represented in the `errno` variable (of type `int`)[2] on POSIX-conforming platforms, but in `HRESULT`[3] objects on Microsoft Windows. Thus, a function that contains separate inner implementations, but wishes to expose the same interface to clients could rely on a weak type alias `using SysError = ...;` in its signature: `void logError(SysError E);`.

3.2 Strong `typedef`

The easiest, most verbose, and most trivial solution that ensures that values representing different concepts do not intermingle is the use of the ***strong type alias*** idiom.

A **wrapper** type is created for each distinct value concept, as shown in Listing 4. This wrapper type only contains – in program memory – the original value, and thus is zero-cost at runtime if compiler optimisations are enabled. However, the moment some calculation needs to be done on the value, e.g., checking whether the submission happened before 12 o'clock, the wrapper must be **unboxed**, as shown in Listing 5.

There are examples for this technique being supported on the language level outside of C++: in **Haskell**, the `\emph {newtype}` declaration creates a type with exactly one "type constructor" – in C++ terms, the boxing/unboxing method – and exactly one data field, the wrapped value. The `Hour` example in Listing 4 would be expressed in Haskell as simply as "`newtype Hour = Hour Int`".

---

[2] Common error values are pre-defined by POSIX to have distinct representations. The entire value identifies an error "category", e.g., `ENOENT`– "no such file or directory" – has the value 2. `int` is the type of 4 byte (32 bit) signed integers.

[3] `HRESULT` is a type alias to `int` which is also defined as the 4-byte (32 bit) signed integer type on Windows. However, understanding `HRESULT` objects is a more involved process, as the representation is split into several values of smaller width. For example, the lower 16 bits (0 to 15) represent the error code, then 11 bits (16 to 26) represent the error reporting subsystem of the OS, and the 5 remaining higher-order bits (27 to 31) are used as one-bit flags. The error for non-existent files is `STG_E_FILENOTFOUND`, formatted as `0x8003002` in hexadecimal. Note that the "actual" *error code* part also contains 2.

```
struct Hour {
  int Value;

  explicit Hour(int V) : Value(V) {}
  explicit operator int() const    { return Value; }
          int       get() const    { return Value; }
};
/* Analogously for the other types like Minute... */

void submit_homework(double forGPA,
                     int year, int month, int day,
                     Hour H, Minute M);

// Does not compile!
submit_homework(8.0, 2022, 7, 5, 12, 0);
// Does not compile!
submit_homework(8.0, 2022, 7, 5, Minute{12}, Hour{0});

// This works and is accepted:
submit_homework(8.0, 2022, 7, 5, Hour{12}, Minute{0});
```

Listing 4. The example in Listing 2 *partially* refactored to strong type *aliases* Hour and Minute.

```
Hour H = 11;
if (H < 12) // Does not compile, no such 'operator<'!

// Instead:
if (H.get() < 12) { ... }
// or:
int Hour = H;
if (Hour    < 12) { ... }
```

Listing 5. Usage of values contained in *strong type aliases* require a manual retrieval.

3.3   Strong Interface

We can get one step further in type safety if we introduce the set of allowed operations into the type system itself, as shown in Listing 6. This way, the user does not have to unbox, transform, and then re-box the values. This corresponds to the tenet of *encapsulation* in object-oriented programming: it should not be the client's – the code that uses the created type – responsibility to understand how an "`Hour`" can be compared with another one – if the type is comparable, it should expose that interface! This further prevents errors that might result from a mistake during programming.

The technique makes the client better readable and understandable by removing the boxing boilerplate, however, there are still no **semantic** validations behind the expressed concepts.

Another useful feature of *strong interfaces* is the ability to define transitions in domain or dimensionality. Suppose a `Distance` and `Time` type. We know from physics that the simplest form of velocity is "$v = s/t$". This relation can be directly   expressed   by   an "`Velocity operator/(Distance, Time)`" function, which, on the level of strong interfaces, would "`return Velocity{S.value()/T.value()};`".

3.4   Strong Type

True *strong types* employ semantic checks in their operations, have more than "trivial" representation and encode meaningful invariants. This way, they combine the features of compile-time checks done by the type system with potential error-checking run-time behaviour. An example for such cases, strengthening the previous case's type safety further, is shown in Listing 7, where we show one possibility to encode the invariant of "Hour $\in [0, 24]$" and to deal with representation issues when the granularity changes.

The most famous example for such strong types in C++ is the *Chrono* library – `##include <chrono>` – which deals with date, time, and clocks. Chrono was originally created as part of the *Boost Libraries* [21] but since C++ 11, it is part of the official *Standard Template Library*. The user can create a "time point", and shift it around with durations, or calculate the duration between two time points. These representations are tied to "clock types" – represented in the type system with the use of `templates` and generative programming [3, 31] – of which there are several in the library, giving the ability to distinguish what kind of clock produced a time point. Commonly available are `std::chrono::system_clock`, that shows the "wall time" with potential adjustments, such as the user changing time, or leap seconds; and `std::chrono::steady_clock`, that must never be adjusted.

The ability to *overload operators* in C++ allows for a sort-of natural expression of time-related concepts, e.g., "`2022y / July / 4d`" expresses the date *2022. July 4*. This expression is the result of calling `operator/` between a *year* literal and a *month* object, and then this intermittent object being `operator/`ed with a *day* literal.

```
struct Hour
{
  int Value;

  explicit Hour(int V) : Value(V) {}
  explicit operator int() const    { return Value; }
           int       get() const   { return Value; }

  bool operator <(Hour H) const    { return Value <  H.Value; }
  bool operator==(Hour H) const    { return Value == H.Value; }
  /* ... */

  // Note this makes possible to have 23 + 8 = 31 as "hour"!
  Hour operator +(int  N) const    { return Hour{Value + N};  }
}
```

Listing 6. Exposing operations and a meaningful interface through a wrapper type removes the need of the user dealing with the representation, enhancing *encapsulation*. Compare with Listing 5.

Chrono is strongly typed because the defined types not only express meaning on their interface, but the implementation has non-trivial business logic and invariants added: shifting a `time_point` forward by the `duration` "`0.5h`" will result in a `time_point` half an hour into the future.

A subset of strong typing, namely *safe arithmetics* or *constrained numerals* can be achieved by libraries akin to Chrono in C++, but is a feature supported on the language-level in **Ada**: `Type Hour Is Range 0..24` will create a constrained integer, with overflows checked during compilation and execution and causing errors.

4   The *"Contact List"* Exercise

In this section, we will detail how to achieve a stronger type safety through a practical exercise that is small and familiar enough to be consumed easily by the Reader, but complex enough to allow the asking of meaningful design and implementation questions. This exercise will consist of a contact list or "phone book", in which the details (such as the phone number or e-mail address) of contacts of an individual can be stored. For simplicity, the program is implemented as an interactive text application running in a terminal, with the following features, each corresponding to the menu action. After the execution of each action, the next action is requested from the user in an interactive loop until the user decides to "`quit`".

- "**add**" a contact, in which the details will be interactively asked from the user
- "**lookup**" the details of a named contact
- "**remove**" a contact

```cpp
struct Hour
{
  int Value;

  explicit Hour(int V) : Value(V) {}
  explicit operator int() const   { return Value; }
          int       get() const   { return Value; }
  /* ... */

  Hour   operator +(Hour   H) const {
    Hour R{Value + H.Value};
    if (R.Value < 0 || R.Value > 24)
      // Semantic check that prohibits going over a day's worth of hours.
      throw day_overflow_exception{};
    return R;
  }

  // "Hours + Minutes" is often not wholly representable on the
  // granularity of Hours.
  Minute operator +(Minute M) const { return {60 * Value + M.Value}; }
}

/* ... */

void test() {
  Hour h1{8}, h2{10};
  Minute m{30};

  static_assert( std::is_same_v<decltype(h1 + m), Minute> );
  assert( (h1        + m).get() ==   510 );  // 60 *  8 =  480; +30 =   510
  assert(   (h1 + h2)   .get() ==    18 );
  assert( ((h1 + h2) + m).get() == 1110 );  // 60 * 18 = 1080; +30 = 1110
}
```

Listing 7. Implementing semantically appropriate *behaviour* through the type system. Compare with Listing 6.

- "`load`" and "`save`" the contacts to a file
    - The "*load*" action is automatically executed at program start-up, requiring the user to specify the name of the file to work with.
    - The "*save*" action is **not** *executed automatically*, allowing the user to discard pending changes without affecting the data file.
    - All changes by the "*add*" and "*remove*" actions are done primarily in system memory.

The full source code of the exercise is available, verbatim, in Section A. The program is designed to be compiled trivially against the C++ 17 standard with any reasonably modern compiler.[4] Starting the resulting `contacts` binary is the only entry point to the application. An example interaction with the program can be seen in Listing 8. It is advised that the Reader tries the application out on their own computer prior to starting the refactoring exercise in order to observe how the program should behave.

4.1   Architectural and Design Overview

The *Phonebook* program is implemented as three translation units, each implementing one "class" of the functionality. First, the `contacts.hpp` header file and `contacts.cpp` source file define and implement `class Contacts`, which is a wrapper over the data structure that keeps the contacts (`class Contacts::Entry`) **in memory**. The methods which index (`lookup()`) or mutate (`add()`, `remove()`) – corresponding to the user-facing action verbs – the list of entries are also on this class's interface. This class exposes the factory function `load()` that creates the data from the contents of a file, and a member method `save()` that drives saving the in-memory data to the file. However, the only data resource in `Contacts` is the list of entries.   `class ContactStorage` implements the actual access to the data files used by the exercise program. This class manages the native resource of opened files and performs the read and write operations. `Contacts` and `ContactStorage` have no interface dependencies on each other: the `Contacts::load()` function takes the file path as a `std::string`. `ContactStorage` is a sort-of *helper* class that is used to implement the `save()` and `load()` functions internally. `ContactStorage`'s input/output methods consume or generate the records in the underlying native file sequentially, transcoding the contacts to and from an `XML`-like format. `ContactStorage::readEntry()` reads the next – in terms of the read "head" of the opened file – entry from the file and populates its **output** parameters, while `ContactStorage::writeEntry()` writes the entry comprised of its **input** parameters to the end of the file. These methods take only the parameters that correspond to values which are **persisted** in the file, and to keep the interfaces separate, `Contacts::Entry` does not appear in their interface! This allows for a hypothetical extension to the `Entry` type with additional fields – such as `Age` – that are not saved to disk, usually because they can be calculated from

---

[4]   On contemporary Linux distributions, "`g++ -std=c++17 main.cpp contacts.cpp contact_storage.cpp -o contacts`" is a sufficient one-liner to compile all parts of the project.

```
Phonebook v1
------------


File to open and use? test.dat
test.dat: 0 entries loaded!
Action? help
add            Add (or overwrite) an entry, step-by-step
remove         Delete an entry based on name
lookup         Get the details of a contact
list           List all names in the phonebook

save           Save changes to the disk
quit           Close the program

Action? add
Adding a new contact...
Name? Test Entry
Reading the details of a contact...
Phone number (without spaces!)? 12345
Fax number (without spaces!)? 67890
E-mail address? aa@aa.a
Physical address? 11111 BBBBBB
Birthday? 2000.01.01.

Action? save
Action? quit
Goodbye!
```

```
Phonebook v1
------------


File to open and use? test.dat
test.dat: 1 entries loaded!
Action? list
 * Test Entry
```

Listing 8. Interaction with the *Contact list* example, first by adding a new contact, then after program restart, observing that it was saved to the file. The lines containing a question (e.g., **Action?**) are prompts for user input.

the persisted fields and the current system state. (`class ContactStorage` is also reusable by third-party clients without ever having to import `class Contacts` to their code.) Although hypothetical due to the small scale of the exercise, this separation of the interface will add several moments where the Reader can question **interface perimeters** during the steps of this Tutorial. For example, dealing with the changes we will describe in Sect. 6 is much more different if `ContactStorage` would be from a third-party software package that we have no technical means or legal rights to modify [26].

The third translation unit implements the user-facing `main()` function and its implementation helpers. This code deals with handling the standard input-output with the user, including also the main loop of the program that draws the menu and requests an action.

5   Finding and Improving Weakly Typed Interfaces

As this is a **controlled** exercise, the first action to do is to find interfaces, and in general program elements, which are weakly typed, e.g., by searching for patterns expressed in Listing 2. This can be achieved by manually reading the original exercise's source code in Section A – doing so is tractable, as it is only approximately 250 lines of code long – or by employing *static analysis*. Further reading about using and developing static analysis in C++ is available in [11].[5]

Because we are hunting for weak interfaces primarily, it is enough to read the *header files* – that have the file extension `.hpp` – and immediately we bump into the functions `ContactStorage::readEntry` and `ContactStorage::writeEntry` which takes several `string` parameters next to one another. Does this mean that `Contacts`, the main business logic library, is type-safe? To answer this question, we need to carefully think about *why* the `readEntry` and `writeEntry` functions have the interface they do. If we look at the definition for `Contacts::Entry`– see Listing 9 – we can observe that the storage input-output interface was purposefully designed to accommodate the members of `Entry`.

The natural refactoring would be to change the functions to take and return instances of `Entry`, however, this is problematic for two reasons. First, note that the `ContactStorage` *does not depend on* `Contacts`, which is a benevolent design decision to not introduce circular dependencies in the interfaces of the project and also allow for different storage methods to be used by the business logic, if needed. As software is not carved out of a single slab of marble in the modern world, we should note that there are always *boundaries* beyond which a specific development action **must not reach**. As it is generally expected that convenience programs are backwards compatible, we should not change the *format* of the serialisation, i.e., what is actually written to the file when the user **saves** the contact list!

---

[5] Using *Clang-Tidy* and the `bugprone-easily-swappable-parameters` analysis routine ("checker") is a verbose but definite way of getting all problematic interfaces highlighted in a C or C++ project.

———— `contacts.hpp` (excerpt) ————

```cpp
class Contacts
{
  public:
    struct Entry
    {
        long Phone;
        long Fax;
        std::string Email;
        std::string Address;
        std::string Birthday;
    };
    /* ... */
};
```

Listing 9. The definition of `Contacts::Entry` in the original exercise code.

What is more important, however, is that there is a *subtle* weak interface that is in fact not written in the code. C++ compilers automatically generate *special member functions* [24] – such as constructors – when the user had not provided one. A constructor with the signature "`Entry::Entry(std::string, long, long, std::string, std::string, std::string)`" is created automatically from `Entry`'s layout, which itself constitutes a weak and swap-prone interface.

5.1   Strong `typedef` for `Phone` and `Fax`

Due to the concerns laid out in the previous section, we will start with reworking the representation of `Entry`. Values of `long` represent "large enough" numbers[6] and thus seem like a good first candidate to just wrap into a strong type alias as discussed in Sect. 3.2. After performing this refactoring and replacing the two `long` fields to be of type `PhoneNumber`, we run into a problem. Now, the `Entry` class – and its associated `Storage` interface – again exhibit a type weakness, but between the objects of type `PhoneNumber`. Unfortunately, both land-lines, mobile phones, and fax systems are wired to the same *public switched telephone network*, and their numbering sceheme uses the same representation. Because of this, this version of the interface that uses a more generic `PhoneNumber` type would still allow a programming misuse, e.g., by letting the developers create a module which accidentally passes the "mobile phone number" to a location expecting a "fax number". In the real world, such a call would result in weird robotic signalling noises being heard by the called person, which is clearly unintended.

Further fixing this problem presents a prime candidate to practice abstraction: we can use the object-oriented programming paradigm to create sub-

---

[6] The *International Telecommunication Union* specifies phone numbers to be at most 15 digits long, which fits into the 18 decimal digit expression capability of an 8-byte `long`; except on Microsoft Windows, where `sizeof(long) == sizeof(int) == 4`, and, thus, could express only 9 decimal digits in 4 bytes.

classes `MobilePhoneNumber` and `FaxNumber` that both inherit from `PhoneNumber`. While inheritance will properly make available both the reading conversion `operator long()` and the `get()` functions, constructors cannot be inherited. Thus, we will have to suffer the increase in verbosity and create an `explicit` constructor that constructs – explicitly – the base class. The version of the project after this refactoring is seen in Listing 10. Note that we did not change the implementation code yet, so this version, as is, will not compile!

5.2   Strong `typedef` for names, `Email`, `Address`, and `Birthday`

Looking at the remaining types used for the fields of `Entry`, we see that there are 3 `strings` that are still mistakable between one another. While it could be discussed whether an "`AbstractAddress`" type is worth to be the superclass of `Address` and `Email`, these two ways of communication differ so much that it is not worth bikeshedding for such details. Instead, for now, it is good to add the strong type aliases, as shown in Listing 4, for all four string-backed values.

It might be surprising to see why 4 was mentioned, when there are only 3 fields remaining. To understand this, note the detail that the "name" of the contact is not stored in the `Entry` object, but instead as a key to the `Entries` mapping, and `std::string` appears in the interface of `add()`, `lookup()` and `remove()` in `Contacts`. Also, the "name" field is explicitly spelled out as the first parameter of both `readEntry()` and `writeEntry()`.

Hereinafter, this Tutorial will assume that the types `PersonName`, `EmailAddress`, `PhysicalAddress`, and `Birthday`, all strong type aliases over `std::string` had been created. Please replace the signatures appropriately, including the change that `getNames()` should return a `std::vector<PersonName>`. The full signature of
`ContactStorage::writeEntry` should be as depicted in Listing 11.

Note that, appropriately, the concept of "filenames" could also be refactored to a stronger type than just `string`, but because file-system APIs are complex, we will not introduce such in this Tutorial.

6   Reworking the Implementation to Use the Strong Type Aliases

First, we will need to make sure that all the function *definitions* in the `.cpp` files match the function *declarations* – signatures – in the `.hpp` files. Without changing any function body, let's try and compile the code in its current state. We will observe an enormous amount of errors emitted by the compiler. When dealing with C++ error messages, it is recommended to always only consume the **first** error. Due to language intricacies, the compiler often emits several *notes* for each observed *error*, which contain useful additional information for understanding the compiler's reasoning. These lengthy details, although, can be scary for someone with only beginner-level experience.

```
———————————————————— types.hpp (new file!) ————————————————————
#ifndef TYPES_HPP
#define TYPES_HPP

struct PhoneNumber
{
  long Value;

  explicit PhoneNumber(long V) : Value(V) {}
  explicit operator   long()  const      { return Value; }
          long        get()   const      { return Value; }
};

struct MobilePhoneNumber : public PhoneNumber
{
  explicit MobilePhoneNumber(long V) : PhoneNumber{V} {}
};

struct FaxNumber        : public PhoneNumber
{
  explicit FaxNumber(long V)             : PhoneNumber{V} {}
};

#endif /* TYPES_HPP */
```

```
———————————————————— contacts.hpp (excerpt) ————————————————————
#include "types.hpp"

class Contacts
{
  public:
    struct Entry
    {
        MobilePhoneNumber Phone;
        FaxNumber         Fax;
        std::string       Email;
        std::string       Address;
        std::string       Birthday;
    };
  /* ... */
};
```

Listing 10. Refactoring of types after introducing strong type aliases over `long`.

```
─────────────────── contact_storage.hpp (excerpt) ───────────────────
class ContactStorage
{
  public:
    /* ... */
    void writeEntry(PersonName      Name,     MobilePhoneNumber Phone,
                    FaxNumber       Fax,      EmailAddress      Email,
                    PhysicalAddress Address, Birthday          BirthDay);
};
```

Listing 11. Refactoring of the storage interface after introducing **all** strong type aliases.

Table 3. CC errors encountered during transformation to use the strong interface

§	Short name	Error message (example)
6.1	Invalid-BinOp	"invalid operands to binary **operator** $\oplus$"
6.2	NoConv-Asg	"no viable overloaded "=", no known conversion"
6.3	NoConv-1Arg	"constructor candidate requires single argument"
6.4	ImpDelete-Ctor	"call to implicitly-deleted constructor of "**Type**""
6.5	ImpDelete-Ctor-Tmp	"*[. . .]* in instantiation"
6.6	NoOverload	"no matching function to call for "**func()**""

During the code transformations performed as part of this exercise, there will be several classes of errors that the compiler will report. The individual cases are summarised in Table 3, with their details expanded upon further in this section. The errors are listed in the order they are likely to appear during the transformation. Resolving one error might transform the code in a way that a different error appears due to the transformation in a different location.

6.1   Invalid-BinOp: *"invalid operands to binary* operator $\oplus$"

This error is emitted when there is an expression "a $\oplus$ b;", but there is no corresponding "operator $\oplus$(A, B)" function defined in an appropriate scope.

"operator <<" and "operator >>".
    are frequent causes of this error in places where the data is read from or output to a stream, such as the standard output or the backing file. In this exercise, this is caused because variables that were of built-in types before the refactoring – e.g., int or std::string – were transitioned to the strong interface. The standard library does not define what it means to output a, e.g., PersonName to the standard output. **This demonstrates hitting a refactoring barrier, as we, the programmers, have no right to alter the definitions in the Standard.**

```
————————— types.hpp (excerpt) —————————
struct PersonName
{
  std::string Value;

  explicit PersonName( std::string V) : Value(std::move(V)) {}
  explicit operator    std::string()  const  { return Value; }
          std::string get()           const  { return Value; }

  bool operator  < (const PersonName& RHS) const
  { return Value < RHS.Value; }
};
```

Listing 12. Trivially exposing the ordering relation of the inner `std::string` through the strong type alias.

Using `.get()` on the variables in the now erroneous usage points makes the casting – e.g., from `PersonName` to `string` – explicit, and resolves this error. An alternative would be to implement the appropriate operator functions, such as "`std::ostream& operator <<(std::ostream&, PersonName)`".

"`operator <`".

is another frequent cause for the same class of error messages, and this error often appears from a piece of code in a non-trivial location deep in the *Standard Template Library (STL)* implementation. Most associative containers, such as `std::map`, are implemented internally using an ordered *search tree*, and this requires an ordering operation to be implemented for the type of the key in the container. By convention and by requirements of the Standard, if the ordering is not explicit specified, the container will try to do the comparison using "`k1 < k2`", which requires the presence of an "`operator <(T, T)`" function. The original version of the code used `std::string` as the "key type" of the mapping, which defines this comparison, but the new `PersonName` class does not. By adding our own `operator <` into the wrapper class – as shown in Listing 12 – this error, and the more dauntingly looking "*static assertion failed due to requirement* "`std::__is_invocable< std::less< PersonName > &, ... >`"" caused for the same reason with newer *STL* implementations that support the "Concepts" language element both disappear.

6.2 NoConv-Asg: *"no viable overloaded* "`=`"*, no known conversion*"

This error message appears when using the "`operator =`" assignment operator in an expression such as "`v1 = v2;`, let v2 be copied into v1". Although looking like a binary operation, and thus expected to fall into the previous category discussed in Sect. 6.1, assignments are special, as a user-defined *assignment operator* always has to be implemented as a **member function**: "`T& operator =(const T& Other)`". Due to this, the left-hand operand of the assignment is never *converted* and must always be known, but the right-hand operand might undergo conversions. (The trivial case is when there are no conversions, and the assigned expression's type matches the recipient variable's type exactly.)

This was the case pre-transformation, e.g., in the `readEntry()` method of `ContactStorage`, where the `Out`-parameters were assigned from the exactly type-matching local variables. However, after the refactoring, the parameter type changed, e.g., `OutName` became `PersonName`, but the local variable is still of type `string`, and there is no "`PersonName::operator =(std::string)`" function. One solution is to implement this function, creating a *converting assignment operator* (and its corresponding *converting constructor*) [25].

If that is not possible, or not desired, we can make the conversion explicit before the assignment takes place, by constructing a temporary "`PersonName{std::move(*Name)}`", and assigning this to the variable, which is performed trivially by the compiler via the implicitly generated constructors and assignment operators, as now both the left-hand variable and the right-hand expression have the same type. In full, the code becomes "`OutName = PersonName{std::move(*Name)};`". All other use cases where the more weakly typed values are assigned to the more strongly typed variables can be fixed using this pattern.

6.3   NoConv1Arg: *"no matching constructor for initialisation – candidate constructor not viable, requires single argument"*

These errors are caused by trying to *default-initialise* an instance of the strong type – e.g., by declaring a local variable "`Strong V;`" – with the "strong type alias" or "strong interface" – see Sects. 3.2 and 3.3 – which defined an "`explicit Strong(V)`" constructor. The language rules stipulate that if a class contains a user-defined constructor – which that converting constructor is – then the compiler is forbidden from generating a *default* – one that may be called without any arguments – constructor. This causes the error message: the statement in the code, e.g., the definition of a local variable, wanted to call a 0-parameter constructor, but the only viable constructor *must* take one argument.

Perhaps repetitively, there are two ways to solve this issue. Either we can make the strong type aliases *default-constructible* by adding a constructor that takes no arguments, or we could give the *default-initialised* instance of the wrapped type in the initialisations of the local variables. As it is not meaningful in the general sense that a "name" or an "address" exists that is "empty", or that a phone number is 0, we should take the second option as it results in more explicit and thus safer code. The solution is to add the character sequence "`{{}}`" after each variable definition, e.g., `PersonName Name{{}};`. This instructs the program to "initialise `Name` using the single-parameter constructor by passing a default-constructed temporary (of the appropriate type, in this case, `std::string` to it". This makes the fact that we have "zeroed" variables glaringly obvious.

6.4 ImpDelete-Ctor: *"call to implicitly-deleted default constructor of "T""*

When encountered in the context of `class Contacts::Entry`, this is the most curious of them all. In principle, this is connected with the previous case, described in Sect. 6.3. The statement "`Entry E;`" tries to initialise an "empty" `Entry` by using `Entry`'s default constructor. As `Entry` has no user-defined constructors, the language states that the compiler should generate an appropriate *default constructor* when needed. Although the compiler attempts this generation, the Standard-specified default behaviour for a default constructor is to *default-initialise* all members (and base classes, etc.) of the `class`. If the solution of not adding a default constructor to the strong types is considered, as described in Sect. 6.3, then this is not possible, as none of the members will be *default-constructible*. (To trigger this error, it is sufficient for only one member to not be appropriately initialisable, with no consideration for the total number of data members.) This condition is called being *"implicitly-deleted"*, which is reported in the error message.

This error message can be taken care of in three different ways: either making all the members default-constructible by changing the strong type definitions; or by adding a default constructor to the larger record type, e.g., `Entry`; or by changing the initialisation of the variables.

As mentioned in Sect. 5 [24], the `Entry` class has a special "invisible", compiler-generated implicit constructor with which we can assign the data members in the appropriate order.[7] Instead of creating an empty object and then assigning each field one-by-one, rewrite the line as follows: "`Entry E{Phone, Fax, Email, Address, Birthday}; `", which will pass each of the read variables into the `Entry`.

6.5 ImpDelete-Ctor-Tmp: *"call to implicitly-deleted default constructor of 'T' in instantiation of a function template specialisation"*

This warning is a subclass of the previous one described in Sect. 6.4, if the issue of trying to *default-initialise* an instance is performed through a chain of `template` instantiations. Thankfully, the compiler eventually – through the use of explanatory *notes* attached to the *error* – points us to culprit code that belong to our exercise project, and not the *STL*: "`Entries[Name] = E; `" in `contacts.cpp`.

The behaviour of `operator []` of a `std::map` is to first construct an "empty" (default) element if one with the key does not exist, and then return the reference to this element. If the element already exists, the reference is returned

---

[7] This is called *aggregate initialisation*. **Aggregates** are `classes` that are simple enough not to contain user-defined constructors, `virtual` or `private` inheritance, or `private` data members. This is the modern abstraction over what was once called *"plain old data (POD)"*: objects that only "hold together" smaller objects, but offer no customisation or semantics on top. If the type is an aggregate, since C++ 17, the syntax suggested in this section is always implicitly available for client code.

```
———————————————— contacts.cpp (excerpt) ————————————————
void Contacts::add(PersonName Name, Entry E)
{
  auto Iter = Entries.find(Name);
  if (Iter == Entries.end())
    Entries.emplace(Name, E);
  else
    Iter->second = E;
}
```

Listing 13. Hoisted logic of potentially overwriting an element of a `std::map` if the value type is not *default-constructible*.

to the existing element. Then, outside, in our code, we use this obtained reference to assign the new value `E` to it. The "in instantiation of a function template specialisation" part refers to the compiler instantiating the `template` "`V& std::map<K, V>::operator [](K) `" function with the right types.

Unfortunately, resolving this error requires an intermediate understanding of the C++ *STL*, and it is not immediately fixable by a simple change around the usage location. Conversely, changing the behaviour of `std::map` is **strictly forbidden** in application code. Instead of using the accessor `operator []`, we should first try to `find()` the existing element, and if so, do the assignment manually. If not, by using the `emplace()` function, the construction of the `map`'s element can be achieved directly, without having to rely on the existence of a *default-initialised* value first. The appropriately rewritten `Contacts::add()` method is shown in Listing 13.

6.6 NoOverload: *"no matching function for call to* "`func()`"""

This error message constitutes the most generic "overload resolution error" indications, but the fundamental issue is the same as was described in Sect. 6.1 and 6.2. The compiler tells us that a call to " `func()`" can not be matched with any of the available definitions for that function. Often, the compiler will give additional *notes* for the attempted function signatures, and more modern compilers can also elaborate on why a candidate could not be called [10]. Unless there is precisely one matching overload of "`func()` ", this error message will appear.

For example, the `std::getline()` calls in `main.cpp` trigger this error. As neither `Entry` nor the member strong type aliases are constructible with default values (see Sect. 6.4 and 6.5), the user-facing input-output operation and the saving of the value into the `Entry` object must be split apart: we have to create temporary variables of the raw I/O types `long` and `std::string`, and do the read into these instead. Once all the values are read into the temporary variables, we can construct the `Entry` by wrapping the raw local variables in the appropriate strong type aliases first:

```
"return Contacts::Entry {
MobilePhoneNumber{RawPhoneNumber}, FaxNumber{RawFaxNumber},
EmailAddress{RawEmail}, PhysicalAddress{RawAddress},
Birthday{RawBirthday} };".
```

6.7 Finishing the Exercise

After the changes described in this Tutorial had been made, the program should compile normally again. In the setting of a lesson, this can be achieved by verifying that everything works as intended. It is paramount with every refactoring that the code not only compiles, but exhibits the same behaviour as before. The best way to achieve this is to create a data file with the old version and reading from it and modifying it with the new version, and then vice-versa.

The complete solution post-refactoring of what was discussed in Sects. 5 and 6 is available in Section B.

7 Conclusion

Type systems and the practice of strong typing are powerful tools to increase the guarantee that developer mistakes are caught in time before causing serious damage. Allowing built-in and widely used generic types only a small footprint in the business logic code, i.e., preventing *primitive obsession*, ensures the expressed logic to be more explicit. As large software systems evolve, their internal quality tends to erode. Combating this entropic trend is usually attributed to a few developers per project for whom it is a personal, internalised goal to do so. For this reason, it is important that the next generation of practitioners and researchers are proactively taught the importance of type safety from the very start. By incorporating the education of strong typing into the material of advanced programming subjects, we ensure that students have the chance to expose themselves to actual, existing and widely used bad code and to obtain hands-on experience in fixing it. This also grows their ability to navigate and understand codebases created by others, something that is also a valuable skill for life in this field.

Acknowledgements. The authors acknowledge the indispensable discussion with their colleagues in the aforementioned project at the *Teachers' Training* event in Nijmegen, the Netherlands, which further helped to refine the educational material of this Tutorial, originally presented in [28].

A   Initial Exercise Code

A.1   Contacts business logic library

```
                                        contacts.hpp
 1   #ifndef CONTACTS_HPP
 2   #define CONTACTS_HPP
 3   #include <map>
 4   #include <string>
 5   #include <vector>
 6
 7   class Contacts
 8   {
 9     public:
10       struct Entry
11       {
12         long Phone;
13         long Fax;
14         std::string Email;
15         std::string Address;
16         std::string Birthday;
17       };
18
19            void    add(    std::string Name, Entry E);
20       const Entry* lookup(std::string Name)              const;
21            bool    remove(std::string Name);
22
23       std::vector<std::string> getNames() const;
24
25            void      save(const std::string& File) const;
26       static Contacts load(const std::string& File);
27
28       std::size_t count() const;
29
30     private:
31       std::map<std::string, Entry> Entries;
32   };
33
34   #endif /* CONTACTS_HPP */
```

```
                                        contacts.cpp
 1   #include "contacts.hpp"
 2   #include "contact_storage.hpp"
 3
 4   std::size_t Contacts::count() const { return Entries.size(); }
 5
 6   void Contacts::add(std::string Name, Entry E)
 7   {
 8     Entries[Name] = E;
 9   }
```

```cpp
const Contacts::Entry* Contacts::lookup(std::string Name) const
{
  auto Iter = Entries.find(Name);
  if (Iter == Entries.end())
    return nullptr;
  return &Iter->second;
}

bool Contacts::remove(std::string Name)
{
  auto Iter = Entries.find(Name);
  if (Iter == Entries.end())
    return false;
  Entries.erase(Iter);
  return true;
}

std::vector<std::string> Contacts::getNames() const
{
  std::vector<std::string> Names;
  for (const std::pair<std::string, Entry>& Contact : Entries)
    Names.push_back(Contact.first);
  return Names;
}

Contacts Contacts::load(const std::string& File)
{
  Contacts Return;
  ContactStorage Storage{File};

  std::string Name, Email, Address, Birthday;
  long Phone, Fax;
  while (Storage.readEntry(Name, Phone, Fax, Email, Address, Birthday))
  {
    Entry E;
    E.Phone = Phone;
    E.Fax = Fax;
    E.Email = Email;
    E.Address = Address;
    E.Birthday = Birthday;

    Return.add(Name, E);
  }

  return Return;
}

void Contacts::save(const std::string& File) const
{
```

```
60    ContactStorage Storage{File};
61    Storage.clear();
62
63    for (const std::pair<std::string, Entry>& Contact : Entries)
64      Storage.writeEntry(Contact.first, Contact.second.Phone,
65                         Contact.second.Fax, Contact.second.Email,
66                         Contact.second.Address, Contact.second.Birthday)
67  }
```

A.2   Contact-Storage persistency library

──────────── contact_storage.hpp ────────────

```
1   #ifndef CONTACT_STORAGE_HPP
2   #define CONTACT_STORAGE_HPP
3   #include <fstream>
4   #include <string>
5
6   class ContactStorage
7   {
8     public:
9       ContactStorage(std::string File);
10      ~ContactStorage();
11
12      // Delete the contents of the backing storage.
13      void clear();
14
15      // Ask whether the backing storage reached its end.
16      bool end() const;
17
18      // Read one (the next, sequentially) record from the backing file.
19      // If the file has emptied, does not modify the output variables.
20      // Returns TRUE if a record was read.
21      bool readEntry(std::string& OutName, long& OutPhone, long& OutFax,
22                     std::string& OutEmail, std::string& OutAddress,
23                     std::string& OutBirthday);
24
25      // Write one record into the backing file.
26      void writeEntry(std::string Name, long Phone, long Fax,
27                      std::string Email, std::string Address,
28                      std::string Birthday);
29
30    private:
31      std::string   Filename;
32      std::fstream* FileOnDisk = nullptr;
33  };
34
35  #endif /* CONTACT_STORAGE_HPP */
```

```cpp
                        ──── contact_storage.cpp ────
 1  #include <cassert>
 2  #include <cstring>
 3  #include <iostream>
 4  #include <optional>
 5  #include <utility>
 6
 7  #include "contact_storage.hpp"
 8
 9  ContactStorage::ContactStorage(std::string File)
10      : Filename(std::move(File)),
11        FileOnDisk(new std::fstream(
12              Filename.c_str(), std::ios_base::in | std::ios_base::out))
13  {
14    if (!FileOnDisk->is_open())
15    {
16      delete FileOnDisk;
17      FileOnDisk = new std::fstream(Filename.c_str(), std::ios_base::out);
18    }
19  }
20
21  ContactStorage::~ContactStorage()
22  {
23    if (!FileOnDisk)
24      return;
25
26    FileOnDisk->flush();
27    FileOnDisk->close();
28    delete FileOnDisk;
29  }
30
31  void ContactStorage::clear()
32  {
33    assert(FileOnDisk && "No file opened?");
34    FileOnDisk->close();
35    delete FileOnDisk;
36
37    FileOnDisk = new std::fstream(Filename.c_str(), std::ios_base::in |
38                                                    std::ios_base::out |
39                                                    std::ios_base::trunc);
40  }
41
42  bool ContactStorage::end() const
43  {
44    assert(FileOnDisk && "No file opened?");
45    return FileOnDisk->eof();
46  }
47
48  void ContactStorage::writeEntry(std::string Name, long Phone, long Fax,
```

```
49                                        std::string Email, std::string Address,
50                                        std::string Birthday)
51  {
52    assert(FileOnDisk && "No file opened?");
53    FileOnDisk->seekp(0, std::ios_base::end);
54    *FileOnDisk << "<ENTRY>\n"
55                   << "<Name>" << Name << "</Name>\n"
56                   << "<Phone>" << Phone << "</Phone>\n"
57                   << "<Fax>" << Fax << "</Fax>\n"
58                   << "<EMail>" << Email << "</EMail>\n"
59                   << "<Address>" << Address << "</Address>\n"
60                   << "<Birthday>" << Birthday << "</Birthday>\n"
61               << "</ENTRY>\n\n";
62  }
63
64  bool ContactStorage::readEntry(std::string& OutName, long& OutPhone,
65                                 long& OutFax, std::string& OutEmail,
66                                 std::string& OutAddress,
67                                 std::string& OutBirthday)
68  {
69    auto ParseLineWithTag = [](const std::string& Line,
70                               const std::string& OpenTag,
71                               const std::string& CloseTag)
72    {
73      std::optional<std::string> Result;
74
75      if (Line.size() >= OpenTag.size() &&
76          Line.substr(0, OpenTag.size()) == OpenTag)
77      {
78        Result.emplace(Line.substr(OpenTag.size()));
79        Result = Result->substr(0, Result->size() - CloseTag.size());
80      }
81
82      return Result;
83    };
84
85    std::string Line;
86    do
87    {
88      std::getline(*FileOnDisk, Line);
89      if (FileOnDisk->eof())
90        return false;
91
92      if (auto Name = ParseLineWithTag(Line, "<Name>", "</Name>"))
93        OutName = std::move(*Name);
94      else if (auto Phone = ParseLineWithTag(Line, "<Phone>", "</Phone>"))
95        OutPhone = std::stol(*Phone);
96      else if (auto Fax = ParseLineWithTag(Line, "<Fax>", "</Fax>"))
97        OutFax = std::stol(*Fax);
98      else if (auto Email = ParseLineWithTag(Line, "<EMail>", "</EMail>"))
```

```cpp
 99        OutEmail = std::move(*Email);
100      else if (auto Address =
101          ParseLineWithTag(Line,"<Address>", "</Address>"))
102        OutAddress = std::move(*Address);
103      else if (auto Birthday =
104          ParseLineWithTag(Line, "<Birthday>", "</Birthday>"))
105        OutBirthday = std::move(*Birthday);
106    } while (Line != "</ENTRY>");
107
108    return true;
109  }
```

A.3   The User-Facing main

```cpp
/* ------------------------------ main.cpp ------------------------------ */
 1  #include <iostream>
 2  #include <limits>
 3  #include <string>
 4
 5  #include "contacts.hpp"
 6
 7  void             printMenu();
 8  Contacts::Entry readEntryFromUser();
 9  void             printEntry(std::string Name, Contacts::Entry E);
10
11  int main()
12  {
13    std::cout << "Phonebook v1\n";
14    std::cout << "------------\n\n";
15    std::cout << "File to open and use? ";
16
17    std::string Filename;
18    std::cin >> Filename;
19    std::cin.ignore(std::numeric_limits<std::streamsize>::max(), '\n');
20
21    Contacts Phonebook = Contacts::load(Filename);
22    std::cout << Filename << ": "
23             << Phonebook.count() << " entries loaded!\n";
24
25    std::string UserChoice;
26    while (true)
27    {
28      std::cout << "Action? ";
29      std::getline(std::cin, UserChoice);
30
31      if (UserChoice == "q" || UserChoice == "quit")
32      {
33        std::cout << "Goodbye!" << std::endl;
```

```cpp
34      return EXIT_SUCCESS;
35    }
36    else if (UserChoice == "h" || UserChoice == "help")
37      printMenu();
38    else if (UserChoice == "a" || UserChoice == "add")
39    {
40      std::cout << "Adding a new contact...\n";
41      std::cout << "Name? ";
42      std::getline(std::cin, UserChoice);
43
44      Contacts::Entry E = readEntryFromUser();
45      Phonebook.add(UserChoice, E);
46      std::cout << std::endl;
47    }
48    else if (UserChoice == "r" || UserChoice == "remove")
49    {
50      std::cout << "Removing a contact...\n";
51      std::cout << "Name? ";
52      std::getline(std::cin, UserChoice);
53
54      bool Removed = Phonebook.remove(UserChoice);
55      if (!Removed)
56        std::cout << "No such entry existed. Not removing anything.";
57      std::cout << std::endl;
58    }
59    else if (UserChoice == "l" || UserChoice == "lookup")
60    {
61      std::cout << "Name? ";
62      std::getline(std::cin, UserChoice);
63
64      if (const Contacts::Entry* EP = Phonebook.lookup(UserChoice))
65        printEntry(UserChoice, *EP);
66      else
67        std::cout << "No such entry!";
68      std::cout << std::endl;
69    }
70    else if (UserChoice == "list")
71    {
72      for (const std::string& Name : Phonebook.getNames())
73        std::cout << " * " << Name << '\n';
74      std::cout << std::endl;
75    }
76    else if (UserChoice == "s" || UserChoice == "save")
77      Phonebook.save(Filename);
78    else
79      std::cerr << "Invalid action, try again! ('h' prints menu)"
80               << std::endl;
81  }
82 }
83
```

```cpp
84   void printMenu()
85   {
86     std::cout << "add        Add (or overwrite) an entry, step-by-step\n";
87     std::cout << "remove     Delete an entry based on name\n";
88     std::cout << "lookup     Get the details of a contact\n";
89     std::cout << "list       List all names in the phonebook\n";
90     std::cout << '\n';
91     std::cout << "save       Save changes to the disk\n";
92     std::cout << "quit       Close the program\n";
93     std::cout << std::endl;
94   }
95
96   Contacts::Entry readEntryFromUser()
97   {
98     Contacts::Entry E;
99     std::cout << "Reading the details of a contact...\n";
100
101    std::cout << "Phone number (without spaces!)? ";
102    std::cin >> E.Phone;
103    std::cin.ignore(std::numeric_limits<std::streamsize>::max(), '\n');
104
105    std::cout << "Fax number (without spaces!)? ";
106    std::cin >> E.Fax;
107    std::cin.ignore(std::numeric_limits<std::streamsize>::max(), '\n');
108
109    std::cout << "E-mail address? ";
110    std::getline(std::cin, E.Email);
111
112    std::cout << "Physical address? ";
113    std::getline(std::cin, E.Address);
114
115    std::cout << "Birthday? ";
116    std::getline(std::cin, E.Birthday);
117
118    return E;
119  }
120
121  void printEntry(std::string Name, Contacts::Entry E)
122  {
123    std::cout << "Name: " << Name << '\n'
124              << "Phone number: " << E.Phone << '\n'
125              << "Fax number: " << E.Fax << '\n'
126              << "E-Mail: " << E.Email << '\n'
127              << "Address: " << E.Address << '\n'
128              << "Birthday: " << E.Birthday << std::endl;
129  }
```

B   Exercise Solution, Refactored to Use a Strong Interface

B.1   The Newly Introduced Types

```
                                    types1.hpp
 1   #ifndef TYPES_HPP
 2   #define TYPES_HPP
 3   #include <string>
 4   #include <utility>
 5
 6   struct PhoneNumber
 7   {
 8     long Value;
 9
10     explicit PhoneNumber(long V) : Value (V) {}
11     explicit operator     long()   const       { return Value; }
12              long         get()    const       { return Value; }
13   };
14
15   struct MobilePhoneNumber : public PhoneNumber
16   {
17     explicit MobilePhoneNumber(long V) : PhoneNumber{V} {}
18   };
19
20   struct FaxNumber          : public PhoneNumber
21   {
22     explicit FaxNumber(long V)           : PhoneNumber{V} {}
23   };
24
25   struct PersonName
26   {
27     std::string Value;
28
29     explicit PersonName( std::string V) : Value(std::move(V)) {}
30     explicit operator    std::string()  const  { return Value; }
31              std::string get()          const  { return Value; }
32
33     bool operator  < (const PersonName& RHS) const
34     { return Value < RHS.Value; }
35   };
36
37   struct EmailAddress
38   {
39     std::string Value;
40
41     explicit EmailAddress(std::string V) : Value(std::move(V)) {}
42     explicit operator     std::string()  const  { return Value; }
43              std::string  get()           const  { return Value; }
44   };
45
```

```cpp
46   struct PhysicalAddress
47   {
48     std::string Value;
49
50     explicit PhysicalAddress(std::string V) : Value(std::move(V)) {}
51     explicit operator        std::string()  const  { return Value; }
52              std::string     get()           const  { return Value; }
53   };
54
55   struct Birthday
56   {
57     std::string Value;
58
59     explicit Birthday(   std::string V) : Value(std::move(V)) {}
60     explicit operator    std::string()  const  { return Value; }
61              std::string get()           const  { return Value; }
62   };
63
64   #endif /* TYPES_HPP */
```

B.2 Contacts business logic library

```cpp
                          ———————— contacts1.hpp ————————
1    #ifndef CONTACTS_HPP
2    #define CONTACTS_HPP
3    #include <map>
4    #include <string>
5    #include <vector>
6
7    #include "types.hpp"
8
9    class Contacts
10   {
11     public:
12       struct Entry
13       {
14         MobilePhoneNumber Phone;
15         FaxNumber         Fax;
16         EmailAddress      Email;
17         PhysicalAddress   Address;
18         class Birthday    Birthday;
19       };
20
21            void    add(   PersonName Name, Entry E);
22     const Entry* lookup(PersonName Name)              const;
23            bool    remove(PersonName Name);
24
25     std::vector<PersonName> getNames() const;
```

```
26
27                   void       save(const std::string& File) const;
28        static Contacts load(const std::string& File);
29
30        std::size_t count() const;
31
32      private:
33        std::map<PersonName, Entry> Entries;
34    };
35
36    #endif  /* CONTACTS_HPP */
```

```
                                 contacts1.cpp
1    #include  "contacts.hpp"
2    #include  "contact_storage.hpp"
3    #include  "types.hpp"
4
5    std::size_t Contacts::count() const { return Entries.size(); }
6
7    void Contacts::add(PersonName Name, Entry E)
8    {
9      auto Iter = Entries.find(Name);
10     if (Iter == Entries.end())
11       Entries.emplace(Name, E);
12     else
13       Iter->second = E;
14   }
15
16   const Contacts::Entry* Contacts::lookup(PersonName Name) const
17   {
18     auto Iter = Entries.find(Name);
19     if (Iter == Entries.end())
20       return nullptr;
21     return &Iter->second;
22   }
23
24   bool Contacts::remove(PersonName Name)
25   {
26     auto Iter = Entries.find(Name);
27     if (Iter == Entries.end())
28       return false;
29     Entries.erase(Iter);
30     return true;
31   }
32
33   std::vector<PersonName> Contacts::getNames() const
34   {
35     std::vector<PersonName> Names;
36     for (const std::pair<PersonName, Entry>& Contact : Entries)
37       Names.push_back(Contact.first);
```

```cpp
38      return Names;
39    }
40
41  Contacts Contacts::load(const std::string& File)
42  {
43    Contacts Return;
44    ContactStorage Storage{File};
45
46    PersonName Name{{}};
47    MobilePhoneNumber Phone{{}};
48    FaxNumber Fax{{}};
49    EmailAddress Email{{}};
50    PhysicalAddress Address{{}};
51    Birthday Birthday{{}};
52    while (Storage.readEntry(Name, Phone, Fax, Email, Address, Birthday))
53    {
54      Entry E{Phone, Fax, Email, Address, Birthday};
55      Return.add(Name, E);
56    }
57
58    return Return;
59  }
60
61  void Contacts::save(const std::string& File) const
62  {
63    ContactStorage Storage{File};
64    Storage.clear();
65
66    for (const std::pair<PersonName, Entry>& Contact : Entries)
67      Storage.writeEntry(Contact.first, Contact.second.Phone,
68                         Contact.second.Fax, Contact.second.Email,
69                         Contact.second.Address, Contact.second.Birthday);
70  }
```

B.3   Contact-Storage persistency library

```cpp
                          ———— contact_storage1.hpp ————
1   #ifndef CONTACT_STORAGE_HPP
2   #define CONTACT_STORAGE_HPP
3   #include <fstream>
4   #include <string>
5
6   #include "types.hpp"
7
8   class ContactStorage
9   {
10    public:
11       ContactStorage(std::string File);
```

```cpp
12          ~ContactStorage();
13
14          // Delete the contents of the backing storage.
15          void clear();
16
17          // Ask whether the backing storage reached its end.
18          bool end() const;
19
20          // Read one (the next, sequentially) record from the backing file.
21          // If the file has emptied, does not modify the output variables.
22          // Returns TRUE if a record was read.
23          bool readEntry(PersonName&         OutName,
24                         MobilePhoneNumber&  OutPhone,
25                         FaxNumber&          OutFax,
26                         EmailAddress&       OutEmail,
27                         PhysicalAddress&    OutAddress,
28                         Birthday&           OutBirthday);
29
30          // Write one record into the backing file.
31          void writeEntry(PersonName       Name,     MobilePhoneNumber Phone,
32                          FaxNumber        Fax,      EmailAddress      Email,
33                          PhysicalAddress Address,  Birthday          Birthday);
34
35      private:
36        std::string   Filename;
37        std::fstream* FileOnDisk = nullptr;
38    };
39
40    #endif /* CONTACT_STORAGE_HPP */
```

```cpp
                        contact_storage1.cpp
1    #include <cassert>
2    #include <cstring>
3    #include <iostream>
4    #include <optional>
5    #include <utility>
6
7    #include "contact_storage.hpp"
8
9    ContactStorage::ContactStorage(std::string File)
10       : Filename(std::move(File)),
11         FileOnDisk(new std::fstream(
12               Filename.c_str(), std::ios_base::in | std::ios_base::out))
13    {
14      if (!FileOnDisk->is_open())
15      {
16        delete FileOnDisk;
17        FileOnDisk = new std::fstream(Filename.c_str(), std::ios_base::out);
18      }
19    }
```

```cpp
ContactStorage::~ContactStorage()
{
  if (!FileOnDisk)
    return;

  FileOnDisk->flush();
  FileOnDisk->close();
  delete FileOnDisk;
}

void ContactStorage::clear()
{
  assert(FileOnDisk && "No file opened?");
  FileOnDisk->close();
  delete FileOnDisk;

  FileOnDisk = new std::fstream(Filename.c_str(), std::ios_base::in |
                                                  std::ios_base::out |
                                                  std::ios_base::trunc);
}

bool ContactStorage::end() const
{
  assert(FileOnDisk && "No file opened?");
  return FileOnDisk->eof();
}

void ContactStorage::writeEntry(PersonName Name, MobilePhoneNumber Phone,
                                FaxNumber Fax, EmailAddress Email,
                                PhysicalAddress Address,
                                Birthday Birthday)
{
  assert(FileOnDisk && "No file opened?");
  FileOnDisk->seekp(0, std::ios_base::end);
  *FileOnDisk << "<ENTRY>\n"
              << "<Name>" << Name.get() << "</Name>\n"
              << "<Phone>" << Phone.get() << "</Phone>\n"
              << "<Fax>" << Fax.get() << "</Fax>\n"
              << "<EMail>" << Email.get() << "</EMail>\n"
              << "<Address>" << Address.get() << "</Address>\n"
              << "<Birthday>" << Birthday.get() << "</Birthday>\n"
           << "</ENTRY>\n\n";
}

bool ContactStorage::readEntry(PersonName& OutName,
                               MobilePhoneNumber& OutPhone,
                               FaxNumber& OutFax, EmailAddress& OutEmail,
                               PhysicalAddress& OutAddress,
                               Birthday& OutBirthday)
```

```cpp
 70  {
 71    auto ParseLineWithTag = [](const std::string& Line,
 72                               const std::string& OpenTag,
 73                               const std::string& CloseTag)
 74    {
 75      std::optional<std::string> Result;
 76
 77      if (Line.size() >= OpenTag.size() &&
 78          Line.substr(0, OpenTag.size()) == OpenTag)
 79      {
 80        Result.emplace(Line.substr(OpenTag.size()));
 81        Result = Result->substr(0, Result->size() - CloseTag.size());
 82      }
 83
 84      return Result;
 85    };
 86
 87    std::string Line;
 88    do
 89    {
 90      std::getline(*FileOnDisk, Line);
 91      if (FileOnDisk->eof())
 92        return false;
 93
 94      if (auto Name = ParseLineWithTag(Line, "<Name>", "</Name>"))
 95        OutName = PersonName{std::move(*Name)};
 96      else if (auto Phone = ParseLineWithTag(Line, "<Phone>", "</Phone>"))
 97        OutPhone = MobilePhoneNumber{std::stol(*Phone)};
 98      else if (auto Fax = ParseLineWithTag(Line, "<Fax>", "</Fax>"))
 99        OutFax = FaxNumber{std::stol(*Fax)};
100      else if (auto Email = ParseLineWithTag(Line, "<EMail>", "</EMail>"))
101        OutEmail = EmailAddress{std::move(*Email)};
102      else if (auto Address =
103          ParseLineWithTag(Line,"<Address>", "</Address>"))
104        OutAddress = PhysicalAddress{std::move(*Address)};
105      else if (auto Birthday =
106          ParseLineWithTag(Line, "<Birthday>", "</Birthday>"))
107        OutBirthday = ::Birthday{std::move(*Birthday)};
108    } while (Line != "</ENTRY>");
109
110    return true;
111  }
```

B.4   The User-Facing `main`

```cpp
―――――――――――――――――――――――― main1.cpp ――――――――――――――――
 1  #include <iostream>
 2  #include <limits>
```

```cpp
 3   #include <string>
 4
 5   #include "contacts.hpp"
 6   #include "types.hpp"
 7
 8   void              printMenu();
 9   Contacts::Entry readEntryFromUser();
10   void              printEntry(PersonName Name, Contacts::Entry E);
11
12   int main()
13   {
14     std::cout << "Phonebook v1\n";
15     std::cout << "------------\n\n";
16     std::cout << "File to open and use? ";
17
18     std::string Filename;
19     std::cin >> Filename;
20     std::cin.ignore(std::numeric_limits<std::streamsize>::max(), '\n');
21
22     Contacts Phonebook = Contacts::load(Filename);
23     std::cout << Filename << ": "
24               << Phonebook.count() << " entries loaded!\n";
25
26     std::string UserChoice;
27     while (true)
28     {
29       std::cout << "Action? ";
30       std::getline(std::cin, UserChoice);
31
32       if (UserChoice == "q" || UserChoice == "quit")
33       {
34         std::cout << "Goodbye!" << std::endl;
35         return EXIT_SUCCESS;
36       }
37       else if (UserChoice == "h" || UserChoice == "help")
38         printMenu();
39       else if (UserChoice == "a" || UserChoice == "add")
40       {
41         std::cout << "Adding a new contact...\n";
42         std::cout << "Name? ";
43         std::getline(std::cin, UserChoice);
44
45         Contacts::Entry E = readEntryFromUser();
46         Phonebook.add(PersonName{UserChoice}, E);
47         std::cout << std::endl;
48       }
49       else if (UserChoice == "r" || UserChoice == "remove")
50       {
51         std::cout << "Removing a contact...\n";
52         std::cout << "Name? ";
```

```cpp
53          std::getline(std::cin, UserChoice);
54
55          bool Removed = Phonebook.remove(PersonName{UserChoice});
56          if (!Removed)
57            std::cout << "No such entry existed. Not removing anything.";
58          std::cout << std::endl;
59        }
60      else if (UserChoice == "l" || UserChoice == "lookup")
61        {
62          std::cout << "Name? ";
63          std::getline(std::cin, UserChoice);
64
65          if (const Contacts::Entry* EP =
66                  Phonebook.lookup(PersonName{UserChoice}))
67            printEntry(PersonName{UserChoice}, *EP);
68          else
69            std::cout << "No such entry!";
70          std::cout << std::endl;
71        }
72      else if (UserChoice == "list")
73        {
74          for (const PersonName& Name : Phonebook.getNames())
75            std::cout << " * " << Name.get() << '\n';
76          std::cout << std::endl;
77        }
78      else if (UserChoice == "s" || UserChoice == "save")
79        Phonebook.save(Filename);
80      else
81        std::cerr << "Invalid action, try again! ('h' prints menu)"
82                  << std::endl;
83    }
84  }
85
86  void printMenu()
87  {
88    std::cout << "add        Add (or overwrite) an entry, step-by-step\n
89    std::cout << "remove     Delete an entry based on name\n";
90    std::cout << "lookup     Get the details of a contact\n";
91    std::cout << "list       List all names in the phonebook\n";
92    std::cout << '\n';
93    std::cout << "save       Save changes to the disk\n";
94    std::cout << "quit       Close the program\n";
95    std::cout << std::endl;
96  }
97
98  Contacts::Entry readEntryFromUser()
99  {
100    std::cout << "Reading the details of a contact...\n";
101
102    std::cout << "Phone number (without spaces!)? ";
```

```cpp
103      long RawPhoneNumber;
104      std::cin >> RawPhoneNumber;
105      std::cin.ignore(std::numeric_limits<std::streamsize>::max(), '\n');
106
107      std::cout << "Fax number (without spaces!)? ";
108      long RawFaxNumber;
109      std::cin >> RawFaxNumber;
110      std::cin.ignore(std::numeric_limits<std::streamsize>::max(), '\n');
111
112      std::cout << "E-mail address? ";
113      std::string RawEmail;
114      std::getline(std::cin, RawEmail);
115
116      std::cout << "Physical address? ";
117      std::string RawAddress;
118      std::getline(std::cin, RawAddress);
119
120      std::cout << "Birthday? ";
121      std::string RawBirthday;
122      std::getline(std::cin, RawBirthday);
123
124      return Contacts::Entry{MobilePhoneNumber{RawPhoneNumber},
125                             FaxNumber{RawFaxNumber},
126                             EmailAddress{std::move(RawEmail)},
127                             PhysicalAddress{std::move(RawAddress)},
128                             Birthday{std::move(RawBirthday)}};
129  }
130
131  void printEntry(PersonName Name, Contacts::Entry E)
132  {
133    std::cout << "Name: " << Name.get() << '\n'
134              << "Phone number: " << E.Phone.get() << '\n'
135              << "Fax number: " << E.Fax.get() << '\n'
136              << "E-Mail: " << E.Email.get() << '\n'
137              << "Address: " << E.Address.get() << '\n'
138              << "Birthday: " << E.Birthday.get() << std::endl;
139  }
```

References

1. Berners-Lee, T., Masinter, L.M., McCahill, M.P.: Uniform Resource Locators (URL). RFC 1738 (1994). https://doi.org/10.17487/RFC1738, http://datatracker.ietf.org/doc/html/rfc1738
2. Butler, S., Wermelinger, M., Yu, Y., Sharp, H.: Exploring the influence of identifier names on code quality: an empirical study. In: 2010 14th European Conference on Software Maintenance and Reengineering, pp. 156–165 (2010). https://doi.org/10.1109/CSMR.2010.27, http://ieeexplore.ieee.org/document/5714430

3. Czarnecki, K., Eisenecker, U.W.: Generative programming: methods, tools, and applications. ACM Press/Addison-Wesley Publishing Co., 1515 Broadway, 17th Floor, New York(2000)

4. Educational Authority of Hungary: Enrolment statistics of previous years, number of enrolments per curricula per institution. http://www.felvi.hu/felveteli/ponthatarok_statisztikak/elmult_evek/!ElmultEvek/index.php/elmult_evek_statisztikai/ponthatarok?filters%5Bsta_int_id%5D=12&filters%5Bsta_kar_id%5D=334&filters%5Bsta_ev%5D=2020%2F%C3%81, accessed 2021.12.27

5. European Commission, E.A.C.E.A., Eurydice: The European Higher Education Area in,: Bologna Process Implementation Report. European Union, European Union, Brussels. (2020). https://doi.org/10.2797/756192, http://op.europa.eu/en/publication-detail/-/publication/c90aaf32-4fce-11eb-b59f-01aa75ed71a1/language-en/format-PDF/source-183354043

6. Fekete, A., Porkoláb, Z.: Report on a field experiment of the comprehension strategies of computer science MSc students. In: Steingartner, W., Korečko, Š., Szakál, A. (eds.) 2022 IEEE 16th International Scientific Conference on Informatics (Informatics), pp. 73–81 (2022). https://doi.org/10.1109/Informatics57926.2022.10083413, http://ieeexplore.ieee.org/document/10083413

7. Fekete, A., Porkoláb, Z.: Field experiment of the memory retention of programmers regarding source code. Studia Universitatis Babeş-Bolyai Informatica **68**(1), 71–82 (2023). https://doi.org/10.24193/subbi.2023.1.05, http://cs.ubbcluj.ro/~studia-i/journal/journal/article/view/89

8. Fredkin, E.: Trie memory. Commun. ACM **3**(9), 490–499 (1960). https://doi.org/10.1145/367390.367400

9. van Heesch, D., et al.: Doxygen: the de facto standard tool for generating documentation from annotated C++ sources (1997-2021). http://doxygen.nl, Accessed 28 Dec 2021

10. Horváth, B.I.: C++ Overload inspector: a tool for analyzing and profiling overloaded function calls. In: Steagall, B., Levi, I. (eds.) C++ Now 2024. In press

11. Horváth, G., Kovács, R.N., Szalay, R., Porkoláb, Z.: Implementing and executing static analysis using LLVM and CodeChecker. In: Grelck, C., Szabó, C. (eds.) Springer LNCS. vol. 11916 (2018), (in press, accepted paper to journal volume of the 3COWS Winter School)

12. ISO/IEC JTC 1/SC 22: ISO/IEC 14882:2017 Information technology — Programming languages — C++, version (C++ 17). International Organization for Standardization, Geneva, Switzerland (2017). http://iso.org/standard/68564.html

13. Kovács, A., Szabados, K.: Internal quality evolution of a large test system - an industrial study. Acta Universitatis Sapientiae, Informatica **8**(2), 216–240 (2016). https://doi.org/10.1515/ausi-2016-0010, http://acta.sapientia.ro/acta-info/C8-2/info82-04.pdf

14. Madsen, O.L., Magnusson, B., Mølier-Pedersen, B.: Strong typing of object-oriented languages revisited. In: Proceedings of the European Conference on Object-Oriented Programming on Object-Oriented Programming Systems, Languages, and Applications, pp. 140–150. OOPSLA/ECOOP '90, Association for Computing Machinery, New York (1990). https://doi.org/10.1145/97945.97964

15. Meyer, B.: Ensuring strong typing in an object-oriented language (abstract). In: Conference Proceedings on Object-Oriented Programming Systems, Languages, and Applications, pp. 89–90. OOPSLA '92, Association for Computing Machinery, New York (1992). https://doi.org/10.1145/141936.290558

16. Peters, R., Zaidman, A.: Evaluating the lifespan of code smells using software repository mining. In: Ferenc, R., Mens, T., Cleve, A. (eds.) 2012 16th European Conference on Software Maintenance and Reengineering, pp. 411–416 (2012). https://doi.org/10.1109/CSMR.2012.79, http://ieeexplore.ieee.org/document/6178888
17. Porkoláb, Z., Brunner, T.: The CodeCompass comprehension framework. In: Proceedings of the 26th Conference on Program Comprehension, pp. 393–396. ICPC '18, Association for Computing Machinery, New York (2018). https://doi.org/10.1145/3196321.3196352
18. Porkoláb, Z., Brunner, T., Krupp, D., Csordás, M.: CodeCompass: an open software comprehension framework for industrial usage. In: Proceedings of the 26th Conference on Program Comprehension, pp. 361–369. ICPC '18, Association for Computing Machinery, New York(2018). https://doi.org/10.1145/3196321.3197546
19. Pusz, M.: Implementing physical units library for C++. C++ Now presentation (2019). http://youtube.com/watch?v=wKchCktZPHU, Accessed 27 Dec 2019
20. Rice, A., Aftandilian, E., Jaspan, C., Johnston, E., Pradel, M., Arroyo-Paredes, Y.: Detecting argument selection defects. Proc. ACM Program. Lang. 1(OOPSLA), 104:1–104:22 (Oct 2017). https://doi.org/10.1145/3133928
21. Schling, B.: The Boost C++ Libraries. XML Press (2011). http://theboostcpplibraries.com/, Accessed 09 Dec 2022
22. Stephenson, A.G., et al.: Mars climate orbiter – mishap investigation report. Tech. rep., National Aeronautics and Space Administration (NASA) (1999). http://llis.nasa.gov/llis_lib/pdf/1009464main1_0641-mr.pdf
23. Szalay, R.: Towards decoupling nullability semantics from indirect access in pointer use. In: Kovásznai, G., Fazekas, I., Tómács, T. (eds.) Proceedings of the 11th International Conference on Applied Informatics, pp. 328–340. Eszterházy Károly University, Eger (2020). http://ceur-ws.org/Vol-2650/paper34.pdf
24. Szalay, R., Porkoláb, Z.: Visualising compiler-generated special member functions of types. In: Heričko, M. (ed.) Proceedings of the 21st International Multiconference on Information Society: Collaboration, Software and Services in Information Society. vol. G, pp. 55–58. University of Maribor, Faculty of Electrical Engineering and Computer Science (2018). http://is.ijs.si/wp-content/uploads/2019/02/Zbornik-G.pdf
25. Szalay, R., Sinkovics, Á., Porkoláb, Z.: Practical heuristics to improve precision for erroneous function argument swapping detection in C and C++. J. Syst. Softw. 181C(111048), 11 (2021). https://doi.org/10.1016/j.jss.2021.111048, http://www.sciencedirect.com/science/article/pii/S016412122100145X
26. Szalay, R.: Interactive tooling support for the migration to Strong Types. In: Steagall, B., Levi, I., Ryan, C. (eds.) C++ Now (2023). http://youtube.com/watch?v=rcXf1VCA1Uc, Accessed 19 July 2023
27. Szalay, R., Porkoláb, Z.: Flexible semi-automatic support for type migration of primitives for C/C++ programs. In: Proceedings of the 2022 IEEE International Conference on Software Analysis, Evolution, and Reengineering, pp. 878–889 (2022). https://doi.org/10.1109/SANER53432.2022.00106, http://ieeexplore.ieee.org/document/9825812/
28. Szalay, R., Porkoláb, Z.: Using strong types in C++ for long-term code management. In: Koopman, P., Lubbers, M., Fernandes, J.P. (eds.) SusTrainable: Promoting Sustainability as a Fundamental Driver in Software Development Training and Education – Teacher Training Revised lecture notes, pp. 36–47. arXiv (2022). https://doi.org/10.48550/ARXIV.2204.13993, https://arxiv.org/abs/2204.13993

29. Takács, R., Kárász, J.T., Takács, S., Horváth, Z., Oláh, A.: Applying the rasch model to analyze the effectiveness of education reform in order to decrease Computer Science students' dropout. Hum. Soc. Sci. Commun. **8**(1), 1–8 (2021). https://doi.org/10.1057/s41599-021-00725-w, http://nature.com/articles/s41599-021-00725-w
30. Velavan, T.P., Meyer, C.G.: The COVID-19 epidemic. Tropical Med. Int. Health (TMIH) **25**, 278–280 (2020). https://doi.org/10.1111/tmi.13383, http://ncbi.nlm.nih.gov/pmc/articles/PMC7169770
31. van de Voorde, D., Josuttis, N.M., Gregor, D.: C++ Templates: The Complete Guide (2nd Edition). Addison-Wesley Professional, 2nd edn. (2017). http://tmplbook.com
32. Wright, H.K.: Incremental type migration using type algebra. In: 2020 IEEE International Conference on Software Maintenance and Evolution (ICSME), pp. 756–765 (2020). https://doi.org/10.1109/ICSME46990.2020.00085, http://ieeexplore.ieee.org/document/9240711
33. Yamashita, A., Moonen, L.: Do developers care about code smells? an exploratory survey. In: 2013 20th Working Conference on Reverse Engineering (WCRE), pp. 242–251 (2013). https://doi.org/10.1109/WCRE.2013.6671299, http://ieeexplore.ieee.org/document/6671299/

Teaching by Example: Temporal-Logic Automated Solutions for Sustainable Educational Progress

Mihail Petrov and Vladimir Valkanov

Plovdiv University "Paisii Hilendarski", Tzar Asen 24, 4000 Plovdiv, Bulgaria
{mihailpetrov,vvalkanov}@uni-plovdiv.bg

Abstract. The process of teaching and learning new knowledge can rarely be classified as standard. Teaching curriculum and the initial professor assessment of the audience are often indirectly related to the end result. The achievements which may vary based on discriminating factors such as gender, age, social background, past experience, learning through techniques such as listening, reading, et al. can be used as the basis of an analysis with the goal of identifying the factors allowing the students to achieve better results in the learning process.

Keywords: eLearning · Intelligent agents · ITL · Behavior analysis · Blockchain · Education

1   Introduction

At the basis of the digital systems for assisting the learning process is the idea of organizing a database which provides the learners a set of resources directed towards learning the subject matter in a structured way, testing new competences as well as providing instant feedback for the progress and the learned competences.

The modern approach towards the various subjects aims to:

- Build competences;
- Focus on inter-subject skills;

This work received financial support through the Erasmus+ Strategic Partnership for Higher Education *SusTrainable—Promoting Sustainability as a Fundamental Driver in Software Development Training and Education* (project number 2020-1-PT01-KA203-078646), funded by the European Union and coordinated by the University of Coimbra, Portugal.

The information and views set out in this publication are those of the authors and do not necessarily reflect the official opinion of the European Union. Neither the European Union institutions and bodies nor any person acting on their behalf may be held responsible for the use which may be made of the information contained therein.

– Be organized around the needs, speed and approach of the learner.

To meet those demands, science is working towards creating software solutions for assisting the learning process through strong integration in the modern classroom. At the basis of all systems for assisting the teaching process the term artificial intelligence (**AI**) [1] [2] can be found.

In the context of systems for assisting the learning process two main types of systems can be considered which serve this idea:

– Adaptive systems for managing learning content;
– Intelligent assistants.

1.1   Directions in Which the Systems for Managing Learning Process are Developing

As we already witnessed, classifying *those systems* [4] gives the possibility to direct our effort to researching various micro aspects of the complete improvement of the process of their integration in the processes related to acquiring new knowledge and skills. We will classify the researches in 4 categories, focusing on the model the researches are united around:

– Cognitive-behavioral models;
– Subject-specific models;
– Models optimizing understanding of the subject matter;
– Models organized around user interaction.

1.2   Instruments for Modeling Temporal Aspects

The technical implementation of the researched concepts would enable us to model real scenarios from the existing problems of the software development with the help of the toolset of temporal logic. At the current moment there are several outstanding examples of products vying to solve this problem:

– Prolog-based projects;
– C -Tempura and Ana Tempura;
– Temporal-based platforms such as **YAP, SWI, SICStus, XSB, tuProlog** and **LPA Prolog**.

The presented temporal dialects show the natural development of the technical implementation in modeling temporal formalisms which however contain a number of obvious weaknesses, including but not only:

– Complex temporal syntax
– Lack of integration interfaces
– Difficulty in expanding implementations

2 Cognitive-Cumulative Processes

At the basis of the adaptive process is the idea of following an independently defined and highly specific path to acquiring knowledge and skills [9]. In the current chapter we want to look at not only the mechanisms for controlling the adaptation processes but also to specifically present the ideas of the processes related to profiling the users of educational services.

2.1 Formal Models for Presenting Knowledge

We formalized the main components of an adaptive educational system by placing at its basis the mechanisms for storing information. The main question we need to answer before going further is what are the popular processes for presenting knowledge. We will use as the basis one of the most popular models – the so-called **naïve Bayes algorithm**.

At its basis the model defines knowledge as a multitude of binary states responding to the question whether a given skill or knowledge have been mastered. Owing to its strictly binary nature, this model does not allow any nuances.

One of the main problems of this model is ignoring the possibility certain knowledge to be forgotten or not completely mastered:

- the model does not allow nuances; it considers certain knowledge either mastered or not which suggests as well as ignores many gray areas in the fine analysis of how certain knowledge is acquired;
- the model ignores the time factor – certain knowledge can be forgotten and/or pushed back by minimizing the efforts for its use in practical and theoretical scenarios.

We will try to suggest an alternative approach in reviewing these problems.

2.2 Architecture of the Cognitive-Cumulative Process (CcP)

The cognitive-cumulative process [5] [6] is a formal description of the necessary steps to acquire a certain set of knowledge in the context of a random educational medium. Any representation of the CcP consists of three analytical structures:

- Element – contains a unit of data which can be consumed by the user of the educational service;
- Condition – characteristic of the element at a given moment in time;
- Event –action altering the state of a certain element.

The elements of the analytical structures are combined in four categories describing the specialized processes:

- **Cognitive-oriented components** – contain information having direct or supplementary relation to the reviewed matter;

- **Problem-oriented components** – contain definitions and states describing scenarios with the goal of confirming or applying newly accumulated knowledge by solving specific problems;
- **Validation-oriented components** – contain mechanisms for checking pre-developed problems. Their main role is assessing or acting as an evaluating mechanism to a specific problem or set of problems;
- **Progress-oriented components** – contain messages or information which attempt to address changes in the condition of the currently-reviewed system.
- **Communication-oriented components** – represent mechanisms showing movement between the conditions of a specific object.

We will formally represent the CcP architecture [13] as a graph whose vertices are the constituent elements and its edges define an algorithmic transition describing the conditions necessary for moving on to the next information element. In Fig. 1 the architecture of a single CcP element is graphically represented as well as the elements sequentially connected to it.

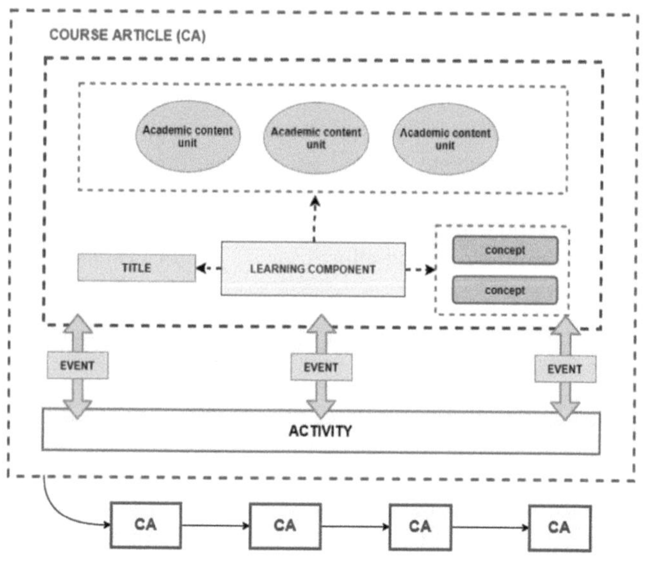

Fig. 1. Architecture of a Cognitive-cumulative process.

3   Modeling Temporal Aspects

The main obstacle for describing a dynamic system is related to the ever-changing environment. The classic logic loses its power [12] when it has to describe a condition of the observed system where time is a factor. From the previously reviewed models it became apparent that the classic plan for modeling knowledge excludes the factor of time and is based only on empirical statistical proofs of the occurrence of a certain condition [14].

Implementing **virtual educational space application** is an example of a system where each agent changes its state in time and space and acquires different features in certain moments. Those features impact its reactivity towards the system (*particularly the platform towards which the interaction is directed*) and constantly change the expected result. In this current chapter we will review the factor of time and how it can be formally defined and practically implemented in systems for modeling knowledge.

3.1   Formal Representation of the Temporal Aspect

As is well known, the actions related to acquiring and analyzing information and the consequent learning knowledge from it takes time and can be classified on the basis of this time-consumption with terms such as – quick or slow, easy or hard. The use of temporal quantifiers in reflecting on educational processes can be of help in predicting or detecting inconsistencies with our initial expectations. Based on this idea, for the needs of this dissertation we will define the term temporal aspect as – *any action or inaction causing change in the condition observed in a certain digital system starting from a set moment in the past up to a set moment in the future.*

We need to point out that the changes in a digital system can be observed only if events are taking place in it which allow for changes made from the user or from the system itself. Unlike nature, digital objects do not alter their features just under the influence of the factor of time (*aging*).

The reason for the occurrence of changes in a system are the events taking place in it. As a main unit for determining the temporal terminology we will use the term temporal event. We will formally define a **temporal event** as an ordered foursome with the following characteristics:

FORMAL REPRESENTATION

$e_t = <title, \{param\}, init_{timestam}, end_{timestamp}>$

title $\qquad\qquad \rightarrow$ *Identifier of the current component*

$\{param\} \qquad \rightarrow$ *Set of parameters*

$init_{timestam} \rightarrow$ *Temporal stamp for initiation of the event*

$end_{timestamp} \rightarrow$ *Temporal stamp for end of the event*

There are various mechanisms for representing the temporal aspects in a digital medium, some of which include:

– Use of real time;
– Use of relative time;
– Use of processing time.

We will define the term empty temporal event, through which we will register the lack of events in a monitored system, as – *the distance from the end of one temporal event to the start of a new one. Despite its name, empty temporal events can carry information for the context in which they are monitored. We will formally designate events of this type with the term* **empty***.*

Temporal events need to be reviewed as a mediator for detecting a relation between two objects in an arbitrary system. An event which exists in isolation and does not address a certain problem can be characterized only with dubious usefulness. There are four user groups in the process of exchanging messages (temporal events) which we will address as:

– Emitters;
– Emittents;
– Listeners;
– Executors.

3.2   Specific Temporal Aspects

Interval. From the used definition for temporal aspect it becomes apparent that the observations we will make on the changes in a given system are clearly defined start and end points. Within this timeframe in the observed system **consecutive** or **parallel** temporal events can arise which can change the condition of various objects. At any moment observations can be made on a specific timeframe located somewhere between the start and end point. For more clarification let us for the needs of this current research give the following definition for an interval – *the amount of temporal events registered in the constraints of a specific timeframe. The endpoints of the interval are called respectively start and end of the interval. A valid interval will be designated with the symbol I.*

For the goals of this research we will call an interval valid if it contains at least one event. A valid temporal event is any event in the multitude of temporal events. There are several special intervals:

– System session;
– User session.

Inactivity. The lack of activity can be registered both within an active process – while executing an action which stops before resuming after a certain period, but also within the context of systems where no activity is being observed [15]. The inactivity is a feature only for the system users as an indicator for the presence of a feature supposing distancing or impeding the normal course of action.

Delay. We will designate one specific type of inactivity which will be classified in its own category.

A typical example of a delay is activating a complex system process which interrupts the natural interaction between the user and the system.

4   Modeling Analytical-Temporal Processes

In this current chapter we will look at the architecture of an adaptive educational system as well as the fundamental analytical processes for profiling [18] [19], analyzing and adapting learning content belonging to it.

When speaking of such a complex concept one inevitably reaches the point where they need to consider the way to communicate this information to the audience in order to turn it into knowledge. It is necessary to account for features of the target group such as background, age, gender and educational capacity. The given abstract characteristics of the audience define the target group to which to deliver the information which we want to transform into an educational product and respectively, knowledge. After looking at the described characteristics we can define **three processes** which need to be considered to maximize the results:

- Profiling the target group – determining the initial characteristics based on which the teaching curriculum will be prepared;
- Analyzing reactive and proactive behavior – within the frames of an observed learning process the users of the educational service initiate actions based on which the learning process' support system responds;
- Adapting the teaching curriculum – all following steps we need to take in order to optimize or maximize the implementation of expected knowledge and skills being tested.

As the basis of looking at these concepts we will present the principle architecture of an adaptive educational system providing the necessary analytical tools for controlling the described processes.

In Fig. 2 is presented a 4-layer basic structure of an adaptive educational system as well as the communication aspects between the separate layers. The architecture consists of:

- Expert layer;
- Adaptive layer;
- Layer of states;
- Communication interfaces.

The described layers and their characteristics do not have concrete technical values but serve as the basis on which a special implementation of the reviewed concepts can be built.

Profiling. Gaining new knowledge is an incremental process suggesting constant building on the basis we established at any given moment. This initial idea stems from the administrative divisions of any school subject. Some courses do not meet the needs of certain learners and often have to define restrictive requirements in order to filter or limit the audience. Some of these are:

- Presence of initial knowledge and skills;
- Meeting certain institutional qualifications;

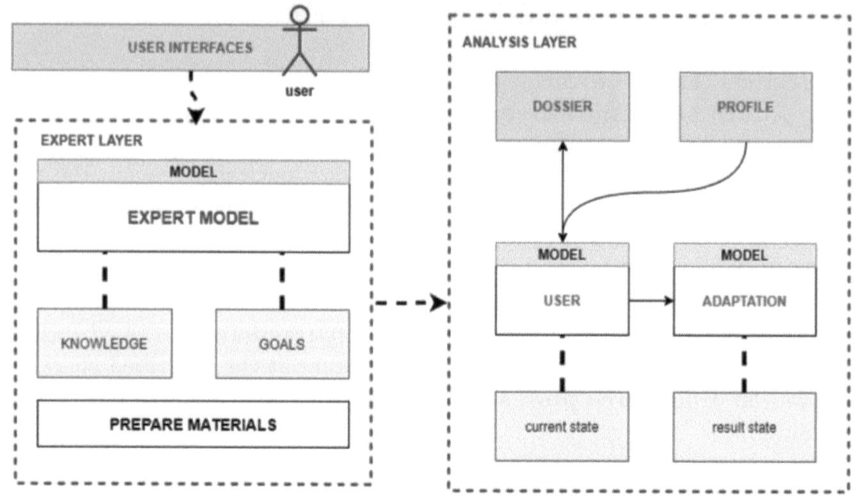

Fig. 2. Adaptive System Architecture.

- Gender or age restrictions.

Some of the instruments we can use to correctly determine the target group as well as categorize it would be:

- Clearly defined matrices and criteria for assessing initial knowledge and skills;
- Putting the learners in situations resembling real-life scenarios;
- Integrating predictive models.

We can determine **three types of characteristics** which we will use in creating a preliminary profile:

- Long-term life characteristics;
- Short-term emotional characteristics;
- Personal characteristics.

Analysis. We can answer the question to what extent a certain learning unit has been mastered by monitoring the process of working with the learning curriculum and the following activities related to the accompanying control and self-control [6]. This process is called analysis of the learning activity. As the basis of this analysis we use the multitude of stable conditions of an arbitrary analytical problem developed within the learning process. For the analysis we will use the so-called:

- Correct states – the solution process was successfully completed
- Incorrect states – an error is observed which in combination with the problem to be solved can assist in the analysis of the learner

We can define several incorrect states:

- wrong result;
- error in solving the problem;
- incorrect understanding of the subject matter.

In order to obtain detailed information from the current state of the error we need to have the following characteristics at our expense:

- The type of error according to the given classification;
- The available primary code at the moment when the error was made;
- Past system states;
- All code states from registering the error to the next correct state.

Adaptation. Adaptation of the learning process is an incremental global process controlling all aspects of the cognitive-cumulative process in the monitored system. Due to its global character the direction of adaptation is related to one of the three following directions:

- Skipping learning units;
- Repeating learning units;
- Expanding learning units.

Depending on the **subject of adaptation** we can distinguish:

- Adaptation of the learning curriculum;
- Adaptation of the individual activities.

Any one of the adaptation characteristics can be implemented alone or as a combination with any of the other two. The starting point in any adaptation activity are the results from the ongoing profiling and of the characteristics of the complete problem graph obtained in the **analysis** phase.

Let us look in detail at the algorithmic setup in any adaptation scenario, given in Fig. 3.

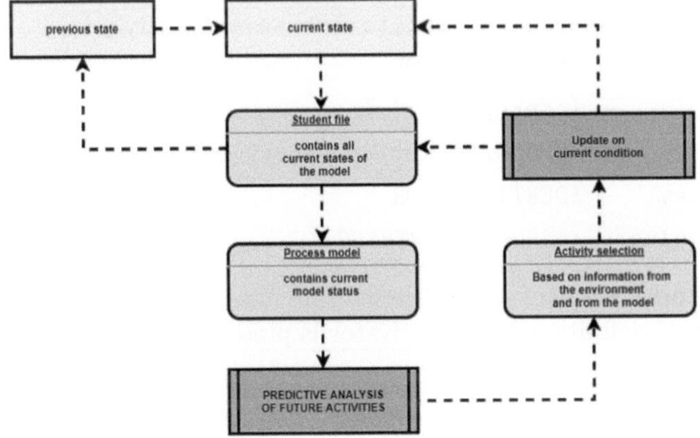

Fig. 3. Graphic representation of the algorithm for determining the current state of modeled knowledge.

5   SystemTempura

SystemTempura is an attempt for this current research to present a mechanism for solving the described problems by providing a certain interpretable protocol. The result is a software implementation of a protocol language for server comunication. Some of the basic aspects we will consider in specifying the protocol will be directed towards:

- Implementing syntactic rules describing programming languages similar to the ones most widely used currently;
- Providing a multi-platform medium; the language should not provide direct implementation characteristics but should try to be independent.

Some of the goals we set in realizing this language:

- Ability for integrating the dialect in media managing the events' context;
- Ability for integrating specialized tools;
- Multi-platform integration;
- Easy integration in the business processes.

5.1   Syntax and Specifications

SystemTempura defines specific syntax constructs describing temporal logic aspects. Some of them include:

Temporal Event. The basic communication mechanism in **SystemTempura** is called event. In its essence it models the formal representation of a temporal event providing functionality in two directions:

- Activating – Initializes the start of activities;
- Transport – Carries a certain information resource.

Syntax	
`event_header`	$\Rightarrow$ `event <event_title> <event_body>`
`event_title`	$\rightarrow$ `IDENTIFICATOR`
`event_body`	$\rightarrow$ `{ <property>? }`
`property`	$\rightarrow$ `<property_key><property_value>?;`
`property_key`	$\rightarrow$ `IDENTIFICATOR`
`property_value`	$\rightarrow$ `: (NUMBER | STRING)`

Listeners and Executors. The listeners are mechanisms for initiating an activity. The formal definition for an activator is presented as follows:

Syntax	
`listener`	$\Rightarrow$ `WHEN <listener_trigger> => ACTIVITY -> EVENT?`
`listener_trigger`	$\rightarrow$ `EVENT	FORMULA`
`{Ec}`	$\rightarrow$ `Set of academic articles`

The execution of a registered listener is always asynchronous and depends on the occurrence of a specific event or the activation of a specific formula. Let us look at some examples of valid listeners and their characteristics.

Monitoring event contexts is a never-ending process and due to that in describing them we should always account for the possibility that a given event might occur multiple times. Of course there are scenarios where we would try to limit the execution of a certain activity a certain number of times independent on the consequences occurring after it.

```
1   - Formula registering the occurrence of an event 5 times
2   IN PROBLEM_CONTEXT WHEN @Error => MessageX
3
4   -- Formula registering the occurrence of an event 10 times
5   IN PROBLEM_CONTEXT WHEN 5_TIME_ERROR => MessageY
6
7   -- Formula registering the occurrence of an event 10 times
8   IN PROBLEM_CONTEXT WHEN 10_TIME_ERROR => MessageZ
```

Temporal Intervals. According to the specification of SystemTempura the interval is a mechanism for defining a specific frame within which certain events take place. The intervals can be registered individually, as their own event but can also be produced as the consequence of a registered event. They are specified in the following way:

Syntax
INTERVAL $\Rightarrow$ <interval_header> <interval_body>
interval_header $\rightarrow$ <session_modifier>? interval <ID>
session_modifier $\rightarrow$ user | system
interval_body $\rightarrow$ { <declaration> }
declaration $\rightarrow$ key: value;
Key $\rightarrow$ ID
Value $\rightarrow$ TEMPORAL | FORMULA;

Temporal formulas. The formula is a mechanism for controlling complex conditions in which the occurrence of a group of events needs to be assessed before a certain activity actually takes place. The formal definition for a formula is presented as follows:

Syntax
formula $\Rightarrow$ FORMULA <ID> = <EXPRESSION>
ID $\rightarrow$ EVENT | FORMULA
EXPRESSION $\rightarrow$ *Set of academic articles*

Formulas are defined statically but their execution is always within the context of an active listener. The result of a given formula at an arbitrary moment can be **TRUE** or **FALSE**.

```
1   - Formula equaling a registered event
2   FORMULA FORMULA_EVENT      = @SampleEvent
```

5.2   Architecture of System Tempura

The system implementing the SystemTempura protocol consists of a protocol definition and a SystemTempura server controlling the process of delivering and receiving messages (Fig. 5).

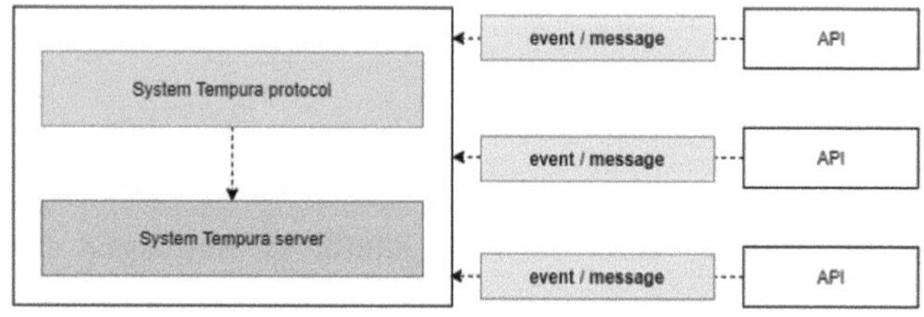

Fig. 4. System tempura work medium.

The architecture of SystemTempura consists of three layers presented in Fig. 4:

– Initiating layer – defines all processes related to the initial administration of registered event artifacts, organization of context in the application and management of the registered listeners in the context of external services. It contains all necessary mechanisms for controlling the process of interpreting the system constructs of SystemTempura;
– Analyzing layer – contains the basic mechanisms for controlling the syntax analysis of SystemTempura. This layer is open to the process of continuous assessment of the registered formulas as well as controlling the system clock.
– Management layer – contains the analytic apparatus for controlling temporal logic aspects which are implemented in the current system such as formulas, events, intervals and working with operators.

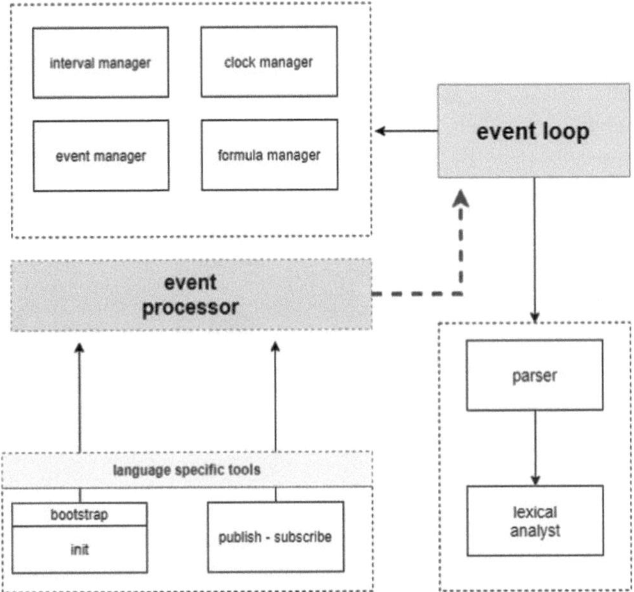

Fig. 5. Principle structure of a SystemTempura interpreter.

6 Architecture of a Platform for Adaptive Learning Based on Temporal Aspects

In this current chapter we will look at the overall architecture of an adaptive system based on tools using temporal concepts for the needs of **virtual education space application** [7].

6.1 Architecture of UniPlayground

We can split the system architecture into 4 basic layers and one control layer:

- Interface layer;
- Service layer;
- Translator layer;
- Data storage layer;
- Service manager.

The relations among all these layers are represented in Fig. 6.

The main idea of the application is for it to be distributed and not exist only on one specific platform or device. Due to this, all solutions for its planning and building were based around the idea of micro services [8] (Fig. 7).

Translators. Adaptive platforms based on tools have the goal of simulating a work medium close to the real one while acquiring a specific skill. Within

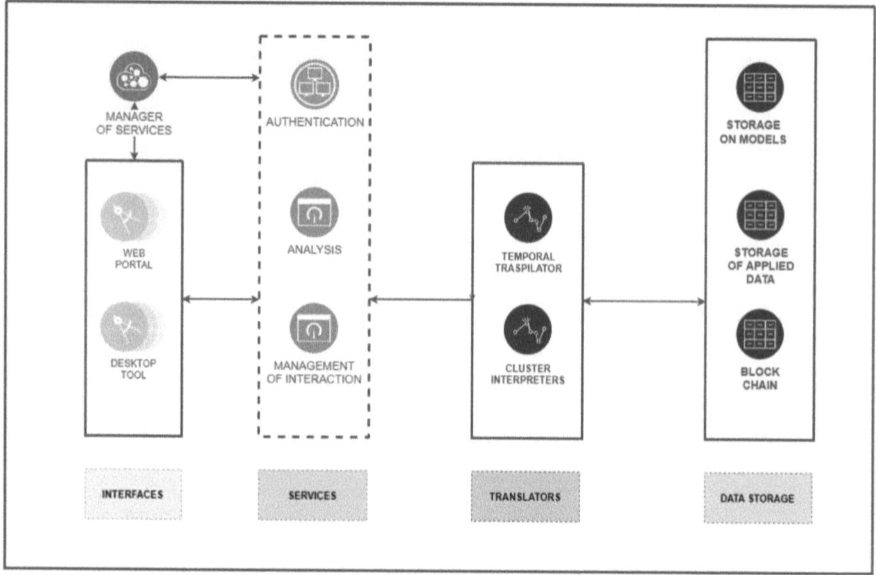

Fig. 6. Basic characteristics of the components comprising the UniPlayground system.

the UniPlayground context the translator layer provides abstracts of controlling and analyzing the communication between the user interface and the service for practicing skills related to the development of software systems [10]. A principle model of communication with the translator layer is presented in Fig. 6.

Let us look in detail at the communication connection between the system users and the translator layer.

The communication cycle begins with sending a command to the **translator layer**. Each command consists of a message containing several parameters:

- The language construct which will be interpreted – This is a set of instructions defined in an artificial programming language;
- Command context, also known as language medium – gives a reference point in interpreting the command in question;
- User identifier based on which the adaptation process is performed [11];
- Exact moment of the occurrence of the event;
- Event – to be able to adequately analyze and build the user graph we need to monitor each one of the user activities within the system. This means we cannot afford to interpret the following only in the context of visible events observed by the user. This is how the system distinguishes events of evident nature.

Services. In the context of principle architecture, the service layer is a set of independent program logical units for controlling all functional aspects within UniPlayground. For the goal of this research they are divided into several categories depending on the context they secure.

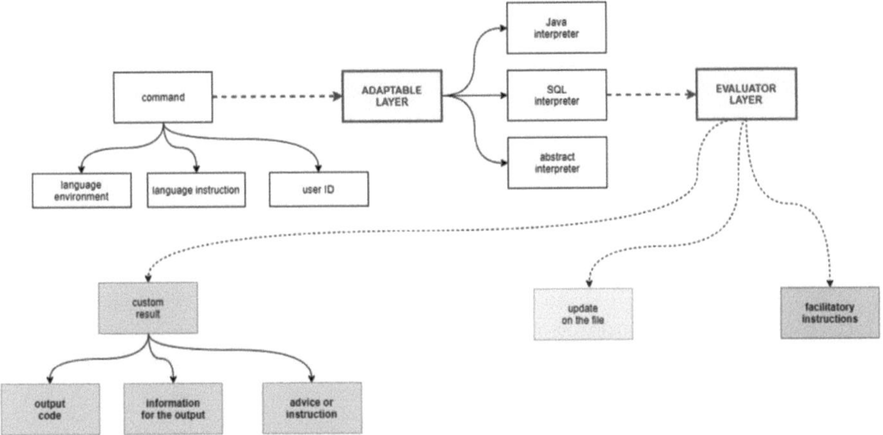

Fig. 7. Communication cycle of client – translator communication.

- System services – they secure processes dealing with assuring all aspects of
  the platform functionalities;
- Services for control of the learning process – their goal is to assist the present
  interfaces in the platform and improve their functionality by adding capabilities based on adaptive context;
- Services for analyzing and adapting the learning process – these are never-
  ending processes for active adapting and profiling the results from working
  with the provided tools.

The various services perform constant internal communication among themselves with the goal of exchanging messages and complementing their independent performance [16].

6.2 Work Cycle of Independent Activity

Much more interesting and more complex is the interaction during the independent activity cycle. Having in mind that the basic direction for the adaptive
systems is providing educational service which changes in time, we can without
a doubt address the activities in the ongoing process as multi-layered.

The work cycle involves all layers from the presented architecture as well as
all active services which were described in the service layer [17].

The process registers the following activities from the educational service
users:

- Authentication and profiling;
- Activating an assigned/generated problem;
- Active work on the assigned problem;
- Receiving and analyzing feedback

7 Development and Testing

The theoretical concepts, architectural models and software processes reviewed so far meet in the practical applied realization of the software system **UniPlayground** whose basic components were described in detail within the dissertation paper. The practical realization of the software product allows us to answer several important questions such as:

- Whether the project target group needs the specific realization?
- Whether the theoretical apparatus can be adequately applied in the frame of the practical realization?
- Whether there are limitations present in integrating the solutions in the context of the contemporary needs of the medium?

7.1 Road Map for Implementing Prototypes

Work on the **UniPlayground** project began in **2017** before the actual start of the activities related to this current scientific research. The main goal was ideally defined as – *Building a system for assisting teaching programming which can successfully be applied to educational institutions as well as to private professional establishments focusing on the development of software products.*
The main problem that was observed is related to the impossibility for the facilitator of the learning discipline to **monitor in real time** what problems the learners encounter; this in turn creates an impossibility to give timely recommendations or evaluation for changes in the learning curriculum or assigning further materials. Within this initial research several important directions were marked:

- monitoring processes in real time;
- analyzing the assigned program problems;
- historical reference for all decisions made for a given time interval;
- flexible adaptation of the assigned exercises based on their level;
- analytical evaluation of all additional activities;
- giving information to the learner for topics related to their development, addressing certain syntax constructs which are used in solving the problem as well as their semantic wholeness.

7.2 Used Software Products and Solutions

For the realization of the software prototypes the following software products, programming languages and technologies were used:

- JavaScript and TypeScript;
- NodeJS and Electron;
- Oracle Database and Orient Db.

As further external services the following software tools were integrated:

- SQL interpreter;
- SystemTempura interpreter;
- Python SDK/R SDK [1]/Java SDK;
- Amazon Web Service [2].

7.3   Testing

The system for dynamic control of the learning process **UniPlayground** has been in operation since **2018** and in its development went through several easily distinguishable stages. Each one has been accompanied by feedback and statistics for its usage, both from participating lectors and from the students/learners who were actively placed in a situation suggesting working with the presented system [20] [21].

All performed experiments followed a unified approach in organizing learning sessions with the only caveat that the total number of academic hours was chosen in such a way that it fit the curriculum assigned for the specific subject:

- Learning sessions with an average duration of 3 academic hours (120 minutes)
- Combination of gaining knowledge and exercising

Research methodology. During the selection of participants several mechanisms for preliminary profiling were used based on:

- long-term life characteristics;
- motivation.

Each of the profiled participants in the research received a modifier for their internal and external motivation defined as either strong or weak. For the goals of our mathematical model the strong internal motivation is viewed as the qualifier carrying the most weight. We used the three standard stereotypes:

- beginners;
- actively engaged;
- professionals.

Let us look at some of the aspects related to the developed prototypes within the testing activities.

First Prototype. The goal of the first prototype is to validate the basic concepts related to teaching languages and programming media in an arbitrary education institution, namely whether the needed set of tools is intuitive enough and well presented for the learners so that it presents a medium which is flexible enough and easy to use in the context of the conducted courses. It was necessary to answer two basic questions.

1. Which pilot technology should be used and which tools it should be built around?

2. Which is the best model for interactive communication between the platform and the agents engaged with it?

Within the testing process a number of problems were identified which were related to the system's technical capabilities and overall level of resistance of the learners towards some of the features of the educational platform.

Second Prototype. The development of the second prototype is entirely based on the feedback during the first iteration. Some of the obtained results were validated as well as their correlation with the errors made during the first development stage.

Work Prototype. Contains a fully configured version of the medium for software development based on the initial choice of concept we made in integrating a functional set of tools. All of the following prototypes build on the concepts defined in the work prototype which is a fundament for the following processes. In the frame of this testing stage strict rules were implemented for profiling the learners by observing and analyzing the development of the tested groups by stereotypes.

Current Prototype. The final stage in the platform development has to do with adding tools for assisting the process of adapting the user experience based on their behavior. The results we obtained based on the previous exhausting tests gave us enough confidence to stop the addition of functional capabilities in the application toolset and concentrate our efforts on improving the tools for analyzing temporal aspects.

Results Summary. The obtained data from the conducted learning as well as the synthesized feedback give us the basic information for the tendencies we can follow in the development of a certain educational practice in a digital environment. Let us look at the summarized data for some of the remarkable aspects of the activities:

- Characteristics of the conducted subjects;
- Demographics of the research;
- Presence and consumption of video content;
- Interaction and independent activities;
- Experimental and autonomous testing.

8   Conclusion and Obtained Results

In this current scientific research, we examined problems related to adapting and analyzing user behavior based on measurables related to time intervals in the concept of systems which are part of the virtual educational space. The focus of this research is modeling knowledge through practical and experimental activities.

8.1   Perspectives for Future Development

Some of the perspectives for future development include:

- Increasing the existing toolset capacity – through integration of mechanisms close to the standardized processes which are currently popular in the software industry;
- Integrating specialized external tools and laboratory processes – related to niche or exotic technologies, including – blockchain and hardware verification;
- Integrating supplementary software tools such as – debuggers, mobile emulators and virtual machines.

References

1. Casalino, G., Vessio, G., Casalino, G.: Exploiting time in adaptive learning from educational data. Bridges and Mediation in Higher Distance Education, ISBN: 978-3-030-67434-2, Springer International Publishing, (2021)
2. Hoy, M.: Alexa, Siri, Cortana, and more: an introduction to voice assistants, pp. 81–88 (2018)
3. Pressey, S.: A simple apparatus which gives tests and scores - and teaches. Sch. Soc. **586**, 373–376 (1923)
4. Skinner, B.: Teaching machines. Science **128**, 969–977 (1958)
5. Ugander, J., Karrer, B., Backstrom, L., Marlow, C.: The Anatomy of the Facebook Social Graph
6. Park, M., Naaman, M., Jonah, B.: A data-driven study of view duration on youtube. In: Tenth International AAAI Conference on Web and Social Media (2016)
7. Sterling, L.: On composing concurrent logic processes. In: Logic Programming, MIT, pp. 531–545 (1995)
8. Vaisman, A.: An introduction to business process modeling. In: Business Intelligence, pp. 29–61 (2012)
9. Brusilovsky, P.: Methods and Techniques of Adaptive Hypermedia (1996)
10. Hildebrand, C., Schlager, T., Herrmann, A., Häubl, G.: Product gamification. Association for Consumer Research, vol. 42, pp. 664–665
11. Dicheva, D., Dichev, C., Agre, G., Angelova, G.: Gamification in education: a systematic mapping study. Educ. Technol. Soc. (2005)
12. Domìnguez, A., Saenz-de-Navarrete, J., de-Marcos, L.: Gamifying learning experiences: practical implications and outcomes. Comput. Educ. **63**, 380–392 (2013)
13. Guillory, A., Nguyen, H.: Learning executable agent behaviors from observation. In: 5th International Joint Conference on Autonomous Agents and Multiagent Systems (2006)
14. Nah, F., Zeng, Q., Telaprolu, V.: Gamification of education: a review of literature. In: International Conference on HCI in Business (2014)
15. Vikramaditya, J., Fotis, L.: An online virtual learning environment for higher education. In: Games and Virtual Worlds for Serious Applications (VS-GAMES) (2011)
16. Toskov, B.: A system of guards in the virtual educational space. Ph.D. thesis, Plovdiv university "Paisii Hilendarski" (2020)

17. Stoyanov, S.: Theoretical model of a virtual educational space. In: International Conference "From DeLC to VelSpace", Plovdiv, pp. 26–28 (2014)
18. Stoyanov, S., Orozova, D., Popchev, I.: Virtual education space – present and future. In Jubilee International Research Conference "The New Idea in Education", Burgas (2016)
19. Todorov, J., Krasteva, I.: Personal assistant to support the learning process in secondary schools. In: Tech-Co Lovech (2019)
20. Kyurkchiev, V.: Tools for adaptive e-learning. Ph.D. thesis, Plovdiv university "Paisii Hilendarski" (2019)
21. Atanasova, M.: Model of adaptive user interface for Business Information Systems. Ph.D. thesis, Plovdiv University "Paisii Hilendarski" (2018)

Energy-Driven Software Engineering: A Tutorial

Ana Oprescu[1(✉)], Kyrian Maat[1], Lukas Koedijk[2], Sander van Oostveen[1], and Stephan Kok[2]

[1] University of Amsterdam, Amsterdam, The Netherlands
{a.m.oprescu,k.maat}@uva.nl
[2] KPMG, Amsterdam, The Netherlands
{Koedijk.Lukas,Kok.Stephan}@kpmg.nl

Abstract. The EU Climate Law targets climate neutrality by 2050. At the same time, the Knowledge Economy is deemed crucial to EU prosperity. Thus, the energy footprint of software services remains an important topic of research and education. In these lecture notes, we set out to inspire students and fellow lecturers on energy-driven software engineering education. We describe relevant background theory, and suggest further reading material. Students can organise their own self-paced learning activities, while lecturers can adapt some of our proposed activities to their existing teaching material.

We show how to assess whether established software smells (anti-patterns), such as 'Long Function', 'Large Class', and 'Duplicate Code', are also energy-related smells. We illustrate two approaches: *removing* code smells from the software project under analysis and *introducing* code smells in the software project under analysis. We summarise research findings on several software projects.

Based on research on programming languages and software smells, we design a tutorial for students and practitioners alike that allows them to discuss and reason about these topics. In this tutorial, students showcase an energy-aware mindset while programming and broaden their knowledge regarding energy consumption and the choice of tooling. In particular, we teach them how to establish whether some programming languages are inherently more energy-friendly, and whether *how* a computer program is written has any impact on the energy consumption. We summarise previous research findings on Java, JavaScript, Python,

---

This work received financial support through the Erasmus+ Strategic Partnership for Higher Education *SusTrainable—Promoting Sustainability as a Fundamental Driver in Software Development Training and Education* (project number 2020-1-PT01-KA203-078646), funded by the European Union and coordinated by the University of Coimbra, Portugal.
The information and views set out in this publication are those of the authors and do not necessarily reflect the official opinion of the European Union. Neither the European Union institutions and bodies nor any person acting on their behalf may be held responsible for the use which may be made of the information contained therein.

PHP, Ruby, C, C++ and C#. We make a selection of programming languages and programs from The Computer Language Benchmarks Game and illustrate how the experimental set-up can be used to identify the programming languages that consume the least amount of energy over all selected problems.

Following the tutorial, we identify interesting future research-based educational directions which include investigating the impact of the difference at the level of (special) programming language constructs on the energy consumption.

Keywords: green programming languages · energy-related software smells · energy footprint of software services

1   Introduction

Global warming is partly the result of the emission of greenhouse gases during energy generation of conventional energy options [28]. One solution for this problem is the renewable energy approach. However, this would still fall short on addressing climate change issues[1]. A complementary solution is to decrease the energy consumption, which is both good for the environment, and could help save significantly on the energy bill.

The energy consumption of communication networks, personal computers and data centers worldwide is increasing every year [29]. This occurs at a growth rate of 10%, 5%, and 4%, respectively [33], therefore, it is important to research ways to decrease energy consumption. One approach relies on manufacturing more energy efficient hardware: Koomey's law [16] describes a doubling of the number of computations per Joule every 1.57 years, a trend that slowed down to every 2.6 after the year 2000. However, the hardware efficiency approach is insufficient because software is increasingly slower, an observation formulated by Wirth in 1995 [35] and later revisited in an industrial setting [34]. Therefore, we need to investigate opportunities to decrease the energy consumption from a software perspective [30]. The software perspective can include a diverse range of views, such as developers awareness and knowledge about software energy consumption [8,23,26], tooling problems [10], production pipelines [7] and maintenance requirements [21].

In the context of a Knowledge Economy that relies on digital services, meeting the sustainability goals of the European Union should be a key aspect of software development[2]. While there is a growing body of related knowledge in both academia and industry, and textbook efforts are beginning to emerge [17], many academic level educational programs are in need of developing a stronger sustainability trajectory in their curricula [31]. In particular, energy-aware software

---

[1] https://www.forbes.com/sites/michaelshellenberger/2019/09/04/why-renewables-cant-save-the-climate/.

[2] https://www.greendigitalcoalition.eu/assets/uploads/2022/02/EGDC-declaration-to-sign_v2.pdf.

engineering could be introduced as one of the default topics in ICT education. This competency does not have to be necessarily graded in terms of how 'green' the final product is, but rather on how in-depth the analysis of the development options in terms of sustainability is [27].

There are various areas of attention for energy-aware software engineering, such as monitoring the application behaviour, software maintainability, and programming language choices. In these lecture notes, we focus on programming languages (tooling) choices. By investigating the effects of different techniques on energy consumption, one will be able to find potential trade-offs to lower their total energy consumption to the required needs.

1.1   Energy Footprints of Programming Languages

Based on research conducted by Koedijk et al. [14], we address the educational gap and introduce students to the concept of energy-aware programming by exploring together two aspects of energy-aware choices regarding programming languages. Ideally, this activity should familiarize the students with the methodology to determine whether it is the programming language or the application that dictates the magnitude of energy consumption; of course, they could both be equally dominant.

To this end, students are guided by the following questions during the creation of software projects in their courses in terms of making a conscious decision about which language and structures their software projects would have. We discuss these questions at length in Sect. 3.

Q1.1 Is there a difference in the energy consumption of software projects in different programming languages that have the same functionality?

Q1.2 Is there a difference in the energy consumption of different software projects (using the same programming language) that have the same functionality?

1.2   Language Constructs and Their Impact on Energy Consumption

An interesting research angle is to find a correlation between the energy consumption of a program and the language constructs used in this, such as for-loops and while-loops. This concept follows from the investigations on different programming languages conducted by Pereira et al. [25] and Koedijk et al. [14]. For instance, in some cases such as Java 8's javac (1.8.0_65), the code is not the same for the conditional operator and the if-else construction. Students can investigate such language constructs to further build their analysis-based thinking on energy consumption. We formulate the following question for our tutorial and discussed at length in Sect. 3:

Q2 Is there a significant impact on the usage of the if-construction over the ternary construction in terms of energy consumption?

1.3 Energy Smells: Code Smells with Respect to Energy Consumption

Code smells are established as undesirable code patterns identified several decades ago by Martin Fowler [9], with a focus on readability and extensibility. The term has become generally applicable to undesirable behaviour, and we apply it to code patterns that would lead to a more energy hungry behaviour as'energy smells'. We thus investigate the impact of code smells on energy consumption.

Based on research conducted by Kok [15] and Oostveen [20], we introduce the students to two approaches to 'energy smells' analysis: *removing* established undesirable code patterns from the software project under analysis and *introducing* established undesirable code patterns in the software project under analysis. We formulate the following questions and discuss them at length in Sect. 4:

Q3.1 What is the impact of refactoring code smells on the energy consumption of Java based open-source software projects?
Q3.2 Is the impact of a code smell on the energy consumption of a software project different between programming languages?

1.4 Intended Learning Objectives

To initiate the integration of energy-aware software engineering in ICT education, and thus address the educational gap, we formulate the following intended learning objectives (ILOs):

ILO_1 Students can investigate whether programming languages have different energy footprints.
ILO_2 Students can conduct an analysis on (special) programming language constructs to identify different energy footprints for different types of statements.
ILO_3 Students can establish whether energy smells exist and whether their impact depends on the programming language under observation.

These three learning objectives encapsulate multiple facets of learning about the energy consumption in software programming. On a higher-level, comparing the energy consumption of the different programming languages that could be used can give a first indication of how much energy the software project will consume. Zooming in several selected programming languages, the analysis of code smells and energy smells will lead students to review their own programming habits to lead to the most energy-aware programming style. Lastly, we can have students scope in on specific language constructs and determine if they would impact the overall cost of their project.

These learning objectives serve as a basis for the content and exercises we present to the students. The following activities are associated with our ILO_s: exercises that allow students to gain insights into the differences between programming languages, such as C and Java, in terms of energy consumption; exercises that require students to make modifications themselves to test the impact of programming language constructions; exercises that require students to conduct an analysis on the impact of different code smells in our tutorial set-up.

We assess how well students achieve the ILO_s by orally discussing how their results compare to those reported by our own research as well as by previous research efforts, such as Pereira et al. [25] and Koedijk et al. [14].

Practical Considerations. The proposed structure of this tutorial comprises a lecture where general concepts are visited and specific research findings are explained. The lecture is followed by lab activities, where specific research findings are leveraged such that students can engage with the hands-on material. Overall, this structure would benefit from a half-day timeslot. Assuming there are sufficient tutorial team members and sufficient hardware available, the structure of this tutorial scales quite easily.

1.5   Outline

These lecture notes are organised as follows. We summarise two energy measurement approaches as well as background on how to process the results towards hypothesis testing in Sect. 2. We provide concise research background on the energy footprint of programming languages in Sect. 3. In Sect. 4 we present a consolidated view of our research efforts to bridge the research gap on the topics related to ILO_3, i.e., the so-called energy smells. We describe the educational activities related to all ILO_s in Sect. 5. We give a concise experience report of the first tutorial ran in practice in Sect. 5.4. We conclude in Sect. 6.

2   Energy Measurement Approaches

Table 1 summarises the energy measurement approaches that we use in our research and educational activities. Both external measurements on the physical device and software model approaches are needed, to understand the sensitivity of the problem, as software models are perceived as coarser-grained, albeit more accessible than external measurement set-ups. A very popular approach to software models is the RAPL-based measurement library which has been shown to estimate power consumption quite accurately [4]. Utilizing direct power distribution unit (PDU) measurements can on the other hand be very useful to

Table 1. Energy measurement approaches used in our research and tutorial

	Current system power	Cumulative
External measurements: accurate, expensive set-up https://github.com/lukaskoedijk/Green-Software Reactivity PDU based framework	✓	✓
Software models: less accurate, more accessible https://github.com/sandervano/GreenCodeSmells RAPL-based C implementation depending on architecture		✓

understand how to gauge the sensitivity of an application to the measurement granularity.

2.1　Software Models

An internal software model is often more easily available but at the cost of accuracy. Such models often make estimations of the energy consumption of the system over a small period of time. While they allow estimates of the consumption of different parts of the system, they may not allow estimates for all parts of the system. As such, the available models and system parts they can measure must be taken into account when using this method of measurement. Extra care can be taken in selecting the code base to only include software projects that make intensive use of only the parts of the system covered by the model. These parts tend to be split between the CPU, memory, and disk. An additional limitation of software models is that they might not always give estimations for all components, such as networking components, or have limited accuracy on aspects such as integrated GPUs.

A common example of a software model is the Running Average Power Limit (RAPL). The hardware performance counters used by RAPL are available on most modern Intel processing architectures, and some modern AMD chipsets use their own version of RAPL that can use potentially different hardware performance counters. Despite these differences, both implementations use hardware counters allowing for the estimation of both the cumulative energy consumed, and the current system power draw. On Linux, there are limited methods for reading these counters that work for both Intel and AMD chipsets. Windows and MacOS are also limited in tools and options for estimations via software models. The estimates are based on readings from a mix of on-chip hardware counters and I/O models [4].

The hardware performance counters used by RAPL are split into several domains to allow for specific readings of the different parts of the processor and memory, as shown in Fig. 1. The top domain is the package, containing readings of the energy and power consumption of all the underlying powerplane domains and any other uncore devices located on the processor. The packages together form the entire processor. Inside the package domain is the core powerplane (PP0). This provides collective readings of the energy and power consumption of the processor's multiple cores. Next is the uncore power plane (PP1). This domain contains readings on the consumption of the integrated graphics (GPU) or other uncore devices, but is not always supported by every chipset, depending on which RAPL-required hardware counters are available; as this can differ for processor models, a user would be required to find which RAPL model-specific registers (MSRs) are present to see if their specific needs are fulfilled. Figure 1 shows that the energy and power consumption of the Package is equal to or greater than the energy and power consumption of PP0 and PP1 combined. Lastly, the DRAM domain provides readings for the memory, separate from the processor and thus also not part of the Package zone. These different components are interesting for student estimations on which parts of the software projects

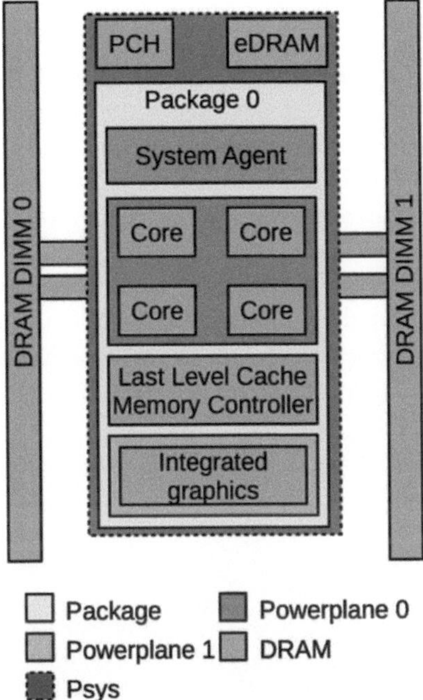

Fig. 1. This diagram illustrates the different domains of RAPL [12]. The Package contains readings of the energy and power consumption of the entire processor, including powerplane 0 (PP0) and powerplane 1 (PP1). Powerplane 0 includes readings for the cores on the processor, and powerplane 1 for an on-core GPU. Psys can be different per platform architecture, but usually comprises the entire system-on-a-chip (SoC). DRAM provides readings for the system memory, and is not included in the package zone.

they should prioritize to improve upon and see what contributes most to the energy consumption of the analysed software projects.

2.2   External Measurements on Device

An external measuring device, such as a power monitor, allows for accurate measuring of the power and cumulative energy consumption of a given system. As these devices are connected through the power plug, they can directly measure the energy consumed by the system without impacting the system itself. The downside of such a method is the need for additional hardware that is reasonably costly and so perhaps available in dedicated research environments, such as universities, and not accessible to the average student/practitioner. Furthermore, using such a device requires an additional setup to integrate its measurements with the framework. Most available external devices allow for the measuring of both the cumulative energy consumed and the current system power, although with varying degrees of accuracy.

2.3 Dealing with Anomalies

In testing with both the external measurements and the software models we encounter many data points, some of which could be classified as anomalies and which need to be addressed appropriately. In this study, we label as anomalies data points which are not intrinsic to the energy consumption behaviour of the observed programming language, such as a potential malfunction in the hardware, or spurious short-lived background processes. The presence of such points can skew the analysis of the results. Clustering algorithms are sometimes suitable for detecting anomalies. By forming clusters of the general data points, any points outside the cluster could be labelled as anomalies.

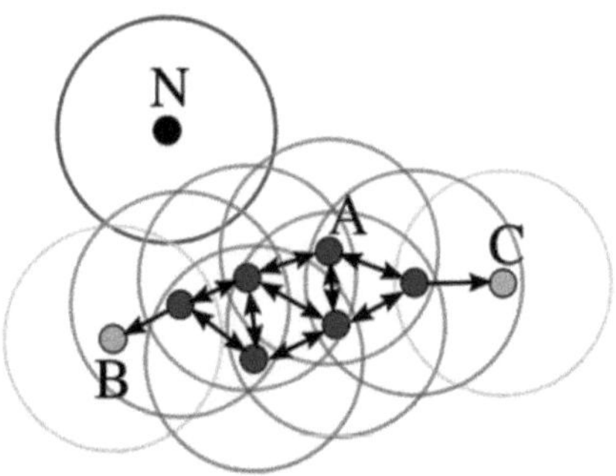

Fig. 2. Cluster labelling by DBSCAN [6,11] with minimal neighbors set to four. Here, the red dots are core points, the yellow dots are border points, and the blue dot is an anomaly. Within the fixed area, core points have at least the minimum amount of neighbors, while border points have less than the minimum amount but do share a border with a core point. Anomalies do not have the minimum amount of neighbors, nor do they border a core point.

One such algorithm is DBSCAN [6], shown in Fig. 2. The algorithm counts the amount of points in a predefined area around the given point. If this amount is equal to or greater than a predefined minimal amount, this point is labelled as a core point. Any point within the predefined area of a core point is a part of the cluster formed by the connected core points. If a point is connected to a core point, but does not have the minimal amount of neighbors, this point is labelled as a border point. A point not connected to any core points is outside of the cluster, and is considered an anomaly. If it does not form any connections with one of these clusters, we deem it a measurement error and remove it.

Figure 3 shows an example of the results that can be obtained using this approach. We observe in the top graph the distribution of measurements for a single program. The red crosses are the measurements DBSCAN labels as an anomaly. The bottom graph shows from every measurement point its distance to its fourth nearest neighbour. This is sorted and the red dot is the choice of the input variable of DBSCAN called *eps*, calculated by finding the first valley.

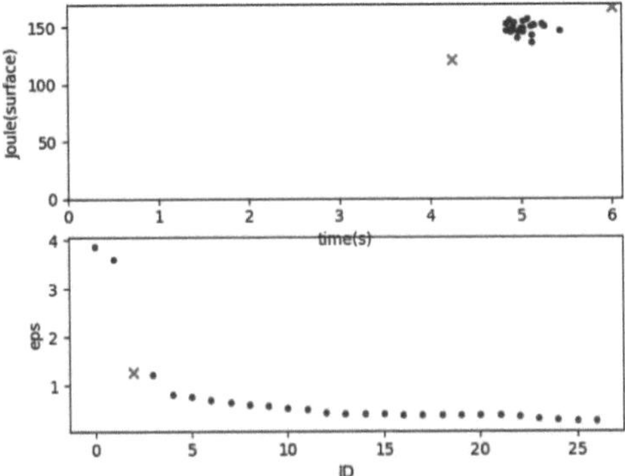

Fig. 3. The top graph shows the distribution of measurements from a program corresponding to a given problem executed on one of the physical nodes. The bottom graph shows the sorted fourth nearest neighbour graph. The top left corner dots are measurements labelled as an anomaly and at the bottom is the choice for the *eps*-value.

Choosing the Parameters for DBSCAN. The DBSCAN algorithm can be used for anomaly detection. The two parameters of the DBSCAN must be chosen based on the data it is used on. According to Ester et al. [6], we can set the minimal number of points neighboring a core point for two-dimensional data, including itself, at four. The radius of the area in which the minimum amount of neighbors must be is called the eps distance. Based on the minimal number of neighbors, we can determine the eps distance by finding the distance to the fourth nearest neighbor for every sample on our data. If we rank these distances from greatest to least, our optimal eps distance can be found as the first knee point in the rankings. The knee point (or elbow point) of the distances is the point where the curve formed by the ranked distances visually bends from a steep slope to a nearly flat slope. Figure 4 and Fig. 5 show an example of choosing the eps distance and respectively the cluster found by DBSCAN for a given set of measurements.

2.4 Statistical Significance

After detecting and removing anomalous data points, the results can be analysed to find answers to the research questions that motivated the energy measurements. For example, we analyse the impact on the energy consumption due to injecting code smells by statistically testing whether the energy consumption distribution observed for the original set of programs and the distribution observed for the set of programs where code smells had been introduced are equal given a problem written in a language. Additionally, in the same example, we may wish to know the direction of the impact: is the original version consuming more energy or less than the smelly one?

The aim of a statistical test is to reject a null hypothesis in favor of an alternative hypothesis. By rejecting the null hypothesis we can claim that the

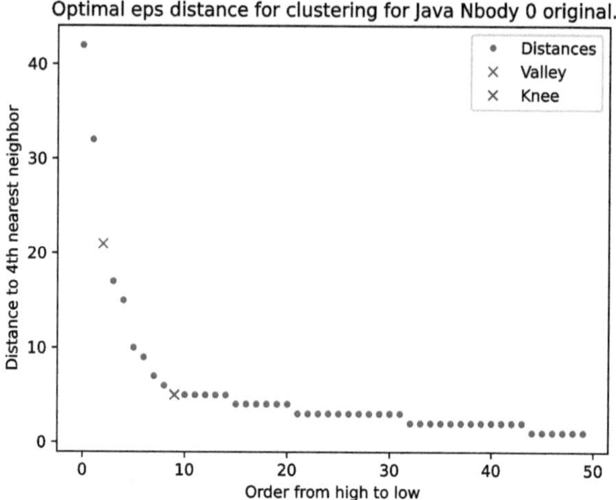

Fig. 4. The ordered distances to the fourth nearest neighbor of a cluster, plotted over its rank in the order. To determine the optimal eps distance, both the first local valley and the first knee point are determined. While the first knee point gives better accuracy, this cannot always be determined from the data, in which case the first local valley is used.

alternative hypothesis is true. For this purpose, a probability called the p-value is computed by the statistical test. If this p-value is smaller than a predetermined significance level ($\alpha - value$), we may reject the null hypothesis. The significance level indicates the probability of rejecting the null hypothesis although it is actually true. Typically, the significance level is set at 0.05.

Thus, when comparing different programs, that are in the same programming language and have the same functionality, the significant difference regarding the energy consumption can be determined by testing the null hypothesis that the two corresponding set of measurements are from the same population. A commonly used test with this null hypothesis is the students t-test. This test, however, has the preconditions that a) the two distributions should follow the normal distribution and b) the variance of the two distributions should be equal. If the data does not match these preconditions we cannot use this test. An alternative is the Mann-Whitney U test.

The Mann-Whitney U Test. The Mann-Whitney U test [18] looks at the probability that a sample from one distribution is greater than a sample from another distribution. If this probability is at 50%, the two samples belong to the same distribution. This test fits the requirements for comparing two programming languages' energy consumption, as we try to see if they are significantly different or not. Because we cannot be certain about the normality of the distributions and the variance, we use the Mann-Whitney U test. There are two Mann-Whitney U tests: The one-sided and the two-sided tests. While the two-

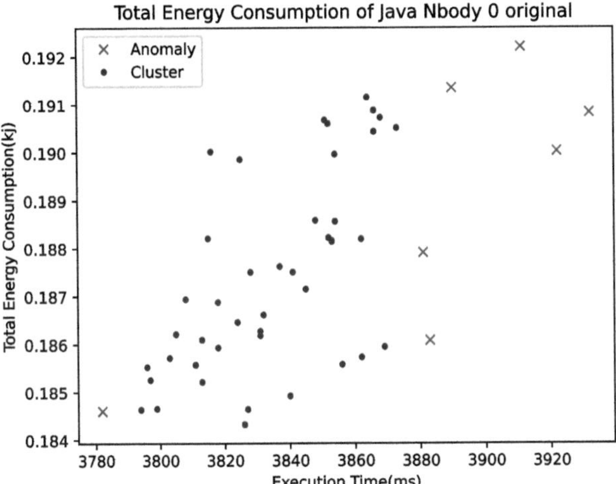

Fig. 5. The cluster found by DBSCAN for a given set of measurements. The blue samples are labeled as part of the cluster while the red crosses are detected anomalies. (Color figure online)

sided test has as an alternative hypothesis claiming that the two samples are from different distributions, the one-sided test has as alternative hypothesis that says the first sample is from a greater distribution [19].

To find the direction of the difference between the distributions, we must use the one-sided test twice. We use a one-sided Mann-Whitney U test with the null hypothesis that a sample from the first distribution is larger than a sample from the second distribution; we refer to this test as *greater*. We also use a one-sided Mann-Whitney U test with the null hypothesis that a sample from the first distribution is smaller than a sample from the second distribution; we refer to this test as *lesser*. If the p-value of the *greater* one-sided test is smaller than our significance level, we can say that the first distribution is statistically significantly *smaller* than the second distribution, and vice versa for the *lesser* one-sided test. If additional guarantees are required, then one can also use the Kolmogorov-Smirnov test adjoint with the Mann-Whitney U test, similar to how Koedijk [14] treated the results of the experiments on the programming languages.

3   Energy Footprints of Programming Languages

The energy footprint of programming languages has been investigated by several efforts, spanning different programming languages and paradigms, and different problem types [13, 25, 34]. In these lecture notes we leverage the study by Koedijk and Oprescu [14].

In the original study, the following programming languages are selected: Java, JavaScript, Python, PHP, Ruby, C, C++ and C#. The target software projects represent several problems, each problem implemented in each programming language in several ways. The Computer Language Benchmarks Game[3] is used to retrieve implementations of selected problems in the selected programming languages. The current system power measurement using specialized hardware[4] is used to perform the energy measurements.

For one of the physical machines (node) in the experimental setup there is a total of 345 anomalies from 7263 measurement points, which is roughly 4.8%. For the other node, there is a total of 305 anomalies from 5918 measurement points, which is roughly 5.1%. Although there are more anomalies found than programs, not all programs yielded anomalies. All the anomalies found are removed from the results before further processing.

Two programming languages are compared by statistically testing the difference in energy consumption between programs solving the same problem in each programming language. The one-sided Mann Whitney U test [19] is used twice on the measurements of the programs. This is calculated for every language with every other language and a table is created for every problem. Table 2 shows the comparison of different languages for the NBody problem. In such a table, a + means that the programming language on the row is performing better than the programming language on the column, i.e. the programming language on the row consumes less energy. The - means the opposite, a *0* means equal and *unknown* means that both one-sided Mann Whitney U tests could not be rejected.

Each such table shows which language is performing better compared to others for a given problem, thus there are seven tables for each physical machine (node) used in the original study [14]. To give a total overview of problems and

Table 2. The comparison of the different languages for the NBody problem on Distributed ASCI Supercomputer (DAS-5) *node28*. A + means that the language on the row consumes less energy than the language on the column, the opposite for -, and the *Unknown* means that we could not reject the null hypothesis. Table obtained from Koedijk and Oprescu [14].

	Java	JavaScript	Python	PHP	C#	Ruby	C-flags	C-noflags	C++-flags	C++-noflags
Java	0	+	+	+	+	+	-	+	-	+
JavaScript	-	0	+	+	+	+	-	+	-	+
Python	-	-	0	-	-	-	-	-	-	-
PHP	-	-	+	0	-	-	-	-	-	-
C#	-	-	+	+	0	+	-	+	-	+
Ruby	-	-	+	+	-	0	-	-	-	-
C-flags	+	+	+	+	+	+	0	+	Unknown	+
C-noflags	-	-	+	+	-	+	-	0	-	+
C++-flags	+	+	+	+	+	+	Unknown	+	0	+
C++-noflags	-	-	+	+	-	+	-	-	-	0

[3] https://benchmarksgame-team.pages.debian.net/benchmarksgame/index.html.
[4] https://github.com/lukaskoedijk/Green-Software.

programming languages across all problems, a score is calculated for every pair
of programming languages on each node as follows:

1. One point is rewarded when there is a plus,
2. One point is subtracted when there is a minus,
3. Nothing is added nor subtracted in the case of a zero or unknown.

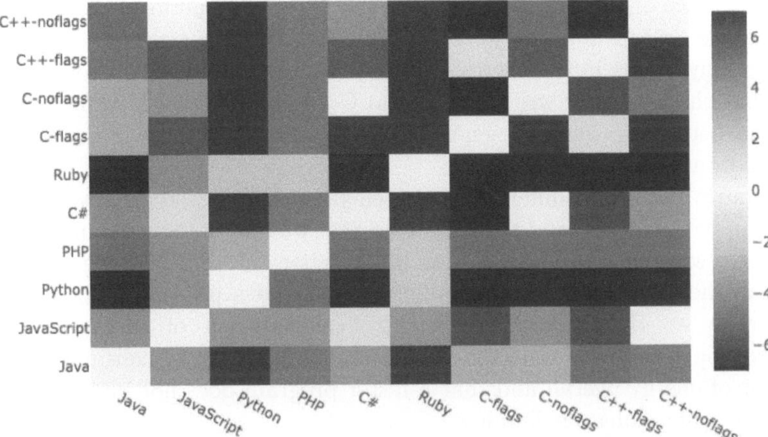

(a) Programming languages comparison heatmap. Green means a lower energy consumption for the row's programming language than the column's.

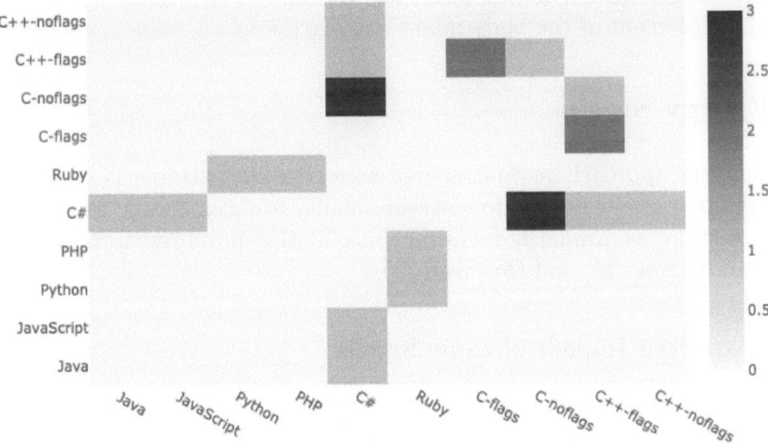

(b) Heatmap of the amount of times the null hypothesis could not be rejected.

Fig. 6. Energy consumption across multiple programming languages and applications.
Plots obtained from Koedijk and Oprescu [14].

A heatmap is created based on these scores, such as in Fig. 6a, where green
means a high score (row's programming language outperforming energy-wise the

column's programming language) and red a low score (underperforming). There is also a companion heatmap indicating for which pair of programming languages *unknown* occurred the most, created by adding a one for every *unknown* in the corresponding cell in each of the seven tables for that node. Thus, Fig. 6b shows the amount of times the Null Hypothesis could not be rejected for each pair of programming languages.

Findings from Koedijk and Oprescu [14]. Comparing the programming languages to each other shows that C-flags and C++-flags perform the best regarding the energy consumption for every problem. We also expect the compilation flags to play a part in the energy consumption of C and C++. The results indicate that the programs written in C and C++ compiled with compilation flags perform better regarding the energy consumption than those obtained without compilation flags.

Because there is a difference between programming languages that have a pre-compilation phase and those that do not, we also analyze the programming languages without a pre-compilation phase separately. JavaScript performs better than the other programming languages without a pre-compilation phase.

Pereira et al. analysed the energy consumption of programming languages [25]. They find that the programming language C consumes the least amount of energy overall and that a faster program does not imply less energy consumed, matching our findings.

When comparing two almost identical programs, we find that a for-loop uses less energy than a while-loop. We also compare the difference between having the body and condition of an if-statement on the same or different lines and find that the placement of the body might not matter.

4   Energy Smells

A promising approach to find energy-hungry code patterns is to investigate whether code smells related to software quality are also energy smells [1–3, 22]. In this section, we primarily focus on consolidating novel research conducted in our team by Kok [15] and Oostveen [20].

4.1   Expected Impact of Code Smells

Based on our expectations [15, 20] regarding the impact refactoring would have on the energy consumption in Java for the 24 code smells identified by Fowler [9], we select the following code smells to investigate in relation to Q3.1 and Q3.2 (Sect. 1.3):

Long Function: Typically refactored using "Extract Function"; the newly extracted function adds an extra function call on the stack. This is expected to increase the energy consumption. However, certain compiler optimisations could remove this overhead by in-lining the code. Thus 'Long Function' would either increase the energy consumption, or would have no impact on compiled languages. Confirming these expectations, Dhaka et al. [5] and Kok [15] for Java, and Park et al. [24] for C++ found that 'Extract Function' increased energy consumption. Inversely, injecting the 'Long Function' code smell to a software project is expected to decrease the energy consumption.

Large Class: Typically refactored using "Extract Class"; when extracting a class we must account for an overhead of an extra class and the overhead for the performance optimisations. As it must be possible to create new objects from this class, it has to stay in memory. As class logic is split, only the required objects must be created, reducing the required memory, however there might be additional message traffic to access data. We expect that this would overall increase the energy consumption.

Duplicate Code: Typically refactored using "Extract Function"; it is expected that extracting the function will have a similar effect to refactoring 'Long Function' and increase the energy consumption, or have no impact. However, as the amount of code is decreased, this also decreases the amount of code needed to be compiled or interpreted. Further compiler optimisations could also lead to the more used function to be cached which could further improve performance. This is expected to decrease the energy consumption.

To determine this impact, a code smell must be selected to be featured in the research. The selection of this code smell is based on its expected impact, and the availability of tools used for detection and refactoring.

4.2 Refactoring Code Smells to Assess Their Energy Footprint

The energy footprint of code smells can be investigated by measuring the impact of established refactorings for the selected code smells on the energy consumption [15]. The impact of refactorings is measured in terms of software metrics via BetterCodeHub[5]. The Java programming language and the JDeodorant code smell detection tool [32] are selected for this study. The energy consumption of code is measured by execution via the JUnit test suite. The selected software projects have a solid testing framework with high code coverage, are open source and medium sized: `FasterXML/java-classmate`, `apache/commons-lang`, `apache/commons-configuration`. The energy measurement approach uses current system power measurement using specialized hardware[6].

There is an increase in energy consumption after refactoring for 'Long Function' and 'Large Class'. Additionally, when refactoring for 'Duplicate Code' there is both an increase and decrease in energy consumption. In a nutshell, developers can safely write quality code while at most negligibly increasing the energy consumption of the code base; sometimes it might even yield a slight decrease in the energy consumption.

---

[5] BetterCodeHub has been replaced by Sigrid – https://bettercodehub.com/.
[6] https://github.com/lukaskoedijk/Green-Software.

Table 3. Overview of relative change in energy consumption, which had a statistically significant impact. The Null Hypothesis – that the sample of refactored measurements and the sample of original measurements are from the same distribution, is rejected if either p-value is less than the α-value of 0.05. Table obtained from Kok [15].

Refactoring	Software Project		
	Classmate	Lang	Configuration
Long Function	+1.06%	+1.52%	+3.29%
Large Class	–	+1.35%	+1.07%
Duplicate Code	–	+2.12%	−1.89%

4.3   Inflicting Code Smells to Assess Their Energy Footprint

Another path to assess the overlap of energy smells and traditional code smells is to introduce the latter in a controlled manner in existing code bases [20].

The following programming languages are selected: Java, Python, C++. In terms of software projects, several problems are selected from the Computer Language Benchmarks Game[7]: Fasta (memory-intensive) and NBody (CPU-intensive), and from Project Euler[8]: Su Doku with a DFS approach (CPU-intensive). Each problem is implemented in each programming language. In each project the same code smell is introduced, namely 'Long Function'. The impact of introduced code smells in terms of software metrics is measured via BetterCodeHub[9].

Figure 7 shows the structure of the software projects implementing the Fasta problem before and after adding the code smell. The 'Make Repeat Fasta' function is removed and its function body is pasted inside the Main function. This increases the length of the 'Main' function and removes one function from the call stack. Any arguments given to the 'Make Repeat Function' are now initialised in the 'Main' Function to prevent issues with the new scope of the function body. None of the functions that are used multiple times are altered, moving these function bodies would result in the 'Duplicate Code' code smell.

A similar method is used to add the code smell to the software projects solving Nbody. As seen in Fig. 8, the 'Main' function's body is removed and added to the 'Make Nbody' function. Additionally, the 'Nbody Advance' function is removed and its function body is added to the 'Make Nbody' function. This increases the length of the 'Make Nbody' function, and removes one function from the call stack. Any arguments given to the 'Nbody Advance' Function are now initialised in the 'Make Nbody' Function. The 'Nbody Energy' function is not altered, as this one is used twice, to prevent adding in the 'Duplicate Code' code smell.

---

[7] https://benchmarksgame-team.pages.debian.net/benchmarksgame/index.html.
[8] https://projecteuler.net.
[9] Sigrid has replaced BetterCodeHub - https://bettercodehub.com/.

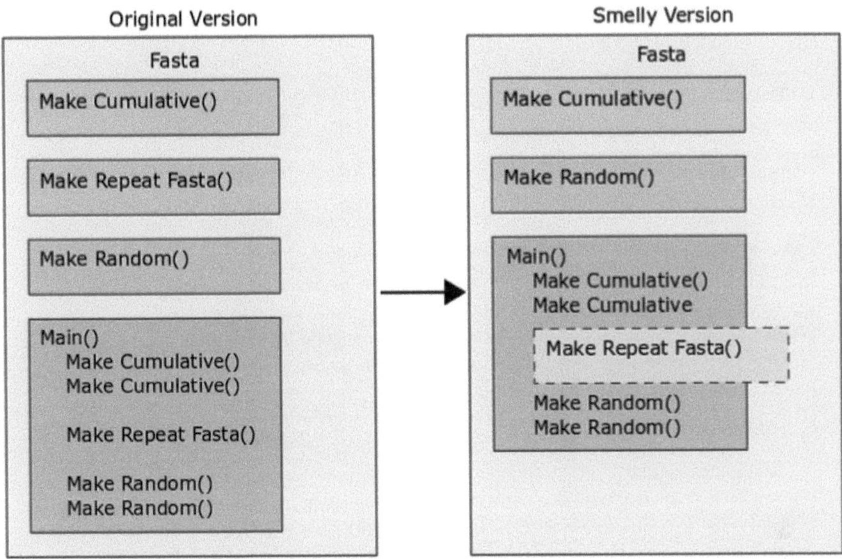

Fig. 7. The Make Repeat Fasta function body is added to the Main function. Diagram obtained from Oostveen [20].

As in the previous cases, 'Long Function' is added to the Su Doku problem by combining several functions into a single, long, function. As shown in Fig. 9, the function body of the 'Is Sudoku Valid' function is moved to the body of the 'Is Sudoku Solved' function. Three hardcoded Su Dokus are added to the 'Main' function, to be solved sequentially by the algorithm. This is done as not all found implementations contained methods to read Su Dokus from user input, and is not changed between the original and smelly counterparts.

For each of the selected languages we investigate the significance of the findings. The one-sided Mann-Whitney U test is used as explained in Sect. 2.4 to determine the probability that a sample from the smelly measurements is greater or lesser than a sample from the original measurements. If the p-value for the comparison is smaller than the significance-value(α-value) of 0.05, then the distribution of smelly samples is significantly greater or smaller than the original samples. If neither p-values are smaller than the α-value, then we cannot reject the Null hypothesis that the two sets of samples are from the same distribution.

The key observations from the experiments can be found in Table 4, where we can see the presence of the 'Long Function' code smell increased the energy consumption of software projects written in different programming languages. The impact of the 'Long Function' code smell on the energy consumption of a software project is different between *different programming languages*, as well as different between different problems written in the *same programming language*. Moreover, the similarity between Java and Python for the Su Doku problem

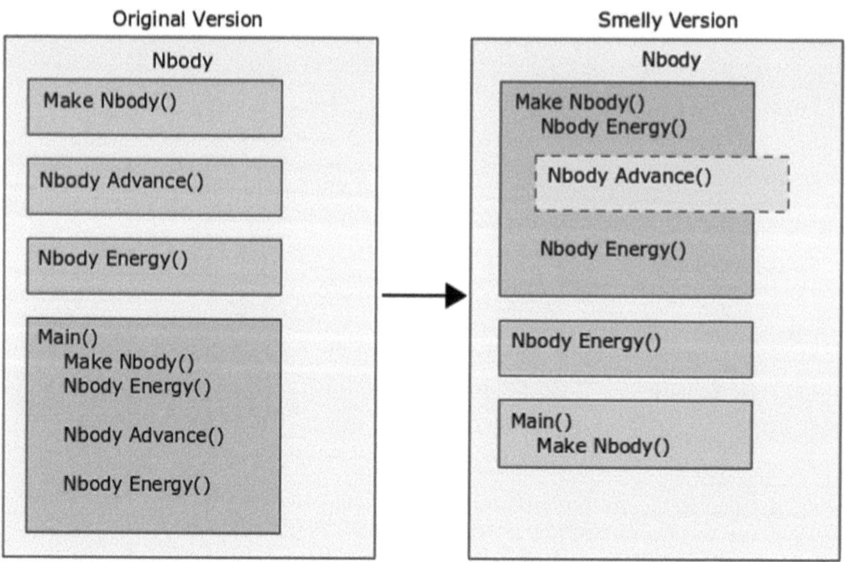

Fig. 8. The Main function body and the Nbody Advance function body are added to the Make Nbody function. Diagram obtained from Oostveen [20].

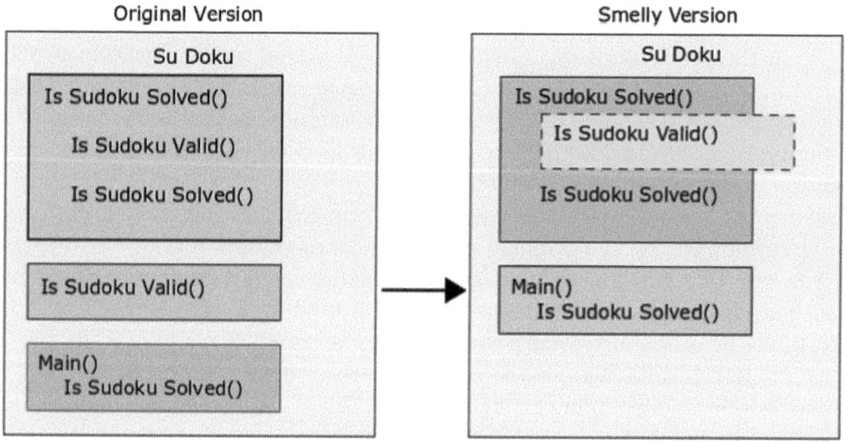

Fig. 9. Diagram of the general structure of the Su Doku problem before and after adding in the 'Long Function' code smell. The dark gray blocks indicate a function in the software project, and the lighter gray dashed box indicates the a function body added to another function. Diagram obtained from Oostveen [20].

warrants further research, particularly in the context of the findings summarized in Table 3.

Table 4. The percentage of significant increase in the mean energy consumption of the smelly version of a software project compared to its original counterpart, for different languages. A negative percentage implies a decrease in the mean energy consumption, while a '-' means that no significant difference was found. The Null Hypothesis is rejected if the p-value is less than the α-value of 0.05. Table obtained from Oostveen [20].

Problem	Language		
	C++	Java	Python
Fasta	-	1.18%	0.22%
Nbody	3.29%	6.14%	−1.71%
Su Doku	-	113%	149%

5 Educational Activities Related to the ILO_s

Section 2 describes the typical work that students should be tackling to properly investigate the effects of different facets of a software project in relation to energy consumption. Not all steps have to be replicated, but the students should be aware and understand the concept and implications of using, for example, a software model over a measuring device. In the following, we describe the goals and objectives for the educational activities.

We group our educational activities organised per learning objective:

ILO_1 investigate whether programming languages have different energy footprints.

ILO_2 investigate whether programming language constructs have different energy footprints.

ILO_3 establish whether energy smells exist and whether their impact depends on the programming language under observation.

ILO_1 is formulated on the basis of common research ground identified in Sect. 3, while the findings related to the for-loop [14] support formulating ILO_2. Section 4 supports formulating ILO_3.

5.1 Lab Exercise Setup for ILO_1

We first briefly explain the different problems to familiarize the students with the concepts and the codebase. The programs could all be easily run on lab computers pre-installed with **perf**, where **perf** is a tool able to use the RAPL energy consumption measurements via kernel events. For their own experiments, students are encouraged to use these pre-installed computers, but could install the set-up on their laptops to take them home.

The starting set-up is as follows:

1. CPU-intensive problem `fannkuchredux` with the following options: C without any optimization flags, C with all optimizations from the Computer Language Benchmarks Game (CLBG), Java and Javascript. For all

of the C optimization flags in this compilation and the following, the flags *-O3*, *-fomit-frame-pointer*, *-march=native* are used, and for C in fannkuchredux *-fopenmp* is also used.

2. Memory-intensive problem `fasta` with the following options: Two C programs, one with openMP parallelization and one without parallelization. With these 2 programs, we have 4 possible C compilation options. Non-Parallel no optimization in C, on-parallel with the CLBG optimizations in C, parallel no optimization in C, parallel with optimizations from the CLBG in C, Java and Javascript.

3. Disk-intensive problem `revcomp` with the following options: C no optimization, C with optimizations from the CLBG, Java and Javascript.

The different compilation options allows the students to get a general overview of how a better compiled program or a better written program in terms of parallelism can affect the energy consumption of the program itself. The students would use this baseline of options to then build upon with experimentation on replacing if-else constructions with ternary operators.

To perform the energy measurements the perf program is used which measures the cores, ram, package domain (pkg) and psys (if available) layers of RAPL. This allows users to see exactly where energy is consumed by reading the RAPL counters. To observe the counters the students run the following command in their terminal:

```
$ perf stat -e power/energy-cores/,power/energy-ram/,
    power/energy-pkg/,power/energy-psys/ ./PROGRAM
```

In this prompt the `perf` energy measurement command is used to allow students to see the energy usage of the `cores/ram/pkg/psys` domains on a given `PROGRAM`. We need to install `perf` on the computers used for the experiment and activate the appropriate kernel_event to allow RAPL measurements. On set-ups without `psys` available, we remove the psys part of the prompt.

To make the experience as smooth as possible for students, all the compilation options and the energy measurement calls are easily accessible from scripts via prompts such as `$ ./run_all_fasta.sh 25000000 1`. This invokes all Fasta programs from C, Java and Javascript and then reports the energy consumption on multiple domains. The command-line arguments in this case specify the problem size for the Fasta execution and the second command-line argument disables parallelism for the C program while testing.

A student can run individual languages and programs for a problem and compare them one-by-one, or run all the programs in one go to get a global overview of performances. The tests could be run on provided equipment but the set-up is also able to be used on their own devices, meaning that the students can see the difference the architecture and hardware made as well. The default executions of the programs are set at a long time (5 s) to observe the power consumption effects, and students are encouraged to run multiple different tweaks and discuss orally with each other their findings. Because noise would most certainly be present on their own laptops, we advise students to measure their idle system energy consumption and take this into account with their findings.

The key aim of the exercise is to familiarize students with the difference in energy consumption between programming languages first-hand, and then try to restructure or rebuild the code themselves to better understand the relations between energy consumption and language and language constructs.

5.2   Educational Activities for ILO_2

A new thought experiment could stem from the discussion between for-loops and while-loops in the previous section, such as whether a ternary operator could differ in energy consumption from the if-else statements that obtain the same result. We propose replicating the methodology for comparison between for-loops and while-loops used by Koedijk et al. [14], for the ternary versus if-else as an educational activity ILO_2.

Students are asked to first measure the initial energy consumption of programs and then modify the code to experimentally find how replacing the if-else statements with ternary operators affected energy consumption. Although the original study by Koedijk used direct energy measurements, these setup required for this approach was not available at the Summer School. We chose to use a method utilizing RAPL via `perf` as it is easy to use and accessible, making it well suited for an experimental session with students.

The original research experimented on multiple languages, however, to account for time constraints, we focus on three languages, namely: Java, Javascript and C. Not only would these languages be quite familiar, they each have very distinct energy consumption patterns and we additionally choose to have a mixture of interpreted and compiled languages to foster more discussion from the students.

The programs to be run are taken from the Computer Language Benchmarks Game[10] and we choose the following: a) CPU-intensive: *fannkuchredux*[11], b) memory-intensive: *fasta*[12] and c) disk-intensive: *revcomp*[13]. We scope the activities to these three programs to keep the time required for the lab reasonable, while still testing different types of programs and their impact on energy.

Lastly, we also instruct students to consider multiple compilation options and explore whether these have a significant difference on the energy consumption.

5.3   Educational Activities ILO_3

Inflicting established code smells in a controlled manner yields interesting results. It is more easily applicable to a wider audience, as refactoring code smells usually

---

[10] https://benchmarksgame-team.pages.debian.net/benchmarksgame/index.html.

[11] https://benchmarksgame-team.pages.debian.net/benchmarksgame/description/fannkuchredux.html.

[12] https://benchmarksgame-team.pages.debian.net/benchmarksgame/description/fasta.html.

[13] https://benchmarksgame-team.pages.debian.net/benchmarksgame/performance/revcomp.html.

requires a deeper familiarity with software development and quality. We instruct students to introduce several established code smells, such as 'Large Class' and 'Duplicate Code' on various software projects. Although the framework and codebase can potentially be reused for multiple code smells, each experiment measures only the impact of a single code smell.

We use the same lab exercise setup as described in Sect. 5.1. If a PDU-based approach is not available on the lab computers, a RAPL-based method can be used.

5.4 Experience Report from the Rijeka Summer School

In the context of the Erasmus+ SusTrainable project[14], this tutorial was given during one of its events, namely the Rijeka Summer School[15]. For a week, students and teachers from all participating universities joined the Rijeka University campus for daily educational activities, consisting mostly of lectures and lab.

While working on our tutorial, students observed noise on their laptops from the measuring environment for energy in RAPL, resulting in inconsistent measurements. However, they did make interesting observations when interchanging the if/else constructs by ternary operators (ILO_2). Some students noted that they could observe a small change in energy consumption, both upwards and downwards, from the original if/else construction, which was mostly tested on non-optimized C. However, they also argued that creating a set-up where a ternary operator differs from an if/else construction is not easy without disabling compiler optimizations and that very specific domain knowledge would be needed to make use of this. Reflecting on this, it would be beneficial for students to make sure that the activities created give a clear indication of the difference in power consumption for these constructs, similar to how the programming languages immediately showcased a difference in energy estimation.

After discussing the need for domain-specific knowledge of the programming languages themselves, the discussion moved on to whether the compiler version, the language version and also implementation-specific bindings could make a difference in energy consumption, which in turn led to the conclusion that conducting an experiment to give a concrete answer would be too difficult for the scope of this introductory training. However, we can say that there is an interesting follow-up research similar to for-loops versus while-loops that can be investigated. In our eyes this would be an interesting exercise for students during a Bachelor's or Master's programme, where the students are encouraged to engage in discussions with each other and experimentally find conclusive results.

One point we could have improved for the tutorial session was releasing the code or pointing the students towards the code for all the different programs we were going to investigate. There was a time sink in understanding the benchmarks that we provided from the Computer Language Benchmark Game that we could not reduce sufficiently even with our introduction at the start of the

---

[14] https://sustrainable.github.io/index.html.
[15] https://sustrainable.uniri.hr/.

tutorial. In a less time-constrained setting, it might be best to have students create and test the energy consumption of a program via an iterative approach, where they keep building on top of past experimentation to not only circumvent the problem of testing on unknown programs, but also be fully aware of the energy expenses of their programs.

For ILO_3, results were quickly found on the CPU-intensive `fannkuchredux` problem when stretching out functions to large proportions to mimic the 'Large Class' code smell, which made the execution time larger and more energy-inefficient. One student interlaced the 'Large Class' with 'Duplicate Code' as well, which similarly showed an increase in energy consumption. However, due to inherent noise in energy measurements, there were instances where the difference was not clear enough. Multiple measurements did show a negative relational trend between the two aforementioned code smells and energy consumption, but the research performed during the seminar was not conclusive on its own. The students did reflect that their coding behaviors could influence the energy consumption of their projects, sometimes surprising themselves as their perceived 'good' coding style gave slightly more energy consumption.

6   Conclusion

Energy-aware programming in software development is a key aspect of meeting the sustainability goals of the European Union, especially in the context of a Knowledge Economy, which relies on digital services. In this work, we consolidate research conducted on relationship between source code smells, *i.e.*, anti-patterns, and the energy consumption of the affected software project, and we revisit research and results related to the energy footprint of programming languages, and find that indeed energy consumption differs across programming languages (Q1.1), and also across different software projects implementing the same functionality in the same programming language (Q1.2). Some language construct choices (such as while-loop versus for-loop) have an impact on energy consumption, while others do not (such as if-construction versus the ternary construction) (Q2).

We analyse the impact of anti-patterns on the energy consumption of Java-based projects (Q3.1), and find that, although most impact is statistically significant, more experimental work is needed to gauge generalizable insights. We also investigate whether the energy consumption of anti-patterns is significantly different across programming languages (Q3.2), and find strong statistical evidence that this is the case.

To support the integration of energy-aware software engineering in modern computer science curricula, we also formulate the following learning objectives:

ILO_1 Students will be able to investigate whether programming languages have different energy footprints

ILO_2 Students will be able to conduct analysis on (special) programming language constructs to identify different energy footprints for different types of statements

ILO$_3$ Students will be able to establish whether energy smells exist and whether their impact depends on the programming language under observation

Based on our collective work, we design and implement educational activities that align with these learning objectives. We achieve the learning goals set out for this tutorial, as students can describe and reason about the different energy consumption of programming languages. We also observe that it is easier for most students to reason about code smells than about the replacement of the ternary operator. Due to the ease of introducing code smells to existing code bases, we conclude that this would suit very well as an assignment for Bachelor students. The ternary operator assignment might be better suited for assignments aimed at master students. The most important improvement on both of these would be allowing for more precise measurements, yet commonly available solutions such as RAPL are also suitable.

Acknowledgments. We would like to thank the SusTrainable project and its inspiring community for their support, in particular Clemens Grelck for his careful, insightful and constructive feedback.

References

1. Carette, A., Younes, M.A.A., Hecht, G., Moha, N., Rouvoy, R.: Investigating the energy impact of Android smells. In: 24th International Conference on Software Analysis, Evolution and Reengineering (SANER), Klagenfurt, Austria. IEEE (2017)
2. Bree, D.C., Cinnéide, M.Ó.: Energy efficiency of the Visitor Pattern: contrasting Java and C++ implementations. Empir. Softw. Eng. **28**(6) (2023)
3. Couto, M., Saraiva, J., Fernandes, J.P.: Energy refactorings for Android in the large and in the wild. In: 27th International Conference on Software Analysis, Evolution and Reengineering (SANER), London, Ontario, Canada. IEEE (2020)
4. Desrochers, S., Paradis, C., Weaver, V.M.: A validation of DRAM RAPL power measurements. In: Proceedings of the 2nd International Symposium on Memory Systems (MEMSYS), Washington DC, USA (2016)
5. Dhaka, G., Singh, P.: An empirical investigation into code smell elimination sequences for energy efficient software. In: 23rd Asia-Pacific Software Engineering Conference (APSEC), Hamilton, New Zealand. IEEE (2016)
6. Ester, M., Kriegel, H.-P., Sander, J., Xu, X., et al.: A density-based algorithm for discovering clusters in large spatial databases with noise. In: Proceedings of the 2nd International Conference on Knowledge Discovery and Data Mining (KDD), Portland, Oregon, USA, vol. 96, p. 34. AAAI (1996)
7. Fagarasan, C., Cristea, C., Cristea, M., Popa, O., Pisla, A.: Integrating sustainability metrics into project and portfolio performance assessment in agile software development: a data-driven scoring model. Sustainability **15**(17) (2023)
8. Fonseca, A., Kazman, R., Lago, P.: A manifesto for energyaware software. IEEE Softw. **36**(6) (2019)
9. Fowler, M.: Refactoring: Improving the Design of Existing Code. Addison-Wesley Professional (2018)

10. Georgiou, S., Rizou, S., Spinellis, D.: Software development lifecycle for energy efficiency: techniques and tools. ACM Comput. Surv. (CSUR) **52**(4) (2019)
11. DBSCAN https://commons.wikimedia.org/wiki/User:Chire (2011). https://en.wikipedia.org/wiki/DBSCAN. Accessed 15 June 2020
12. Khan, K.N., Hirki, M., Niemi, T., Nurminen, J.K., Ou, Z.: RAPL in action: experiences in using RAPL for power measurements. Trans. Model. Perform. Eval. Comput. Syst. (ToMPECS) **3**(2) (2018)
13. Kirkeby, M.H., Santos, B., Fernandes, J.P., Pardo, A.: Compiling Haskell for energy efficiency: empirical analysis of individual transformations. In: Proceedings of the 39th Symposium on Applied Computing (SAC), Avila, Spain (2024)
14. Koedijk, L., Oprescu, A.: Finding significant differences in the energy consumption when comparing programming languages and programs. In: International Conference on ICT for Sustainability, ICT4S, Polvdiv, Bulgaria. IEEE (2022)
15. Kok, S.: The impact of Refactoring Code Smells on the Energy Consumption of Java-based Open-source Software. MSc. Software Engineering. Universiteit van Amsterdam, Netherlands (2019)
16. Koomey, J., Berard, S., Sanchez, M., Wong, H.: Implications of historical trends in the electrical efficiency of computing. IEEE Ann. Hist. Comput. **33**(3) (2010)
17. Kruglov, A., Succi, G.: Developing Sustainable and Energy-Efficient Software Systems. Springer (2023)
18. Mann, H.B., Whitney, D.R.: On a test of whether one of two random variables is stochastically larger than the other. Ann. Math. Stat. (1947)
19. Nachar, N., et al.: The Mann-Whitney U: a test for assessing whether two independent samples come from the same distribution. Tutor. Quant. Methods Psychol. **4**(1) (2008)
20. van Oostveen, S.: Regarding the Impact of Code Smells on the Energy Efficiency of Different Computer Language. BSc. Informatics. Universiteit van Amsterdam, The Netherlands (2020)
21. Ournani, Z., Rouvoy, R., Rust, P., Penhoat, J.: On reducing the energy consumption of software: from hurdles to requirements. In: Proceedings of the 14th International Symposium on Empirical Software Engineering and Measurement (ESEM), Bari, Italy. ACM (2020)
22. Palomba, F., Di Nucci, D., Panichella, A., Zaidman, A., De Lucia, A.: On the impact of code smells on the energy consumption of mobile applications. Inf. Softw. Technol. **105** (2019)
23. Pang, C., Hindle, A., Adams, B., Hassan, A.E.: What do programmers know about software energy consumption? IEEE Softw. **33**(3) (2015)
24. Park, J.J., Hong, J.-E., Lee, S.-H.: Investigation for software power consumption of code refactoring techniques. In: Software Engineering and Knowledge Engineering (SEKE), Vancouver, Canada (2014)
25. Pereira, R., et al.: Energy efficiency across programming languages: how do energy, time, and memory relate? In: Proceedings of the 10th International Conference on Software Language Engineering. ACM SIGPLAN (2017)
26. Pinto, G., Castor, F.: Energy efficiency: a new concern for application software developers. Commun. ACM **60**(12) (2017)
27. Saraiva, J., Zong, Z., Pereira, R.: Bringing green software to computer science curriculum: perspectives from researchers and educators. In: Proceedings of the 26th ACM Conference on Innovation and Technology in Computer Science Education V. 1, ITiCSE 2021, Virtual Event, Germany. ACM (2021)
28. Schneider, S.H.: The greenhouse effect: science and policy. Science **243**(4892) (1989)

29. Andrae, A.S.G.: New perspectives on internet electricity use in 2030. Eng. Appl. Sci. Lett. **3**(2) (2020)
30. Toczé, K., Madon, M., Garcia, M., Lago, P.: The dark side of cloud and edge computing: an exploratory study. In: Workshop on Computing within Limits (LIMITS) (2022)
31. Torre, D., Procaccianti, G., Fucci, D., Lutovac, S., Scanniello, G.: On the presence of green and sustainable software engineering in higher education curricula. In: 1st International Workshop on Software Engineering Curricula for Millennials (SECM), Buenos Aires, Argentina (2017)
32. Tsantalis, N., Chaikalis, T., Chatzigeorgiou, A.: Ten years of JDeodorant: lessons learned from the hunt for smells. In: 25th International Conference on Software Analysis, Evolution and Reengineering (SANER), Campobasso, Italy. IEEE (2018)
33. Van Heddeghem, W., Lambert, S., Lannoo, B., Colle, D., Pickavet, M., Demeester, P.: Trends in worldwide ICT electricity consumption from 2007 to 2012. Comput. Commun. **50** (2014)
34. Verdecchia, R., Procaccianti, G., Malavolta, I., Lago, P., Koedijk, J.: Estimating energy impact of software releases and deployment strategies: the KPMG case study. In: Proceedings of the 11th International Symposium on Empirical Software Engineering and Measurement (ESEM). IEEE Press (2017)
35. Wirth, N.: A plea for lean software. Computer **28**(2) (1995)

UActor: Group Based Actor Model Middleware with UDP Protocol

Jianhao Li and Viktória Zsók[(✉)]

Department of Programming Languages and Compilers, Faculty of Informatics,
Eötvös Loránd University, Pázmány Péter sétány 1/C., Budapest 1117, Hungary
`{lijianhao,zsv}@inf.elte.hu`

Abstract. This tutorial introduces a new approach to distributed communication middleware by using connectionless transport protocols and the actor model. In contrast to traditional connection-oriented protocols like TCP, which are often compared to phone systems due to their reliability and higher bandwidth needs, our approach utilizes connectionless protocols akin to postal services, aligning more naturally with the decentralized nature of the actor model. This shift towards a brokerless system reduces the complexity and overhead associated with message brokers and broker clusters, often with significant deployment and maintenance costs. Moreover, this middleware integrates various communication patterns into a unified framework that enhances usability. In this tutorial, students can learn how to design, implement, and test a distributed communication middleware. Furthermore, by involving energy consumption estimation tests in the evaluation phase, the tutorial inspires students to consider sustainability when designing or implementing their projects.

Keywords: Distributed system · Middleware · Actor model

1 Introduction

In distributed systems, the choice of transport protocol significantly affects performance and system architecture. Connection-oriented protocols like TCP are often compared to phone systems due to their reliable, ordered communication, and higher bandwidth requirements. In contrast, connectionless protocols like

This work received financial support through the Erasmus+ Strategic Partnership for Higher Education *SusTrainable—Promoting Sustainability as a Fundamental Driver in Software Development Training and Education* (project number 2020-1-PT01-KA203-078646), funded by the European Union and coordinated by the University of Coimbra, Portugal.

The information and views set out in this publication are those of the authors and do not necessarily reflect the official opinion of the European Union. Neither the European Union institutions and bodies nor any person acting on their behalf may be held responsible for the use which may be made of the information contained therein.

UDP (User Datagram Protocol) are compared to postal services for their simplicity and lower overhead [27]. HTTP/2 relies on TCP, while HTTP/3 leverages QUIC, a protocol built on UDP, to reduce latency and improve deployment efficiency [20]. This shift towards connectionless protocols reflects a growing trend to handle transport control at the application layer [12]. However, QUIC is primarily designed for web applications and it does not fully address the broader needs of distributed systems requiring diverse communication mechanisms.

The actor model [15] is inherently asynchronous. Each actor maintains a private message queue (mailbox) and processes incoming messages sequentially. Although messages from the same sender are processed in order, no global ordering is enforced across different actors. The actor model itself does not provide strong guarantees about the order in which messages arrive or are processed. Message delivery guarantees, including whether messages are lost or not, depend on the underlying implementation (like the TCP protocol).

The actor model is inherently well suited for parallelism and message-driven computation, often compared to UDP in its decentralized approach. Most actor model implementations are based on TCP, which limits their scalability and introduces complexity related to connection management. Therefore, we build an actor model using UDP to better align with its inherent characteristics. We argue that connectionless transport protocols are more compatible with nowadays actor models.

The UACTOR system introduced here is a new lightweight, distributed and decentralized communication middleware designed to offer versatile communication services for distributed systems. By leveraging a connectionless transport protocol and the actor model, UACTOR provides a more harmonious solution. This design eliminates the need for centralized message brokers, simplifies error handling, and reduces deployment and maintenance overhead. It ensures message delivery and provides versatile messaging mechanisms while maintaining ease of maintenance and upgrades. Once members join a group, they can communicate with all group members in a load-balanced or broadcast manner.

2   Related Work

Various systems have addressed the challenges of distributed communication with different approaches, each with its tradeoffs.

ERLANG can be used to implement robust distributed systems [7]. However, it has scalability limitations because of its fully connected TCP structure. Moreover, ERLANG lacks group communication service constructs with built-in mechanisms and it needs a virtual machine running environment. In contrast, UACTOR leverages a connectionless UDP-based actor model that scales more efficiently and inherently supports decentralized group communication.

RABBITMQ [13] is a mature open-source message broker that implements the AMQP 0-9-1 protocol [36]. However, the message brokers are additional components that increase the deployment and maintenance costs. All the protocols supported by the broker are based on long-lived TCP connections [25], which causes a single point of failure or bottleneck if there is no complex broker cluster service. UACTOR avoids these limitations by eliminating intermediary brokers through its decentralized design, reducing system complexity and overhead.

NSQ [31] supports pub/sub and load-balanced message delivery without centralized brokers. The nsqd daemon is used to receive, queue, and deliver messages [33] while nsqlookupd is used to manage the topology information [34]. Each nsqd has a long-lived TCP connection to nsqlookupd [32]. The high availability of nsqlookupd is achieved by running multiple instances. Clients are connected over TCP to all nsqd daemons. Clients send query requests to the nsqlookupd daemons through the HTTP interfaces to discover the nsqd information. As a result, NSQ has reduced the SPOF while the number of instances, connections, and messages have increased. Thus, system complexity increased. Moreover, the clients are not independent because they still rely on external components (nsqlookupd) to maintain the topology. By contrast, UACTOR integrates group communication and message delivery within a unified framework based on UDP, simplifying the overall architecture.

MANGOS [21] and NANOMSG [24] aim at implementing "scalability protocols" (communication patterns). However, they have usability problems because of their separated communication constructs (sockets) with different communication patterns. NNG [30] is the successor of NANOMSG. Compared to the NANOMSG, NNG aims to improve scalability and usability by engaging multiple cores and implementing more intuitive new APIs [10]. Besides the usability problem of separated communication constructs, NNG provides fewer delivery guarantees (only for some protocols). Additionally, it follows a different model than the actor one, and it does not have unified group communication services and failover mechanisms. By contrast, UACTOR unifies diverse communication patterns under a single actor-based framework, enhancing both reliability and ease of use.

AKKA is a toolkit for building distributed message-driven applications used to build a clustered dynamic actor system [35]. The AKKA cluster is a ring of nodes, where each node is an actor system that listens on the network [38]. The actor systems in AKKA gossip to each other about the current cluster state (containing a list of member nodes) using a Gossip Protocol [9]. By contrast, UACTOR uses UDP and a simpler architecture without a ring. UACTOR group is based on the gossip-based SWIM protocol [11]. However, compared to the traditional Gossip protocol, SWIM optimizes the Gossip method, making it more efficient in large-scale clusters and reducing unnecessary message overhead.

QUIC is a secure general-purpose connection-oriented transport protocol that aims to improve transport performance for HTTPS traffic [17], and it also uses UDP to create a stateful interaction between a client and a server. User Datagram Protocol (UDP) is a lightweight, connectionless transport protocol designed for low-latency communication. It transmits messages without establishing a dedicated connection or guaranteeing delivery [26]. However, it does not follow the actor model, and its focus is not on communication patterns when constructing a decentralized distributed system. ACTIX [37] is the famous actor framework for

RUST [22]. It can handle communication in a local/thread context. However, it can not handle remote communication. Instead, UACTOR offers a comprehensive UDP-based actor model that supports local and distributed messaging.

The paper [6] introduces a design that uses a universal ring to provide a scalable infrastructure. However, it only provides bootstrapping services: indexing, multicasting, code obtaining, and joining service. Furthermore, regarding the communication, there is only one multicast service based on the SCRIBE [5]. By contrast, UACTOR distinguishes itself by integrating point-to-point communication, group broadcast, group load balance sending, and built-in reliability mechanisms into a single middleware solution without a ring structure.

The middleware that this project refers to is not a traditional one that focuses on message queueing [29]. It is a library inside the application layer of the network, and it aims to augment and simplify the facilities provided by the lower-level socket APIs [28]. A classical message-oriented middleware requires an extra component (broker) as a messaging server, which is typically complex and expensive to maintain [18]. The cluster (of brokers) handles the single point of failure (SPOF) problem (i.e. if a component or part fails, it causes the entire system to fail) by connecting multiple machines. The coordination between the brokers in a cluster may need even more extra components.

For example, the KAFKA [19] broker (KAFKA server) cluster needs the centralized service the ZOOKEEPER [16] server provides. The brokers (and the distributed coordination servers of brokers) are preferable to be run on separate machines for high availability and reliability, which consume more energy. Contrary, all the messages do not need to pass through the broker, and the network communication can be reduced in brokerless projects [4], which is more sustainable due to the lower deployment and maintenance costs. Additionally, designing the system in a decentralized way frees the system from a single point of failure (SPOF), and the broker will no longer turn into a bottleneck of the system. Our approach eliminates the need for brokers by utilizing a decentralized design. This brokerless model simplifies deployment and maintenance and eliminates SPOF issues [4].

Overall, UACTOR presents a brokerless, UDP-based, actor model-based middleware that simplifies deployment, enhances scalability, and delivers unified communication services (including point-to-point and group messages) with message delivery guarantee, addressing many of the limitations observed in existing systems.

3   UACTOR System Design

In this section, the design decisions and the architecture of the system are described. To prevent the single point of failure (SPOF) and the single point of bottleneck, we design our system as broker-less, which means there is no central component that is only responsible for interchanging messages. The system implemented following this design is a decentralized transient messaging system.

The actor model is suitable for large-scale concurrent and distributed systems [1]. Therefore, this system is designed following the actor model. The

UActor system design is based on the connectionless communication protocol that defines the actor's address (mail address) and the message (mail) format. The UActor contains two essential components: the `Communicator` and the group communication service.

The `Communicator` defines the message structure, including encoding and decoding, and handles actor-level functionalities. It utilizes features from the underlying transport protocol to achieve scalability by choosing a connectionless protocol to minimize the number of connections. Consequently, the `Communicator` maintains a connectionless nature. It implements mechanisms such as acknowledgments (ACKs), message resending, and local message storage for retrying failed deliveries to enhance message reliability. Additionally, it can use techniques like splitting and reassembling messages to support larger message sizes. The `Communicator` also sets up various model-related structures, such as a mailbox-like buffer for the actor model. The `Communicator` also has a message handling service in which goroutines concurrently receive messages from the port and process them according to the type of the messages.

The group communication service, built on top of the `Communicator` and integrated with a group membership protocol, provides communication methods for nodes within a group. This service typically includes broadcasting and load-balanced sending mechanisms.

4   UActor System Implementation

UActor is implemented in the Go programming language, which is ideally suited for constructing scalable, high-performance networked systems. The Go language has efficient concurrent constructs, like goroutines and channels, which enable the system to execute multiple low-cost tasks concurrently and to communicate with each other with minimal overhead. Furthermore, Go's comprehensive networking libraries and automatic memory management via garbage collection facilitate the development of robust distributed systems.

Figure 1 illustrates the architecture of the UActor implementation. The group communication service deals with group membership and group communication based on the `Communicator`. `CreateNodeAndJoin`, `Broadcast`, and `RoundRobinSend` are the main public functions provided by the group communication service. The `Communicator` handles the mailbox, acknowledgment, retransmission, and storage of messages. The `Communicator` provides functions like `NewCommunicator` (including starting the received messages handler) and `Send` (with the guarantee of message delivery). The connectionless transport utility provides the essential datagram transmission-related functions to the `Communicator`.

4.1   Message Structure

This section introduces the basic message format used in UActor. In our system, each message includes a unique message identifier (MsgId) that ensures

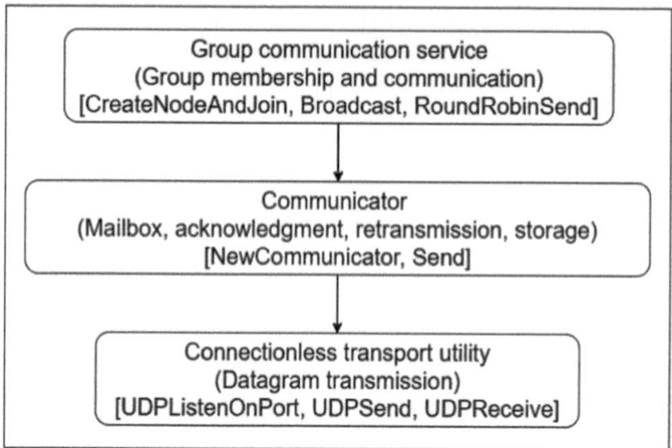

Fig. 1. UActor architecture

every message can be individually tracked, even though the destination is determined separately by the underlying network protocol.

We deliberately omit the destination from the message because the routing information is provided by the underlying transport layer (i.e., the UDP socket address). At the same time, the MsgId uniquely identifies the message for acknowledgment and retransmission purposes.

As shown in Listing 1.1, the message structure contains a `MsgId` message ID, a `MsgType` message type, and a `Body` message body. The `MsgId` is of type `UUID` (16-byte universal unique identifier), which is imported from the `uuid` package. The `MsgType` is of type `uint8` ranging from 0 to 255 (takes one byte). The `Body` is a slice of bytes. Every message carries a message type. To reduce the message size, we simplified the message type to an unsigned integer with 8 bits.

For the current implementations, the type 1 message denotes standard application messages that require acknowledgment, while type 2 is reserved exclusively for acknowledgment signals. There can be more types of messages when this system is enhanced with more features.

```
package ua                                          1
import (                                            2
    "github.com/google/uuid"                        3
)                                                   4
type Msg struct {                                   5
    MsgId    uuid.UUID                               6
    MsgType uint8                                    7
    Body        []byte                              8
}                                                   9
```

Listing 1.1. The message structure

As shown in Listing 1.2, the function `MsgToBytes` converts a `Msg` structure into a byte slice, enabling efficient message encoding. It begins by calculating the length of the message body and creating a byte slice `r` that can accommodate the message ID, message type, and body. The first 16 bytes of `r` are filled with the message ID, followed by the message type byte at position 16. The message body is then copied into the remaining portion of `r`. The function finishes by returning the complete byte slice representing the `Msg` structure.

The `BytesToMsg` function is the message decoding process that transforms a byte slice back into a `Msg` structure. It starts by declaring a new `Msg` variable, m. The function extracts the first 16 bytes of the byte slice b and assigns them to m.`MsgId`, interprets the 17th byte as the message type and assigns it to m.`MsgType`, and sets the remaining bytes as the message body m.`Body`. Finally, the function returns the reconstructed `Msg` structure.

The `GenId` function generates a new universally unique identifier (UUID). It utilizes a UUID generation library to create and return a new UUID, ensuring each message can have a unique identifier. This function is straightforward; it simply calls `uuid.New()` to obtain the UUID and return it. The UUID is crucial for uniquely identifying messages in a distributed system.

```
func MsgToBytes(m Msg)([]byte) {                              1
    l:=len(m.Body)                                            2
    r := make([]byte, 17+1)                                   3
    copy(r[0:16],(m.MsgId)[:])                                4
    r[16] = m.MsgType                                         5
    copy(r[17:],m.Body)                                       6
    return r                                                  7
}                                                             8
func BytesToMsg(b []byte)(Msg) {                              9
    var m Msg                                                10
    m.MsgId = *((*[16]byte)(b[0:16]))                        11
    m.MsgType = b[16]                                        12
    m.Body = b[17:]                                         13
    return m                                                14
}                                                            15
func GenId() uuid.UUID {                                     16
    return uuid.New()                                        17
}                                                            18
```

Listing 1.2. The message functions

4.2   Message Storage

Although correct socket numbers are used, message sending can still fail due to network congestion, packet loss, or transient errors. Therefore, unsent messages are persisted on the disk for later retransmission.

This file-based storage mechanism ensures that no message is permanently lost, enabling the system to achieve reliable communication even under adverse network conditions. The messages that failed to be sent are stored on the hard disk. This implementation utilizes the file storage system of the local operating system to avoid additional third-party dependencies.

As shown in Listing 1.3, the `MsgWithAddr` struct is an extended version of the `Msg` structure, designed to include the IP address and port number along with the message itself. It includes the fields `IP`, a 4-byte array representing the sender's IP address, and `Port`, a 16-bit unsigned integer representing the sender's port number. Additionally, it embeds the `Msg` struct, which contains the message ID, type, and body.

When `MsgWithAddr` stores received messages in memory, because the receiver is the node itself, the fields `IP` and `Port` are used to store the sender's address. When `MsgWithAddr` stores failed sending messages on a hard disk, because the sender is always the node itself, the fields `IP` and `Port` are used to store the receiver's address.

The `MsgWithAddrToBytes` function converts a `MsgWithAddr` structure into a byte slice. It begins by determining the length of the message body and creating a byte slice `r` that can hold the IP address, port number, message ID, message type, and message body. The first 4 bytes of `r` are filled with the IP address, followed by 2 bytes for the port number, encoded in the little endian format. The following 16 bytes are allocated for the message ID, followed by 1 byte for the message type. The remaining bytes are used to store the message body. The function returns the complete byte slice representing the `MsgWithAddr` structure.

The `BytesToMsgWithAddr` function reconstructs a `MsgWithAddr` structure from a byte slice. It starts by declaring a new `MsgWithAddr` variable, m. The function extracts the first 4 bytes of the byte slice b and assigns them to m.`IP`. The next 2 bytes are interpreted as the sender's port number, decoded from the little-endian format, and assigned to m.`Port`. The subsequent 16 bytes are allocated to m.`MsgId`, and the next byte is assigned to m.`MsgType`. The remaining bytes are set as the message body, m.`Body`. Finally, the function returns the reconstructed `MsgWithAddr` structure.

As shown in Listing 1.4, the `SaveMsg` function saves a message to a file on disk. It first creates the directory specified by `msgFolder` if it does not already exist. Then, it creates a new file named after the `msgId` within this directory. The function writes the byte slice `msgB` to the file and ensures the data is flushed to disk with `f.Sync()`. If any step fails, an error and a zero are returned. Otherwise, the function returns the number of bytes successfully written.

The `GetMsg` function retrieves a message from a file. It reads the contents of the specified file into a byte slice and converts this byte slice to a `MsgWithAddr` structure using `BytesToMsgWithAddr`. An appropriate error is returned if any error occurs during file reading, message conversion, or file deletion. If successful, the function returns a pointer to the reconstructed `MsgWithAddr` structure.

The `DiskMsgNum` function returns the number of files in `msgFolder` directory, which indicates the number of saved messages. It reads the directory contents and returns the count of files. If reading the directory fails, it returns -1.

```
type MsgWithAddr struct {                                           1
    IP   [4]byte                                                    2
    Port uint16                                                     3
    Msg                                                             4
}                                                                   5
const msgFolder = "failedSend/"                                     6
func MsgWithAddrToBytes(m MsgWithAddr) []byte {                     7
    l := len(m.Body)                                                8
    r := make([]byte, 23+l)                                         9
    copy(r[0:4], m.IP[:])                                           10
    binary.LittleEndian.PutUint16(r[4:6], m.Port)                   11
    copy(r[6:22], (m.MsgId)[:])                                     12
    r[22] = m.MsgType                                               13
    copy(r[23:], m.Body)                                            14
    return r                                                        15
}                                                                   16
func BytesToMsgWithAddr(b []byte) MsgWithAddr {                     17
    var m MsgWithAddr                                               18
    m.IP = *((*[4]byte)(b[0:4]))                                    19
    m.Port = binary.LittleEndian.Uint16(b[4:6])                     20
    m.MsgId = *((*[16]byte)(b[6:22]))                               21
    m.MsgType = b[22]                                               22
    m.Body = b[23:]                                                 23
    return m                                                        24
}                                                                   25
```

Listing 1.3. The `MsgWithAddr` struct

The `DeleteMsgFolder` function deletes the entire `msgFolder` directory and its contents. It uses `os.RemoveAll` to remove the directory and all its files. If the removal fails, an error is returned. If successful, the function returns `nil`.

4.3   Connectionless Transport Utility

Our middleware (UActor) acts as an abstraction layer over the raw UDP protocol. It encapsulates basic send/receive functions while adding reliability features such as acknowledgments, retransmission mechanisms, persistent storage and group messages. This approach decouples the application logic from low-level network details, simplifying development and enhancing robustness.

The transport protocol utility encapsulates the protocol usage, and it provides functions to listen on a port, and to send and receive messages. The maximum transmission unit (MTU) of the project can be declared in this utility. The send and receive functions are implemented under the MTU restrictions.

As shown in Listing 1.5, the `UDPListenOnPort` function creates a UDP connection bound to the specified address and port. It first checks if the port number is within the valid range (1024 to 49151). If the port number is invalid, it returns an error. Otherwise, it attempts to start listening on the specified UDP address. If successful, it returns the UDP connection; otherwise, it returns an error indicating the failure to listen on the UDP port. There are two important functions related to UDP ports: `UDPSend` and `UDPReceive`.

The `UDPSend` function sends data to a specified UDP address using a given UDP connection. It first checks if the data length exceeds the Maximum Transmission Unit (MTU), which is set to 1400 bytes. If the data is too large, it returns

```go
const msgFolder = "failedSend/"                                       1
func SaveMsg(msgId uuid.UUID,                                         2
    msgB []byte) (int, error) {                                      3
    err := os.MkdirAll(msgFolder, os.ModePerm)                       4
    if err != nil {                                                  5
        return 0, errors.New("failed␣to␣MkdirAll")                   6
    }                                                                7
    f, err := os.Create(msgFolder + msgId.String())                 8
    if err != nil {                                                  9
        fmt.Println(err)                                            10
        return 0, errors.New("failed␣to␣create␣file")               11
    }                                                               12
    defer f.Close()                                                 13
                                                                    14
    n, err := f.Write(msgB)                                         15
    if err != nil {                                                 16
        return 0, errors.New("failed␣to␣write␣file")                17
    }                                                               18
    f.Sync()                                                        19
    return n, nil                                                   20
}                                                                   21
func GetMsg(fileName string) (*MsgWithAddr, error) {               22
    var msg MsgWithAddr                                             23
    b, err := os.ReadFile(fileName)                                 24
    if err != nil {                                                 25
        fmt.Println(err)                                            26
        return nil, errors.New("failed␣to␣ReadFile")                27
    }                                                               28
    msg = BytesToMsgWithAddr(b)                                     29
    err = os.Remove(fileName)                                       30
    if err != nil {                                                 31
        return nil, errors.New("failed␣to␣Remove")}                 32
    return &msg, nil                                                33
}                                                                   34
func DiskMsgNum() int {                                            35
    fs, err := os.ReadDir(msgFolder)                               36
    if err != nil {                                                37
        return -1 }                                                 38
    return len(fs)                                                  39
}                                                                   40
func DeleteMsgFolder() error {                                     41
    err := os.RemoveAll(msgFolder)                                  42
    if err != nil {                                                 43
        return errors.New("failed␣to␣RemoveAll") }                  44
    return nil                                                      45
}                                                                   46
```

Listing 1.4. The message storage operations

an error. Otherwise, it sends the data to the destination address and returns the number of bytes written. If an error occurs during sending, it returns an error message indicating the failure.

The `UDPReceive` function receives data from a UDP connection. It creates a buffer of size MTU to store the incoming data. It then reads data from the UDP connection and puts it into the buffer. If successful, it returns the received data (sliced to the number of bytes read), the address of the sender, and nil for the error. If an error occurs during reading, it returns nil for both the data and the sender address, along with an error message indicating the failure to read from the UDP port.

```go
package ua                                                      1
                                                               2
import (                                                        3
    "errors"                                                    4
    "net"                                                       5
)                                                              6
const MTU = 1400                                               7
                                                               8
func UDPListenOnPort(addr *net.UDPAddr) (                       9
    *net.UDPConn, error) {                                     10
    if addr.Port < 1024 || addr.Port > 49151 {                11
        return nil, errors.New("invalid port number")         12
    }                                                          13
    localConn, e := net.ListenUDP("udp", addr)                14
    if e != nil {                                              15
        return nil, errors.New("failed to Listen UDP")        16
    }                                                          17
    return localConn, nil                                      18
}                                                              19
func UDPSend(c *net.UDPConn,                                   20
    dest *net.UDPAddr, data []byte) (int, error) {            21
    if len(data) > MTU {                                      22
        return 0, errors.New("the data length exceeds MTU")   23
    }                                                          24
    n, e := c.WriteTo(data, dest)                             25
    if e != nil {                                             26
        return 0, errors.New("failed to write")              27
    }                                                          28
    return n, nil                                             29
}                                                              30
func UDPReceive(c *net.UDPConn) (                             31
    []byte, *net.UDPAddr, error) {                           32
    buf := make([]byte, MTU)                                  33
    n, senderAddr, e := c.ReadFromUDP(buf)                   34
    if e != nil {                                            35
        return nil,                                          36
            nil, errors.New("failed to read from udp port")  37
    }                                                         38
    return buf[:n], senderAddr, nil                          39
}                                                             40
```

Listing 1.5. The connectionless transport functions

4.4 Communicator

The `Communicator` handles point-to-point communication based on the functions provided by the connectionless transport utility and provides reliable message sending to the group message service. The `Communicator` acts as an abstraction layer over the raw UDP protocol. It encapsulates basic send/receive functions while adding reliability features such as acknowledgments, retransmission mechanisms, and persistent storage.

As shown in Listing 1.6, the `Config` struct defines the configuration settings for the `Communicator`. It includes the following fields: `ReceiveGoroutineNumber`, which specifies the number of goroutines for receiving data; `MailboxSize`, which determines the size of the mailbox or the capacity of the queue or buffer used for storing incoming messages; `RetryTimes`, which sets the number of retry attempts the system should make in case of failures; and `RetryInitialInterval`, of type `time.Duration`, which specifies the initial interval between retry attempts. This interval helps to manage the delay between retries.

```
type Config struct {                                                1
    ReceiveGoroutineNumber int                                      2
    MailboxSize int                                                 3
    RetryTimes int                                                  4
    RetryInitialInterval  time.Duration                             5
}                                                                   6
```

Listing 1.6. The `Config` struct

As shown in Listing 1.7, the `Communicator` struct defines the core communication component of a system. It includes several fields: `LocalAddr`, a pointer to a `net.UDPAddr` that holds the local UDP address configuration; `LocalConn`, a pointer to a `net.UDPConn` representing the local UDP connection; and `Config`, an instance of the previously defined `Config` struct, which contains various configuration settings.

Additionally, the `Mailbox` is a channel for receiving `MsgRecv` messages, acting as a message queue. The `ackMap` is a map that uses a 16-byte array as a key to map to boolean channels, used for managing acknowledgments.

To ensure concurrent access to `ackMap` is handled safely, the `ackMutex` field employs a `sync.RWMutex` for read-write locking. It uses a `done` channel and a `sync.WaitGroup` to coordinate the shutdown of its receiver goroutines.

```
type Communicator struct {                                          1
    LocalAddr *net.UDPAddr                                          2
    LocalConn *net.UDPConn                                          3
    Config    Config                                                4
    Mailbox   chan MsgWithAddr                                      5
                                                                    6
    ackMap    map[[16]byte]chan bool                                7
    ackMutex  sync.RWMutex                                          8
                                                                    9
    done chan struct{}                                              10
    wg   sync.WaitGroup                                             11
}                                                                   12
```

Listing 1.7. The `Communicator` struct

As shown in Listing 1.8, the `NewCommunicator` function initializes a new `Communicator` instance. It begins by setting default values for configuration parameters if they are not provided.

These defaults include setting the number of receive goroutines to 10, mailbox size to 100, retry times to 5, and initial retry interval to 5 s. It then validates the configuration using the `Validate` function. If the configuration is invalid, it returns an error.

Next, the function establishes a UDP connection on the specified address using `UDPListenOnPort`. If this fails, it logs the error and returns an error. The function then creates a message mailbox channel with the specified size and initializes the `Communicator` struct. Suppose `resendFailedMsgs` is true. In that case, it attempts to resend any previously failed messages by reading from the `msgFolder` directory, retrieving messages, and sending them.

Finally, it spawns the specified number of goroutines for handling incoming messages and returns the initialized `Communicator` instance.

The `receiveHandle` function operates in a loop to continuously receive and process incoming UDP messages. It reads data from the UDP connection and checks for errors. The `select` statement is used to monitor the `done` channel, allowing for an exit when the `Communicator` is closing. Upon successfully receiving a message, it converts the bytes into a `Msg` using `BytesToMsg`. The function then processes the message based on its type. If the message type is 1, it sends an acknowledgment message of type 2 back to the sender and places the message into the mailbox. If the message type is 2, it checks if there is a corresponding acknowledgment channel in the `ackMap` and sends a `true` value if found. This function helps manage incoming messages and acknowledgments within the `Communicator`.

As shown in Listing 1.10, the `Send` method of the `Communicator` struct is a public function that facilitates sending a message to a specified destination. It takes in a `net.UDPAddr` representing the destination address and a byte slice as the message body. It then calls the private `send` method with a message type of 1, which indicates a regular message type. The `Send` method returns the number of bytes sent, a `UUID` of the message, and any error encountered during the process.

The `send` method of the `Communicator` struct is a private function that handles the actual sending of a message and its acknowledgment. It takes in a destination address, a message type, and a byte slice representing the message body. The function generates a unique message ID using `GenId()` and creates a `Msg` struct. It converts this message to bytes using `MsgToBytes`. If the message type is not 1, it directly sends the message using `UDPSend` and returns the number of bytes sent along with the message ID.

For messages of type 1, which require acknowledgment, the function creates a channel for acknowledgment and stores it in `ackMap` under the message ID. It then attempts to send the message up to `RetryTimes` times, waiting for an acknowledgment or a timeout period before retrying. Suppose an acknowledgment is received within the retry attempts. In that case, it removes the acknowledgment channel from `ackMap` and returns the number of bytes sent along with the message ID. Suppose no acknowledgment is received after all retries. In that case, the message is stored in a file using `SaveMsg`, and the function returns the number of bytes saved along with the message ID.

Although advanced techniques such as sliding window protocols and embedded acknowledgments can improve transmission efficiency, they introduce additional complexity.

```
func NewCommunicator(addr *net.UDPAddr,                                    1
    cf Config, resendFailedMsgs bool) (*Communicator, error) {             2
    if cf.ReceiveGoroutineNumber == 0 {                                    3
        cf.ReceiveGoroutineNumber = 10                                     4
    }                                                                      5
    if cf.MailboxSize == 0 {                                               6
        cf.MailboxSize = 100                                               7
    }                                                                      8
    if cf.RetryTimes == 0 {                                                9
        cf.RetryTimes = 5                                                 10
    }                                                                     11
    if cf.RetryInitialInterval == 0*time.Second {                        12
        cf.RetryInitialInterval = 3 * time.Second                        13
    }                                                                     14
    err := Validate(cf)                                                  15
    if err != nil {                                                      16
        return nil, errors.New("invalid Config")                        17
    }                                                                     18
    localConn, e := UDPListenOnPort(addr)                                19
    if e != nil {                                                        20
        log.Println(e)                                                   21
        return nil, errors.New("failed to ListenOnPort")                22
    }                                                                     23
    mailb := make(chan MsgWithAddr, cf.MailboxSize)                      24
    c := Communicator{LocalAddr: addr,                                   25
        LocalConn: localConn, Config: cf, Mailbox: mailb,               26
        ackMap: make(map[[16]byte]chan bool),                           27
        done:   make(chan struct{}),                                    28
    }                                                                     29
                                                                          30
    for i := 0; i < c.Config.ReceiveGoroutineNumber; i++ {              31
        c.wg.Add(1)                                                      32
        go receiveHandle(&c)                                            33
    }                                                                     34
                                                                          35
    if resendFailedMsgs {                                               36
        files, err := os.ReadDir(msgFolder)                            37
        if err != nil {                                                 38
            return nil, errors.New("failed to os.ReadDir")            39
        }                                                                 40
        for _, file := range files {                                   41
            mr, err := GetMsg(msgFolder + file.Name())                 42
            if err != nil {                                            43
                log.Println("Failed to GetMsg", err)                  44
                continue                                                45
            }                                                            46
            addr := &net.UDPAddr{IP: (*mr).IP[:],                      47
                Port: (int)((*mr).Port)}                              48
            _, _, err = c.Send(addr, (*mr).Body)                       49
            if err != nil {                                            50
                log.Println("Failed to Send", err)                   51
            }                                                            52
        }                                                                 53
    }                                                                     54
    return &c, nil                                                       55
}                                                                         56
```

Listing 1.8. The NewCommunicator function

Listing 1.11 shows a new Close method has been introduced to allow a clean
shutdown of the Communicator. This method closes the UDP connection, signals
all receiver goroutines to exit via the done channel, waits for them to finish using
the wait group, and cleans up all channels in the acknowledgment map.

```go
func receiveHandle(c *Communicator) {
    defer c.wg.Done()
    for {
        select {
        case <-c.done:
            return
        default:
            b, senderAddr, err :=
                UDPReceive(c.LocalConn)
            if err != nil {
                if errors.Is(err, net.ErrClosed) {
                    return
                }
                continue
            }
            m := BytesToMsg(b)

            if m.MsgType == 1 {
                c.send(senderAddr, 2, (m.MsgId)[:])
                mRecv := MsgWithAddr{
                    ([4]byte)(senderAddr.IP.To4()),
                    uint16(senderAddr.Port), m}
                c.Mailbox <- mRecv
            }

            if m.MsgType == 2 {
                c.ackMutex.RLock()
                v, ok :=
                    c.ackMap[*(*[16]byte)(m.Body[0:16])]
                c.ackMutex.RUnlock()
                if ok {
                    v <- true
                }
            }
        }
    }
}
```

Listing 1.9. The `receiveHandle` function

4.5 Group Communication Service

In UActor, a group refers to a collection of nodes. This grouping allows efficient broadcast and load-balanced message delivery, enabling coordinated operations among multiple distributed actors.

As shown in Listing 1.12, the `GroupNode` struct is a component in a distributed system that facilitates communication and manages group membership. It includes several key fields for this purpose. The `C` field is a pointer to a `Communicator`, which handles sending and receiving messages over UDP. This component is crucial for network communication within the node. The `ML` field holds a pointer to a `memberlist.Memberlist` from the `hashicorp/memberlist` package [23], which manages cluster membership and member failure detection using a gossip-based protocol. This field manages the list of group members, ensuring that each node is aware of the others and participates in the group membership protocol. The `rrIndex` field is an integer used to keep track of the index for round-robin operations. It can help to distribute tasks or messages evenly among group members. The `rrIndexMutex` field is a mutex that provides

```go
func (c *Communicator) Send(dest *net.UDPAddr,
    body []byte) (int, *uuid.UUID, error) {
    return c.send(dest, 1, body)
}

func (c *Communicator) send(dest *net.UDPAddr,
    mType uint8, body []byte) (int, *uuid.UUID, error) {
    var success bool
    var sent int
    mid := GenId()
    m := Msg{mid, mType, body}
    b := MsgToBytes(m)

    if mType != 1 {
        bn, err := UDPSend(c.LocalConn, dest, b)
        if err != nil {
            return 0, nil, errors.New("failed␣UDPSend")
        }
        return bn, &mid, nil
    }
    ackch := make(chan bool)
    c.ackMutex.Lock()
    c.ackMap[mid] = ackch
    c.ackMutex.Unlock()

L:
    for i := 0; i < c.Config.RetryTimes; i++ {
        bn, err := UDPSend(c.LocalConn, dest, b)
        if err != nil {
            continue
        }
        select {
        case <-ackch:
            c.ackMutex.Lock()
            delete(c.ackMap, mid)
            c.ackMutex.Unlock()
            success = true
            sent = bn
            break L
        case <-time.After(c.Config.RetryInitialInterval):
        }
    }

    if success {
        return sent, &mid, nil
    } else {
        c.ackMutex.Lock()
        delete(c.ackMap, mid)
        c.ackMutex.Unlock()
        rm := MsgWithAddr{IP: [4]byte(dest.IP.To4()),
            Port: uint16(dest.Port), Msg: m}
        storeB := MsgWithAddrToBytes(rm)
        saved, err := SaveMsg(mid, storeB)
        if err != nil {
            log.Println("Send␣failed␣SaveMsg", err)
            return 0, nil, errors.New("failed␣SaveMsg")
        }
        return saved, &mid, nil
    }
}
```

Listing 1.10. The Send function

```
func (c *Communicator) Close() error {                              1
    if err := c.LocalConn.Close(); err != nil {                    2
        return err                                                  3
    }                                                               4
    close(c.done)                                                   5
    c.wg.Wait()                                                     6
    close(c.Mailbox)                                                7
    c.ackMutex.Lock()                                               8
    for k, ackCh := range c.ackMap {                               9
        close(ackCh)                                                10
        delete(c.ackMap, k)                                         11
    }                                                               12
    c.ackMutex.Unlock()                                             13
    return nil                                                      14
}                                                                   15
```

Listing 1.11. The `Close` function

synchronization for access to the `rrIndex` field. It ensures that concurrent modifications and reads of the index do not lead to race conditions or inconsistent states.

```
package ua                                                          1
import (                                                            2
    "errors"                                                        3
    "io"                                                            4
    "log"                                                           5
    "net"                                                           6
    "sync"                                                          7
    "github.com/hashicorp/memberlist"                              8
)                                                                   9
type GroupNode struct {                                            10
    C               *Communicator                                   11
    ML              *memberlist.Memberlist                          12
    rrIndex         int                                             13
    rrIndexMutex    sync.Mutex                                      14
}                                                                   15
```

Listing 1.12. The `GroupNode` struct

As shown in Listing 1.13, the `CreateNodeAndJoin` function is responsible for setting up a new node in a distributed system and integrating it into an existing group of nodes. The function `CreateNodeAndJoin` initiates a new node with the given address, connects it to known members, and optionally handles failed message resending. It begins by validating the port number of the provided address, ensuring that it falls within the acceptable range of 10000 to 24500. If the port number exceeds this range, the function returns an error indicating the invalid port.

Next, it configures a new `memberlist.Memberlist` instance using default settings suitable for a local area network (LAN). The configuration includes setting the node's name, bind address, and port. If the `memberlist.Create` function fails, it logs the error and returns a failure message. If no known members are provided, the node will include itself in the list to ensure it can join the group.

The node attempts to join the specified known members using the `Join` method of the `memberlist.Memberlist`. If this operation encounters an error, the function returns a failure message. Upon successful contact with other nodes, the number of successfully contacted hosts is logged.

The function then sets up the node's communication address, adjusting the port by adding 15000 to the original port. It creates a new `Communicator` instance using this address and the provided configuration, while also handling any errors that arise during initialization.

Finally, the function creates a `GroupNode` struct, which includes the initialized `Communicator` and `memberlist.Memberlist`. It logs the count and details of the members in the group. If all steps are successful, the function returns the newly created `GroupNode` and a nil error.

```go
func CreateNodeAndJoin(selfAddr *net.UDPAddr,                                      1
    knowMember []string,                                                          2
    resendFailedMsgs bool) (*GroupNode, error) {                                  3
    if selfAddr.Port < 10000 || selfAddr.Port > 24500 {                           4
        return nil,                                                               5
            errors.New("port number should be in [10000,24500]")                  6
    }                                                                             7
    mlConfig := memberlist.DefaultLANConfig()                                     8
    mlConfig.Name = selfAddr.String()                                            9
    mlConfig.BindAddr = selfAddr.IP.String()                                     10
    mlConfig.BindPort = selfAddr.Port                                            11
    mlConfig.Logger = log.New(io.Discard, "", 0)                                 12
    list, err := memberlist.Create(mlConfig)                                     13
    if err != nil {                                                              14
        log.Println(err)                                                         15
        return nil, errors.New("failed memberlist.Create")                       16
    }                                                                            17
    if len(knowMember) == 0 {                                                    18
        knowMember = append(knowMember, selfAddr.String())                       19
    }                                                                            20
    _, err = list.Join(knowMember)                                               21
    if err != nil {                                                              22
        return nil, errors.New("failed list.Join")                               23
    }                                                                            24
    selfCommAddr := &net.UDPAddr{IP: selfAddr.IP,                                25
        Port: int((selfAddr.Port) + 15000)}                                      26
    var cf Config                                                                27
    communicator, err := NewCommunicator(selfCommAddr,                           28
        cf, resendFailedMsgs)                                                    29
    if err != nil {                                                              30
        return nil, errors.New("failed NewCommunicator")                         31
    }                                                                            32
    g := &GroupNode{C: communicator, ML: list}                                   33
    return g, nil                                                                34
}                                                                                35
```

Listing 1.13. The `CreateNodeAndJoin` function

As shown in Listing 1.14, the `Broadcast` method in the `GroupNode` struct is used to send a message to all group members. The `Broadcast` function sends a message to each group member. It counts any errors encountered during the process. It takes a **body** parameter, which represents the content of the message to be broadcasted. The function starts by initializing an **errorCount** variable to zero, which will keep track of the number of failed send operations. It then iterates through the list of members in the group obtained from the **memberlist.Memberlist** instance **g.ML.Members()**. For each member, the function constructs a **net.UDPAddr** instance using the member's IP address and port.

The port is adjusted by adding 15000 to match the communication setup. The method then calls the **Send** method of the **Communicator** to send the message to the constructed address. Suppose the **Send** method returns an error. In that case, the function logs the error details, including the address and the message body, and increments the **errorCount** variable. After iterating through all members and attempting to send the message, the function returns the total **errorCount**, indicating how many send operations failed.

```
func (g *GroupNode) Broadcast(body []byte) uint32 {         1
    var errorCount uint32                                   2
    for _, member := range g.ML.Members() {                 3
        addr := &net.UDPAddr{IP: member.Addr,               4
            Port: int((member.Port) + 15000)}               5
        _, _, err := g.C.Send(addr, body)                   6
        if err != nil {                                     7
            log.Println("Failed Send", addr, body, err)     8
            errorCount++                                     9
        }                                                   10
    }                                                       11
    return errorCount                                       12
}                                                           13
```

Listing 1.14. The **Broadcast** function

As shown in Listing 1.15, the **RoundRobinSend** method in the **GroupNode** struct is designed to send a message to group members in a round-robin fashion. The **RoundRobinSend** function is used to send a message to a group member in a round-robin manner, ensuring that each member gets a turn to receive a message sequentially.

The function begins by locking the **rrIndexMutex** to ensure that the round-robin index **g.rrIndex** is accessed thread-safe. This mechanism prevents race conditions if multiple goroutines attempt to send messages concurrently. The method fetches the list of members from the **memberlist.Memberlist** instance. This list contains all the current members of the group. It selects a member based on the current round-robin index **g.rrIndex**. The member at the current index is chosen to receive the message. The method updates the round-robin index. If the current index is the last in the list, it wraps around to the beginning by setting **g.rrIndex** to 0. Otherwise, it increments the index to point to the next member.

After updating the index, the mutex is unlocked, allowing other goroutines to access and modify the round-robin index safely. The function constructs a

net.UDPAddr instance for the selected member by adding 15000 to their port number. It then uses the **Send** method of the **Communicator** to send the message to the selected address.

Suppose that the **Send** method encounters an error. In that case, it logs the details of the failure, including the address and the message body, for troubleshooting purposes.

```go
func (g *GroupNode) RoundRobinSend(body []byte) {          1
    var addr *net.UDPAddr                                  2
    g.rrIndexMutex.Lock()                                  3
    members := g.ML.Members()                              4
    m := members[g.rrIndex]                                5
    if g.rrIndex >= len(members)-1 {                       6
        g.rrIndex = 0                                      7
    } else {                                               8
        g.rrIndex++                                        9
    }                                                      10
    g.rrIndexMutex.Unlock()                                11
    addr = &net.UDPAddr{IP: m.Addr,                        12
        Port: int((m.Port) + 15000)}                       13
    _, _, err := g.C.Send(addr, body)                      14
    if err != nil {                                        15
        log.Println("Failed Send",                         16
            addr, body, err)                               17
    }                                                      18
}                                                          19
```

Listing 1.15. The RoundRobinSend function

As shown in Listing 1.16, the **CloseNode** function enables the shutdown of a node. This function closes the underlying **Communicator** and shuts down the memberlist, ensuring that all resources are properly released.

```go
func (g *GroupNode) CloseNode() error {                    1
    if err := g.C.Close(); err != nil {                    2
        log.Println("Error closing communicator:",         3
            err)                                           4
        return err                                         5
    }                                                      6
    if err := g.ML.Shutdown(); err != nil {                7
        log.Println("Error shutdown memberlist:",          8
            err)                                           9
        return err                                         10
    }                                                      11
    return nil                                             12
}                                                          13
```

Listing 1.16. The CloseNode function

4.6   Usage Examples

This section presents three code snippets demonstrating the use of the
Communicator and group communication functionalities provided by UACTOR.

Listing 1.17 shows how to create two Communicator instances on different
UDP ports. One instance sends a message using the Send method. Another
instance receives the message through its mailbox channel. The complete testing
example can be found in Listing 1.22.

```
var cf ua.Config                                                          1
addr1 := &net.UDPAddr{IP: net.IPv4(127,0,0,1), Port: 10000}               2
c1, _ := ua.NewCommunicator(addr1, cf, false)                             3
addr2 := &net.UDPAddr{IP: net.IPv4(127,0,0,1), Port: 10001}               4
c2, _ := ua.NewCommunicator(addr2, cf, false)                             5
// c1 sends a message                                                     6
p := []byte("Hello")                                                      7
_, _, _ = c1.Send(addr2, p)                                               8
// c2 receives the message                                                9
rm := <-c2.Mailbox                                                        10
_ = rm // process the message as needed                                   11
```

Listing 1.17. Minimal Communicator example

Two nodes are created in the example shown in Listing 1.18. The first
node sends a broadcast message with the Broadcast function. The second node
receives the message from its mailbox. After the message exchange, both nodes
are closed. The complete testing example can be found in Listing 1.27.

```
// Create the first node                                                  1
addrFirst := &net.UDPAddr{IP: net.IPv4(127,0,0,1), Port: 10000}           2
firstNode, _ := ua.CreateNodeAndJoin(addrFirst, []string{}, false)        3
// Create the second node                                                 4
addrSecond := &net.UDPAddr{IP: net.IPv4(127,0,0,1), Port: 10001}          5
secondNode, _ := ua.CreateNodeAndJoin(addrSecond,                         6
[]string{"127.0.0.1:10000"}, false)                                       7
// The first node broadcasts a message                                    8
msg := []byte("123")                                                      9
firstNode.Broadcast(msg)                                                  10
// The second node receives the broadcast message                         11
rm := <-secondNode.C.Mailbox                                              12
_ = rm // process the message as needed                                   13
// Clean up nodes                                                         14
firstNode.CloseNode()                                                     15
secondNode.CloseNode()                                                    16
```

Listing 1.18. Minimal group broadcast example

As shown in Listing 1.19, beside the Broadcast function, the first node can
also send group messages in a round-robin manner using the RoundRobinSend
function. The complete testing example can be found in Listing 1.28.

```
// The first node sends messages in a round-robin manner    1
for i := 0; i < 10; i++ {                                    2
firstNode.RoundRobinSend([]byte(strconv.Itoa(i)))           3
    time.Sleep(time.Second)                                  4
}                                                            5
// The second node retrieves a message from its mailbox     6
msg := <-secondNode.C.Mailbox                                7
_ = msg // process the message as needed                     8
```

Listing 1.19. Minimal group load balance sending example

5 Performance Tests with Energy Consumption Estimation

This section evaluates the performance of UACTOR through various tests, focusing on message reliability, `Communicator` functionality, and group message transmission. Each subsection presents results from specific test cases to validate different aspects of the system.

The UACTOR is designed to run on a distributed network, but the tests are performed on a single machine. The machine used for tests has the following system configuration: AMD Ryzen 7 7840HS processor with Radeon 780M Graphics 3.80 GHz, 32 GB of RAM, and Windows 11 as operating system.

The energy consumption is estimated with the help of AMD uProf [2].

The average power consumption during the profiling period is computed as follows:

$$P_{\text{avg}} = \frac{1}{N} \sum_{i=1}^{N} P_i$$

where P_{avg} represents the average power (in Watts), P_i denotes the instantaneous power sample at time i, and N is the total number of samples recorded. The average power consumption is obtained by summing up all recorded power values and dividing by the total number of samples.

5.1 Message Tests

To evaluate the performance of message encoding and decoding, we conducted benchmark tests using two serialization methods: a custom binary encoding approach (Listing 1.20) and Go's built-in **gob** [14] encoding (Listing 1.21). The tests were executed on a system with an AMD processor, and power consumption was measured using the AMD uProf tool. Each test involved serializing (encoding) and deserializing (decoding) messages of varying sizes, ranging from 1B to 1200B. Each operation was repeated 100,000 times to obtain reliable average latency values. The AMD uProf tool recorded power consumption at 50 ms intervals for 10 s.

Figure 2 illustrates the average encoding and decoding latency for both **Msg** and **Gob** implementations. The custom **Msg** encoding consistently outperformed the Gob approach in terms of latency. For small payloads (1B), Msg encoding latency was approximately 215ns, while Gob required 12,339ns. As the payload size increased to 1200B, Msg encoding latency remained efficient at 268ns, whereas Gob encoding latency reached 12,786ns. Decoding results exhibited a similar trend. The Msg decoding latency was significantly lower, ranging from 6ns to 10ns across all payload sizes. In contrast, Gob decoding latency remained above 12,000ns for all message sizes. In summary, UActor Msg encoding is around 10 times faster than the Gob encoding, while the decoding is around 1200 times faster than the Gob decoding.

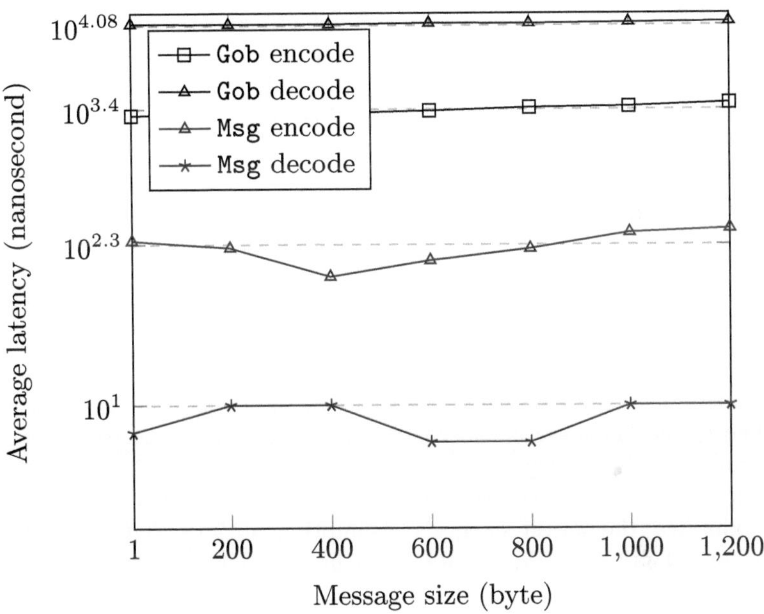

Fig. 2. Encoding and decoding latency for different message sizes

Table 1 presents the estimated energy consumption of both serialization methods for the testing process. The custom Msg implementation demonstrated a substantially lower average power consumption of 2.2 W compared to Gob's 11.9 W, which is approximately 5 times less. The considerable difference suggests that the Msg serialization method provides superior performance and significantly reduces energy usage.

Our tests show that using our custom Msg serialization cuts both latency and power usage considerably compared to Go's Gob encoding. This suggests that a lean binary format can really boost performance and save energy in distributed communication middleware.

Table 1. Comparison of average power consumption estimation of `Gob` and `Msg` tests

	Gob test	Msg test
Average power (Watts)	11.9012	2.21161

5.2   Communicator Tests

We benchmarked UDP, TCP, RABBITMQ, and the `Communicator` to compare their performance. We focused on two primary metrics: the time taken for messages to travel back and forth (latency) and the energy consumed during the test. For each method, we sent messages ranging from 1 byte to 600 bytes across 1000 iterations, using the average round-trip time as our measure of latency. We also tracked power usage with AMD uProf. The AMD uProf tool recorded power consumption at 50 ms intervals for 25 s. The code listings can be found in Sect. B.

The latency results for different message sizes are illustrated in Fig. 3. UDP achieved the lowest latency across all message sizes, with values ranging from approximately 16 μs to 25 μs. TCP exhibited slightly higher latencies than UDP, with an average latency increase of 20–30%. `Communicator` had a moderate increase in latency compared to raw UDP, averaging around 65 μs to 77 μs. It is slower than the pure UDP utility since we added more mechanisms like the message delivery guarantee. RABBITMQ displayed significantly higher latencies, particularly for larger message sizes (at 600B, the latency surged beyond 6600 μs). The latency increase is caused by the broker-based message queuing and additional reliability mechanisms.

Table 2 presents the average power consumption recorded during the tests. UDP exhibited the lowest energy consumption at 2.18 W. `Communicator` is 2.26 W, and TCP is 2.60 W. RABBITMQ has the highest power consumption at 6.10 W.

Table 2. Energy consumption estimation of message sending test

	UDP	`Communicator`	TCP	RABBITMQ
Average power (Watts)	2.1794	2.2646	2.60866	6.09695

The experimental results suggest that UDP is high-performing and has low latency. TCP introduces additional reliability mechanisms but induces a slight increase in latency and energy consumption. The `Communicator` (built on UDP) provides additional reliability while maintaining reasonable performance. RABBITMQ is helpful for reliable message queuing but incurs significant latency and power overhead.

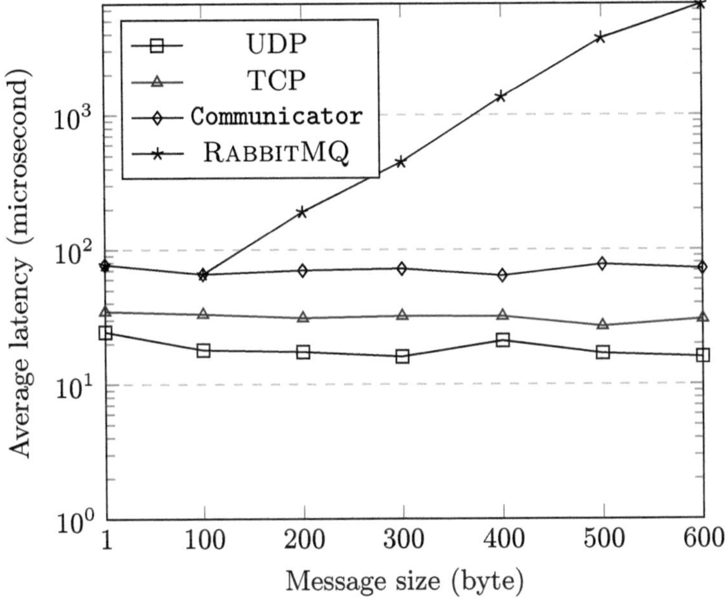

Fig. 3. Communicator tests

5.3   Message Storage Tests

The performance and energy consumption of message storage operations are evaluated in this section. Each test involved 1000 iterations with messages ranging from 1 to 600 bytes. The test code is provided in Listing 1.26. Before each run, we cleared any existing message folders. The AMD uProf tool recorded power consumption at 50 ms intervals for 25 s.

In each iteration, a message was generated and saved to disk using `ua.SaveMsg`. We recorded the time taken for storage and verified that the file size matched the expectations. After storing all messages, we retrieved each from disk using `ua.GetMsg`. We checked the integrity of the retrieved messages by comparing their sizes with the original payloads and measured the read duration.

Figure 4 shows the average latency (in microseconds) for storing and reading across different message sizes. The results indicate that storing messages takes significantly longer than reading them. For example, for a 600-byte message, the average store latency is approximately 1137.508 µs, while the average read latency is about 313.449 µs. Even when message sizes increase from 1 byte to 600 bytes, the average latency for both operations remains relatively constant, showing no apparent linear increase.

The average power consumption during storage tests was 12.9035 W. This is much higher than the power usage in the message send/receive tests (UDP at 2.18 W, `Communicator` at 2.26 W, TCP at 2.60 W, and RabbitMQ at 6.10 W).

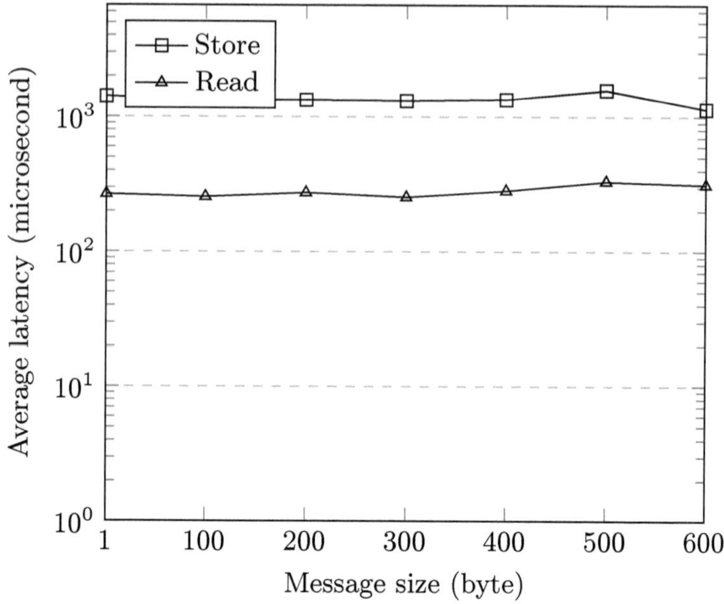

Fig. 4. Message store and read tests

Moreover, within the tested range, storage and read latency are not directly influenced by message size.

5.4 Group Message Sending Tests

The tests were conducted using UACTOR. Two primary message dissemination strategies were evaluated: *broadcast* (BC) and *load-balanced sending* (LB). Each test scenario varied the group size (10, 100, 200, 300, and 400 members) and message transmission interval (1 s and 3 s). The experimental duration was set to 30 s, during which the first node initiated message dissemination to the entire group (BC) or distributed messages in a round-robin fashion (LB). Energy consumption was measured using the AMD uProf tool, collecting power data at 50 ms intervals over a 55 s period. Message reception was monitored to ensure the group members received the expected number of transmissions. The code listings can be found in Sect. D.

Figure 5 presents the average power consumption estimation results. Broadcasting generally showed higher power consumption than load-balanced sending. This result is reasonable since BC involves simultaneous message duplication for all nodes, increasing network and processing overhead. When the interval was set to 1 s, power consumption increased steadily with the number of nodes, peaking at 6.52 W for 400 members. However, the BC 3 s case displayed an unexpected spike at 200 members (5.64 W). Load-balanced sending (LB) maintained a more stable energy profile, with lower power usage across all node configurations than

BC. The irregular power spike observed at 200 nodes in the BC 3 s test may be attributed to transient workload imbalances or network contention.

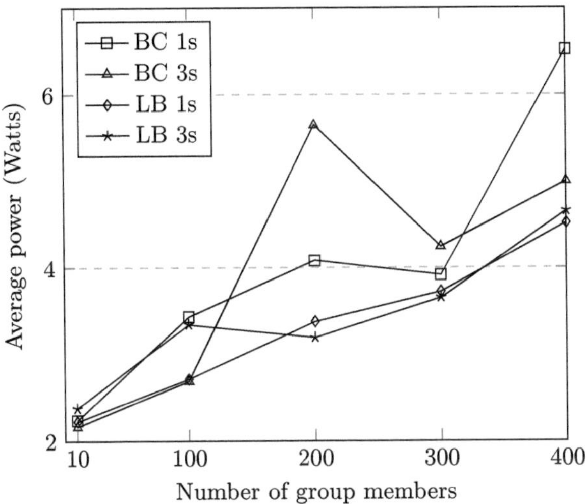

Fig. 5. Group sending tests

The results demonstrate that compared to load-balance sending, the broadcast strategy incurs higher power costs, particularly at lower intervals.

Due to the quadratic communication overhead and constraints from distributed garbage collection, Distributed Erlang groups typically scale up to 100 nodes [8]. In contrast, the tests show that building a group of up to 400 nodes is easy even on one machine using UActor.

The scalability increased with the help of the SWIM protocol. However, during the test period, we found that even though there is no group change, there are still periodic ping messages from each node. The gossip protocol is not particularly robust in dealing with correlated loss patterns and malicious behaviors [3]. Additionally, the periodic ping messages involved in gossip can affect sustainability. Therefore, our future work is to design and validate a more sustainable membership protocol.

6   Conclusion

This tutorial presented UActor, a lightweight and brokerless middleware. UActor uses the actor model with UDP to improve distributed system communication. It overcomes the scalability issues and overhead in traditional TCP-based actor frameworks, resulting in a more efficient and decentralized solution.

Our design shows that UDP works well with the actor model. It delivers low-latency, reliable, and scalable messaging without complex connection management. The `Communicator` module ensures message reliability through acknowledgment mechanisms, retransmission, and persistent storage. The group communication service supports dynamic actor coordination, group broadcasting, and group load balance sending functions using the group membership protocol SWIM.

More importantly, the tutorial shows how to improve the overall sustainability of a distributed system by including energy consumption estimations of different decisions taken during the design and implementation of key components. In the future, we plan to design more sustainable membership protocols that reduce communication overhead and enhance energy efficiency in large-scale distributed systems.

A   Codes of Message Tests

```
package testMsg

import (
    "Uactor/ua"
    "fmt"
    "os/exec"
    "testing"
    "time"
)

const times = 100000

func TestLatency(t *testing.T) {
    var tests = []struct {
        payload int
        encode  int
    }{
        {1, 18},
        {200, 217},
        {400, 417},
        {600, 617},
        {800, 817},
        {1000, 1017},
        {1200, 1217},
    }

    cmd := exec.Command(
        `C:\Program Files\AMD\AMDuProf\bin\AMDuProfCLI.exe`,
        "timechart", "--event", "power", "-o",
        `C:\Temp\Poweroutput`,
        "--interval", "50", "--duration", "10",
    )
    if err := cmd.Start(); err != nil {
        t.Errorf("Failed to start AMD uprof: %v", err)
    }

    for _, tt := range tests {
        testname := fmt.Sprintf("Average latency for %dB msg", tt.payload)
        t.Run(testname, func(t *testing.T) {
            p := make([]byte, tt.payload)
            mid := ua.GenId()
            msg := ua.Msg{MsgId: mid, MsgType: 1, Body: p}
```

```
        startEncode := time.Now()                                    44
        var encoded []byte                                           45
        for i := 0; i < times; i++ {                                 46
            encoded = ua.MsgToBytes(msg)                             47
        }                                                            48
        totalEncodeTime := time.Since(startEncode)                   49
        averageEncode := totalEncodeTime / time.Duration(times)      50
                                                                     51
        startDecode := time.Now()                                    52
        var decodedMsg ua.Msg                                        53
        for i := 0; i < times; i++ {                                 54
            decodedMsg = ua.BytesToMsg(encoded)                      55
        }                                                            56
        totalDecodeTime := time.Since(startDecode)                   57
        averageDecode := totalDecodeTime / time.Duration(times)      58
                                                                     59
        if decodedMsg.MsgId.String() != msg.MsgId.String() {         60
            t.Errorf("Decoded␣MsgId␣mismatch")                       61
        }                                                            62
        if len(decodedMsg.Body) != tt.payload {                      63
            t.Errorf("Decoded␣body␣length␣mismatch")                 64
        }                                                            65
        t.Logf("Average␣encode␣latency␣for␣%dB␣msg:␣%s",             66
        tt.payload, averageEncode)                                   67
        t.Logf("Average␣decode␣latency␣for␣%dB␣msg:␣%s",             68
        tt.payload, averageDecode)                                   69
    })                                                               70
}                                                                    71
                                                                     72
    if err := cmd.Wait(); err != nil {                               73
        t.Errorf("AMD␣uprof␣process␣finished␣with␣error:␣%v", err)   74
    }                                                                75
}                                                                    76
```

Listing 1.20. Message tests

```
package testGob                                                      1
                                                                     2
import (                                                             3
    "Uactor/ua"                                                      4
    "bytes"                                                          5
    "encoding/gob"                                                   6
    "fmt"                                                            7
    "os/exec"                                                        8
    "testing"                                                        9
    "time"                                                          10
)                                                                   11
                                                                    12
const times = 100000                                                13
                                                                    14
func TestGobTime(t *testing.T) {                                    15
    var tests = []struct {                                          16
        payload int                                                 17
    }{                                                              18
        {1},                                                        19
        {200},                                                      20
        {400},                                                      21
        {600},                                                      22
        {800},                                                      23
        {1000},                                                     24
        {1200},                                                     25
    }                                                               26
    cmd := exec.Command(                                            27
        `C:\Program␣Files\AMD\AMDuProf\bin\AMDuProfCLI.exe`,        28
        "timechart", "--event", "power", "-o",                      29
        `C:\Temp\Poweroutput`,                                      30
        "--interval", "50", "--duration", "10",                     31
```

```
    )                                                                             32
    if err := cmd.Start(); err != nil {                                           33
        t.Errorf("Failed␣to␣start␣AMD␣uprof:␣%v", err)                            34
    }                                                                             35
    for _, tt := range tests {                                                    36
        testname := fmt.Sprintf("Average␣latency␣for␣%dB␣msg", tt.payload)        37
        t.Run(testname, func(t *testing.T) {                                      38
            p := make([]byte, tt.payload)                                         39
            mid := ua.GenId()                                                     40
            msg := ua.Msg{MsgId: mid, MsgType: 1, Body: p}                        41
                                                                                  42
            startEncode := time.Now()                                            43
            var encoded []byte                                                   44
            for i := 0; i < times; i++ {                                          45
                var buf bytes.Buffer                                             46
                enc := gob.NewEncoder(&buf)                                      47
                if err := enc.Encode(msg); err != nil {                          48
                    t.Errorf("Iteration␣%d:␣encode␣error:␣%v", i, err)          49
                }                                                                 50
                encoded = buf.Bytes()                                           51
            }                                                                     52
            totalEncodeTime := time.Since(startEncode)                          53
            averageEncode := totalEncodeTime / time.Duration(times)             54
                                                                                  55
            startDecode := time.Now()                                           56
            var decodedMsg ua.Msg                                               57
            for i := 0; i < times; i++ {                                          58
                buf := bytes.NewBuffer(encoded)                                 59
                dec := gob.NewDecoder(buf)                                      60
                if err := dec.Decode(&decodedMsg); err != nil {                 61
                    t.Errorf("Iteration␣%d:␣decode␣error:␣%v", i, err)          62
                }                                                                 63
            }                                                                     64
            totalDecodeTime := time.Since(startDecode)                          65
            averageDecode := totalDecodeTime / time.Duration(times)             66
                                                                                  67
            if decodedMsg.MsgId.String() != msg.MsgId.String() {                68
                t.Errorf("Decoded␣MsgId␣mismatch")                              69
            }                                                                     70
            if len(decodedMsg.Body) != tt.payload {                             71
                t.Errorf("decoded␣body␣length␣%d␣!=␣expected␣%d",               72
                len(decodedMsg.Body), tt.payload)                              73
            }                                                                     74
            t.Logf("Average␣gob␣encode␣latency␣for␣%dB␣msg:␣%s",               75
            tt.payload, averageEncode)                                          76
            t.Logf("Average␣gob␣decode␣latency␣for␣%dB␣msg:␣%s",               77
            tt.payload, averageDecode)                                          78
        })                                                                        79
    }                                                                             80
                                                                                  81
    if err := cmd.Wait(); err != nil {                                            82
        t.Errorf("AMD␣uprof␣process␣finished␣with␣error:␣%v", err)               83
    }                                                                             84
}                                                                                 85
```

Listing 1.21. Gob tests

B Codes of Communicator Tests

```
package testCommunicator                                                          1
                                                                                  2
import (                                                                          3
    "Uactor/ua"                                                                   4
    "fmt"                                                                         5
```

```go
    "net"                                                     6
    "os/exec"                                                 7
    "testing"                                                 8
    "time"                                                    9
)                                                            10
                                                             11
const times = 1000                                           12
                                                             13
func TestSendLatency(t *testing.T) {                         14
    var tests = []struct {                                   15
        senderPort   int                                     16
        receiverPort int                                     17
        payload      int                                     18
        send         int                                     19
        recv         int                                     20
    }{                                                       21
        {10000, 10001, 1, 18, 1},                            22
        {10002, 10003, 100, 117, 100},                       23
        {10004, 10005, 200, 217, 200},                       24
        {10006, 10007, 300, 317, 300},                       25
        {10008, 10009, 400, 417, 400},                       26
        {10010, 10011, 500, 517, 500},                       27
        {10012, 10013, 600, 617, 600},                       28
    }                                                        29
                                                             30
    err := ua.DeleteMsgFolder()                              31
    if err != nil {                                          32
        t.Errorf("Failed to DeleteMsgFolder")               33
    }                                                        34
                                                             35
    cmd := exec.Command(                                     36
        `C:\Program Files\AMD\AMDuProf\bin\AMDuProfCLI.exe`, 37
        "timechart", "--event", "power", "-o",              38
        `C:\Temp\Poweroutput`,                               39
        "--interval", "50", "--duration", "25",             40
    )                                                        41
    if err := cmd.Start(); err != nil {                      42
        t.Errorf("Failed to start AMD uprof: %v", err)       43
    }                                                        44
    for _, tt := range tests {                               45
        testname := fmt.Sprintf("Average latency for %dB msg", 46
            tt.payload)                                      47
        t.Run(testname, func(t *testing.T) {                 48
            var cf ua.Config                                 49
            addr1 := &net.UDPAddr{IP: net.IPv4(127, 0, 0, 1), 50
                Port: tt.senderPort}                         51
            c1, err := ua.NewCommunicator(addr1, cf, false)  52
            if err != nil {                                  53
                t.Errorf("Failed to NewCommunicator")       54
            }                                                55
            addr2 := &net.UDPAddr{IP: net.IPv4(127, 0, 0, 1), 56
                Port: tt.receiverPort}                       57
            c2, err := ua.NewCommunicator(addr2, cf, false)  58
            if err != nil {                                  59
                t.Errorf("Failed to NewCommunicator")       60
            }                                                61
                                                             62
            start := time.Now()                              63
            for i := 0; i < times; i++ {                     64
                p := make([]byte, tt.payload)               65
                                                             66
                s, _, err := c1.Send(addr2, p)              67
                if err != nil {                              68
                    t.Errorf("Failed to Send")              69
                }                                            70
                rm := <-c2.Mailbox                           71
                                                             72
                if s != tt.send {                            73
```

```
                    t.Errorf("s_%d_!=_tt.send_%d", s, tt.send)   74
                }                                                  75
                if len(rm.Body) != tt.recv {                      76
                    t.Errorf("len(rm.Body)_%d_!=_tt.recv_%d",     77
                        len(rm.Body), tt.recv)                    78
                }                                                  79
            }                                                      80
            elapsed := time.Since(start)                          81
                                                                   82
            average := elapsed / time.Duration(times)             83
            t.Logf("Average_latency_for_%dB_msg:_%s",             84
                tt.payload, average)                              85
        })                                                         86
    }                                                              87
    if err := cmd.Wait(); err != nil {                            88
        t.Errorf("AMD_uprof_process_finished_with_error:_%v",     89
            err)                                                  90
    }                                                              91
}                                                                  92
```

Listing 1.22. Communicator tests

```
package testUDP                                                    1
                                                                   2
import (                                                           3
    "Uactor/ua"                                                    4
    "fmt"                                                          5
    "net"                                                          6
    "os/exec"                                                      7
    "testing"                                                      8
    "time"                                                         9
)                                                                 10
                                                                  11
const times = 1000                                                12
                                                                  13
func TestSendReceive(t *testing.T) {                              14
    addr1 := &net.UDPAddr{IP: net.IPv4(127, 0, 0, 1),             15
        Port: 10000}                                              16
    conn1, err := ua.UDPListenOnPort(addr1)                       17
    if err != nil {                                               18
        t.Errorf("Failed_to_UDPListenOnPort")                     19
    }                                                             20
    addr2 := &net.UDPAddr{IP: net.IPv4(127, 0, 0, 1),             21
        Port: 10001}                                              22
    conn2, err := ua.UDPListenOnPort(addr2)                       23
    if err != nil {                                               24
        t.Errorf("Failed_to_UDPListenOnPort")                     25
    }                                                             26
                                                                  27
    var tests = []struct {                                        28
        send int                                                  29
        recv int                                                  30
    }{                                                            31
        {1, 1},                                                   32
        {100, 100},                                               33
        {200, 200},                                               34
        {300, 300},                                               35
        {400, 400},                                               36
        {500, 500},                                               37
        {600, 600},                                               38
    }                                                             39
                                                                  40
    cmd := exec.Command(                                          41
        `C:\Program_Files\AMD\AMDuProf\bin\AMDuProfCLI.exe`,      42
        "timechart", "--event", "power", "-o",                    43
        `C:\Temp\Poweroutput`,                                    44
        "--interval", "50", "--duration", "25",                   45
```

```go
    )                                                                          46
    if err := cmd.Start(); err != nil {                                        47
        t.Errorf("Failed to start AMD uprof: %v", err)                         48
    }                                                                          49
                                                                               50
    for _, tt := range tests {                                                 51
        testname := fmt.Sprintf("Average latency for %dB msg",                 52
            tt.send)                                                           53
        t.Run(testname, func(t *testing.T) {                                   54
            start := time.Now()                                                55
            for i := 0; i < times; i++ {                                       56
                p := make([]byte, tt.send)                                     57
                                                                               58
                sentB, err := ua.UDPSend(conn1, addr2, p)                      59
                if err != nil {                                                60
                    t.Errorf("Failed to UDPSend")                             61
                }                                                              62
                                                                               63
                if sentB != tt.send {                                          64
                    t.Errorf("sentB != tt.send")                               65
                }                                                              66
                                                                               67
                recvB, _, err := ua.UDPReceive(conn2)                          68
                                                                               69
                if err != nil {                                                70
                    t.Errorf("Failed to UDPReceive")                          71
                }                                                              72
                if len(recvB) != tt.recv {                                     73
                    t.Errorf("recvB != tt.recv")                               74
                }                                                              75
            }                                                                  76
            elapsed := time.Since(start)                                       77
            average := elapsed / time.Duration(times)                          78
            t.Logf("Average latency for %dB msg: %s",                          79
                tt.send, average)                                              80
        })                                                                     81
    }                                                                          82
    if err := cmd.Wait(); err != nil {                                         83
        t.Errorf("AMD uprof process finished with error: %v",                  84
            err)                                                               85
    }                                                                          86
}                                                                              87
```

Listing 1.23. UDP tests

```go
package testTCP                                                                1
                                                                               2
import (                                                                       3
    "fmt"                                                                      4
    "net"                                                                      5
    "os/exec"                                                                  6
    "sync"                                                                     7
    "testing"                                                                  8
    "time"                                                                     9
)                                                                             10
const times = 1000                                                           11
func TestSendReceive(t *testing.T) {                                         12
    var tests = []struct {                                                   13
        send int                                                             14
        recv int                                                             15
    }{                                                                       16
        {1, 1},                                                              17
        {100, 100},                                                          18
        {200, 200},                                                          19
        {300, 300},                                                          20
        {400, 400},                                                          21
        {500, 500},                                                          22
```

```go
        {600, 600},
    }
    cmd := exec.Command(
        `C:\Program Files\AMD\AMDuProf\bin\AMDuProfCLI.exe`,
        "timechart", "--event", "power", "-o",
        `C:\Temp\Poweroutput`,
        "--interval", "50", "--duration", "25",
    )
    if err := cmd.Start(); err != nil {
        t.Fatalf("Failed to start AMD uprof: %v", err)
    }

    listener, err := net.Listen("tcp", "127.0.0.1:10000")
    if err != nil {
        t.Fatalf("Failed to start TCP listener: %v", err)
    }
    defer listener.Close()

    for _, tt := range tests {
        testname := fmt.Sprintf("Average latency for %dB msg",
            tt.send)
        t.Run(testname, func(t *testing.T) {
            var wg sync.WaitGroup
            errChan := make(chan error, times)

            wg.Add(1)
            go func() {
                defer wg.Done()

                conn, err := listener.Accept()
                if err != nil {
                    errChan <- fmt.Errorf(
                        "Failed to accept connection: %v", err)
                    return
                }
                defer conn.Close()

                buf := make([]byte, tt.recv)
                start := time.Now()

                for i := 0; i < times; i++ {
                    n, err := conn.Read(buf)
                    if err != nil {
                        errChan <- fmt.Errorf(
                            "Failed to read from connection: %v",
                            err)
                        return
                    }
                    if n != tt.recv {
                        errChan <- fmt.Errorf(
                            "Received bytes %d != expected %d",
                            n, tt.recv)
                        return
                    }
                }

                elapsed := time.Since(start)
                average := elapsed / time.Duration(times)
                t.Logf("Average latency for %dB msg: %s",
                    tt.send, average)
            }()

            clientConn, err := net.Dial("tcp", "127.0.0.1:10000")
            if err != nil {
                t.Fatalf("Failed to connect to TCP server: %v",
                    err)
            }
            defer clientConn.Close()
```

```
                                                                        91
            for i := 0; i < times; i++ {                               92
                p := make([]byte, tt.send)                             93
                _, err := clientConn.Write(p)                          94
                if err != nil {                                         95
                    t.Errorf("Failed to send data: %v", err)           96
                }                                                       97
            }                                                           98
            wg.Wait()                                                   99
            close(errChan)                                             100
            for err := range errChan {                                 101
                t.Errorf(err.Error())                                  102
            }                                                          103
        })                                                            104
    }                                                                 105
    if err := cmd.Wait(); err != nil {                                106
        t.Errorf("AMD uprof error: %v", err)                          107
    }                                                                 108
}                                                                     109
```

Listing 1.24. TCP tests

```
package testRabbitMQ                                                     1
                                                                        2
import (                                                                3
    "fmt"                                                               4
    "os/exec"                                                           5
    "sync"                                                              6
    "testing"                                                           7
    "time"                                                              8
                                                                        9
    "github.com/streadway/amqp"                                        10
)                                                                      11
                                                                       12
const times = 1000                                                    13
                                                                       14
func TestRabbitMQSendReceive(t *testing.T) {                          15
    var tests = []struct {                                            16
        send int                                                      17
        recv int                                                      18
    }{                                                                19
        {1, 1},                                                       20
        {100, 100},                                                   21
        {200, 200},                                                   22
        {300, 300},                                                   23
        {400, 400},                                                   24
        {500, 500},                                                   25
        {600, 600},                                                   26
    }                                                                 27
                                                                      28
    cmd := exec.Command(                                              29
        `C:\Program Files\AMD\AMDuProf\bin\AMDuProfCLI.exe`,          30
        "timechart", "--event", "power", "-o",                       31
        `C:\Temp\Poweroutput`,                                        32
        "--interval", "50", "--duration", "25",                      33
    )                                                                 34
    if err := cmd.Start(); err != nil {                              35
        t.Fatalf("Failed to start AMD uprof: %v", err)               36
    }                                                                 37
                                                                      38
    connPub, err := amqp.Dial("amqp://guest:guest@localhost:5672/")  39
    if err != nil {                                                  40
        t.Fatalf("Failed to connect to RabbitMQ: %v", err)           41
    }                                                                 42
    defer connPub.Close()                                            43
                                                                     44
    connSub, err := amqp.Dial("amqp://guest:guest@localhost:5672/")  45
```

```go
    if err != nil {
        t.Fatalf("Failed to connect to RabbitMQ: %v", err)
    }
    defer connSub.Close()

    pubChan, err := connPub.Channel()
    if err != nil {
        t.Fatalf("Failed to open publisher channel: %v", err)
    }
    defer pubChan.Close()

    subChan, err := connSub.Channel()
    if err != nil {
        t.Fatalf("Failed to open subscriber channel: %v", err)
    }
    defer subChan.Close()

    err = pubChan.ExchangeDeclare("pExchange", "direct",
        false, true, false, false, nil)
    if err != nil {
        t.Fatalf("Failed to declare an exchange: %v", err)
    }

    for _, tt := range tests {
        testname := fmt.Sprintf("RabbitMQ latency for %dB msg",
            tt.send)
        t.Run(testname, func(t *testing.T) {
            var wg sync.WaitGroup
            errChan := make(chan error, times)

            q, err := subChan.QueueDeclare("",
                false, true, false, false, nil)
            if err != nil {
                t.Fatalf("Failed to declare a queue: %v", err)
            }

            err = subChan.QueueBind(q.Name, "key", "pExchange",
                false, nil)
            if err != nil {
                t.Fatalf("Failed to bind a queue: %v", err)
            }

            msgs, err := subChan.Consume(q.Name, "",
                false, false, false, false, nil)
            if err != nil {
                t.Fatalf("Failed to register a consumer: %v", err)
            }

            wg.Add(1)
            go func() {
                defer wg.Done()
                start := time.Now()
                count := 0

                for d := range msgs {
                    if len(d.Body) != tt.recv {
                        errChan <- fmt.Errorf(
                            "Received %d bytes, expected %d",
                            len(d.Body), tt.recv)
                    }

                    count++
                    d.Ack(false)
                    if count >= times {
                        break
                    }
                }
            }
```

```
            elapsed := time.Since(start)                           114
            average := elapsed / time.Duration(times)              115
            t.Logf("Average␣latency␣for␣%dB␣msg:␣%s",              116
                tt.send, average)                                  117
        }()                                                        118
                                                                   119
        for i := 0; i < times; i++ {                               120
            payload := make([]byte, tt.send)                       121
            err := pubChan.Publish("pExchange", "key",             122
                false, false, amqp.Publishing{                     123
                    ContentType: "text/plain",                     124
                    Body:        payload,                          125
                })                                                 126
            if err != nil {                                        127
                t.Errorf("Failed␣to␣publish:␣%v", err)            128
            }                                                      129
        }                                                          130
                                                                   131
        wg.Wait()                                                  132
                                                                   133
        close(errChan)                                             134
        for err := range errChan {                                 135
            t.Errorf(err.Error())                                  136
        }                                                          137
    })                                                             138
    }                                                              139
                                                                   140
    if err := cmd.Wait(); err != nil {                            141
        t.Errorf("AMD␣uprof␣error:␣%v", err)                      142
    }                                                              143
}                                                                  144
```

Listing 1.25. RABBITMQ tests

C   Codes of Message Storage Tests

```
package testStorage                                                1
                                                                   2
import (                                                           3
    "Uactor/ua"                                                    4
    "fmt"                                                          5
    "net"                                                          6
    "os/exec"                                                      7
    "testing"                                                      8
    "time"                                                         9
)                                                                  10
                                                                   11
const times = 1000                                                 12
const msgFolder = "failedSend/"                                    13
                                                                   14
func TestStorage(t *testing.T) {                                  15
    var tests = []struct {                                        16
        payload int                                                17
        stored  int                                                18
    }{                                                             19
        {1, 24},                                                   20
        {100, 123},                                                21
        {200, 223},                                                22
        {300, 323},                                                23
        {400, 423},                                                24
        {500, 523},                                                25
        {600, 623},                                                26
    }                                                              27
                                                                   28
```

```go
if err := ua.DeleteMsgFolder(); err != nil {                          29
    t.Errorf("Failed␣to␣DeleteMsgFolder:␣%v", err)                    30
}                                                                     31
                                                                      32
cmd := exec.Command(                                                  33
    'C:\Program␣Files\AMD\AMDuProf\bin\AMDuProfCLI.exe',              34
    "timechart", "--event", "power", "-o",                           35
    'C:\Temp\Poweroutput',                                           36
    "--interval", "50", "--duration", "25",                          37
)                                                                     38
if err := cmd.Start(); err != nil {                                  39
    t.Errorf("Failed␣to␣start␣AMD␣uprof:␣%v", err)                   40
}                                                                     41
                                                                      42
senderIP := net.IPv4(127, 0, 0, 1)                                  43
senderPort := 10000                                                  44
                                                                      45
for _, tt := range tests {                                          46
    testname := fmt.Sprintf("Storage␣test␣for␣%dB␣msg",             47
        tt.payload)                                                  48
    t.Run(testname, func(t *testing.T) {                            49
        fileNames := make([]string, 0, times)                       50
                                                                      51
        storeStart := time.Now()                                    52
        for i := 0; i < times; i++ {                                53
            mId := ua.GenId()                                       54
            p := make([]byte, tt.payload)                           55
            m := ua.MsgWithAddr{                                    56
                IP:   ([4]byte)(senderIP),                         57
                Port: uint16(senderPort),                          58
                Msg:  ua.Msg{MsgId: mId, MsgType: 1, Body: p},     59
            }                                                       60
            mb := ua.MsgWithAddrToBytes(m)                         61
                                                                      62
            saved, err := ua.SaveMsg(mId, mb)                      63
            if err != nil {                                        64
                t.Errorf("Iteration␣%d:␣Failed␣to␣SaveMsg:␣%v",   65
                    i, err)                                         66
                return                                              67
            }                                                      68
            if saved != tt.stored {                                69
                t.Errorf("Iteration␣%d:␣saved␣%d␣!=␣expected␣%d",  70
                    i, saved, tt.stored)                           71
            }                                                      72
            fileNames = append(fileNames, mId.String())           73
        }                                                          74
        storeTotal := time.Since(storeStart)                      75
        averageStore := storeTotal / time.Duration(times)         76
        t.Logf("Average␣store␣time␣for␣%dB␣msg:␣%s",              77
            tt.payload, averageStore)                              78
                                                                      79
        readStart := time.Now()                                   80
        for _, f := range fileNames {                             81
            msg, err := ua.GetMsg(msgFolder + f)                  82
            if err != nil {                                        83
                t.Errorf("Failed␣to␣GetMsg␣for␣file␣%s:␣%v",     84
                    f, err)                                        85
                return                                             86
            }                                                     87
            if len(msg.Body) != tt.payload {                      88
                t.Errorf("For␣file␣%s:␣got␣msg␣size␣%d,␣expected␣%d", 89
                    f, len(msg.Body), tt.payload)                 90
            }                                                     91
        }                                                         92
        readTotal := time.Since(readStart)                        93
        averageRead := readTotal / time.Duration(times)          94
        t.Logf("Average␣read␣time␣for␣%dB␣msg:␣%s",              95
            tt.payload, averageRead)                              96
```

```
                                                                          97
        if ua.DiskMsgNum() != 0 {                                         98
            t.Errorf("Expected␣%d␣messages␣on␣disk,␣got␣%d",              99
                0, ua.DiskMsgNum())                                       100
        }                                                                 101
                                                                          102
        if err := ua.DeleteMsgFolder(); err != nil {                      103
            t.Errorf("Failed␣to␣DeleteMsgFolder␣after␣test:␣%v",          104
                err)                                                       105
        }                                                                 106
    })                                                                    107
  }                                                                       108
                                                                          109
  if err := cmd.Wait(); err != nil {                                      110
      t.Errorf("AMD␣uprof␣process␣finished␣with␣error:␣%v",               111
          err)                                                            112
  }                                                                       113
}                                                                         114
```

Listing 1.26. Message storage tests

D   Codes of Group Message Sending Tests

```
package group_test                                                        1
                                                                          2
import (                                                                  3
    "fmt"                                                                 4
    "net"                                                                 5
    "os/exec"                                                             6
    "strconv"                                                             7
    "sync"                                                                8
    "testing"                                                             9
    "time"                                                                10
                                                                          11
    "Uactor/ua"                                                           12
)                                                                         13
const testTime = 30                                                       14
                                                                          15
func TestGroupBroadcastMulti(t *testing.T) {                              16
    intervals := []int{1, 3}                                              17
    memberCounts := []int{10, 100, 200, 300, 400}                         18
                                                                          19
    for _, inter := range intervals {                                     20
        for _, mCount := range memberCounts {                             21
            testName := fmt.Sprintf("Interval_%ds_Members_%d",            22
                inter, mCount)                                            23
            t.Run(testName, func(t *testing.T) {                          24
                var nodes []*ua.GroupNode                                 25
                                                                          26
                addrFirst := &net.UDPAddr{IP: net.IPv4(127, 0, 0, 1),     27
                    Port: 10000}                                          28
                firstNode, err := ua.CreateNodeAndJoin(addrFirst,         29
                    []string{}, false)                                    30
                if err != nil {                                           31
                    t.Fatalf("Failed␣to␣CreateNodeAndJoin:␣%v", err)      32
                }                                                         33
                nodes = append(nodes, firstNode)                          34
                msgNum := testTime / inter                                35
                                                                          36
                exist := "127.0.0.1:10000"                                37
                for port := 10001; port < 10001+mCount-1; port++ {        38
                    addr := &net.UDPAddr{IP: net.IPv4(127, 0, 0, 1),      39
                        Port: port}                                       40
                    node, err := ua.CreateNodeAndJoin(addr,               41
```

```go
                        []string{exist}, false)                        42
            if err != nil {                                            43
            t.Errorf("Failed_to_CreateNodeAndJoin_at_port_%d:_%v",     44
                    port, err)                                         45
                continue                                               46
            }                                                          47
            nodes = append(nodes, node)                                48
            time.Sleep(10 * time.Millisecond)                          49
    }                                                                  50
                                                                       51
    cmd := exec.Command(                                               52
        'C:\Program_Files\AMD\AMDuProf\bin\AMDuProfCLI.exe',           53
        "timechart", "--event", "power", "-o",                         54
        'C:\Temp\Poweroutput',                                         55
        "--interval", "50", "--duration", "55",                        56
    )                                                                  57
    if err := cmd.Start(); err != nil {                                58
        t.Errorf("Failed_to_start_AMD_uprof:_%v", err)                 59
    }                                                                  60
                                                                       61
    var wg sync.WaitGroup                                              62
    for idx, node := range nodes {                                     63
        wg.Add(1)                                                      64
        go func(i int, n *ua.GroupNode) {                              65
            defer wg.Done()                                            66
            count := 0                                                 67
            timeout := time.After(time.Duration(testTime+15) *         68
                time.Second)                                           69
        loop:                                                          70
            for {                                                      71
                select {                                               72
                case msg, ok := <-n.C.Mailbox:                         73
                    if !ok {                                           74
                        break loop                                     75
                    }                                                  76
                    r := string(msg.Body)                              77
                    msgVal, err := strconv.Atoi(r)                     78
                    if err != nil {                                    79
                        t.Errorf(                                      80
                "Failed_to_convert_message_body_to_int:_%v",           81
                            err)                                       82
                        continue                                       83
                    }                                                  84
                    if msgVal >= msgNum {                              85
            t.Errorf("Received_message_value_%d_>=_msgNum_%d",         86
                        msgVal, msgNum)                                87
                    }                                                  88
                    count++                                            89
                    if count >= msgNum {                               90
                        break loop                                     91
                    }                                                  92
                case <-timeout:                                        93
                    break loop                                         94
                }                                                      95
            }                                                          96
            if err := n.CloseNode(); err != nil {                      97
                t.Errorf("Error_closing_node_at_%s:_%v",               98
                    n.C.LocalAddr.String(), err)                       99
            }                                                          100
        }(idx, node)                                                   101
    }                                                                  102
                                                                       103
    for i := 0; i < msgNum; i++ {                                      104
        _ = firstNode.Broadcast([]byte(strconv.Itoa(i)))              105
        time.Sleep(time.Duration(inter) * time.Second)                106
    }                                                                  107
                                                                       108
    if err := cmd.Wait(); err != nil {                                109
```

```
            t.Errorf("AMD␣uprof␣process␣finished␣with␣error:␣%v",   110
                err)                                                  111
        }                                                            112
        wg.Wait()                                                    113
    })                                                               114
    }                                                                115
    }                                                                116
}                                                                    117
```

Listing 1.27. Group broadcast tests

```
package TestGroupLB                                                  1
                                                                     2
import (                                                             3
    "fmt"                                                            4
    "net"                                                            5
    "os/exec"                                                        6
    "strconv"                                                        7
    "sync"                                                           8
    "testing"                                                        9
    "time"                                                           10
    "Uactor/ua"                                                      11
)                                                                    12
const testTime = 30                                                 13
                                                                     14
func TestGroupLBMulti(t *testing.T) {                               15
    intervals := []int{1, 3}                                        16
    memberCounts := []int{10, 100, 200, 300, 400}                   17
                                                                     18
    for _, inter := range intervals {                              19
        for _, mCount := range memberCounts {                       20
            testName := fmt.Sprintf("LB_Interval_%ds_Members_%d",    21
                inter, mCount)                                       22
            t.Run(testName, func(t *testing.T) {                     23
                var nodes []*ua.GroupNode                            24
                                                                     25
                addrFirst := &net.UDPAddr{                           26
                    IP:   net.IPv4(127, 0, 0, 1),                    27
                    Port: 10000}                                     28
                sender, err := ua.CreateNodeAndJoin(addrFirst,       29
                    []string{}, false)                               30
                if err != nil {                                      31
                    t.Fatalf("Failed␣to␣CreateNodeAndJoin:␣%v",      32
                        err)                                         33
                }                                                    34
                nodes = append(nodes, sender)                       35
                msgNum := testTime / inter                           36
                                                                     37
                exist := "127.0.0.1:10000"                           38
                for port := 10001; port < 10001+mCount-1; port++ {  39
                    addr := &net.UDPAddr{IP: net.IPv4(127, 0, 0, 1), 40
                    Port: port}                                      41
                    node, err := ua.CreateNodeAndJoin(addr,          42
                        []string{exist}, false)                      43
                    if err != nil {                                  44
                    t.Errorf("Failed␣to␣CreateNodeAndJoin␣at␣port␣%d:␣%v", 45
                        port, err)                                   46
                        continue                                     47
                    }                                                48
                    nodes = append(nodes, node)                     49
                    time.Sleep(10 * time.Millisecond)               50
                }                                                    51
                                                                     52
                cmd := exec.Command(                                 53
                    `C:\Program␣Files\AMD\AMDuProf\bin\AMDuProfCLI.exe`, 54
                    "timechart", "--event", "power", "-o",           55
                    `C:\Temp\Poweroutput`,                           56
```

```go
                "--interval", "50", "--duration", "55",        57
        )                                                       58
        if err := cmd.Start(); err != nil {                     59
            t.Errorf("Failed to start AMD uprof: %v", err)      60
        }                                                       61
                                                                62
        counts := make([]int, len(nodes))                       63
        var wg sync.WaitGroup                                    64
        for i, node := range nodes {                            65
            wg.Add(1)                                            66
            go func(i int, n *ua.GroupNode) {                   67
                defer wg.Done()                                 68
                count := 0                                       69
                timeout := time.After(                          70
                    time.Duration(testTime+15) * time.Second)   71
            loop:                                               72
                for {                                           73
                    select {                                    74
                    case msg, ok := <-n.C.Mailbox:             75
                        if !ok {                                76
                            break loop                          77
                        }                                       78
                        r := string(msg.Body)                   79
                        msgVal, err := strconv.Atoi(r)          80
                        if err != nil {                         81
            t.Errorf("Node %s: failed to convert message: %v",  82
                            n.C.LocalAddr.String(), err)        83
                            continue                            84
                        }                                       85
                        if msgVal >= msgNum {                   86
                    t.Errorf(                                   87
            "Node %s: Received message value %d >= msgNum %d",  88
                            n.C.LocalAddr.String(), msgVal, msgNum)  89
                        }                                       90
                        count++                                 91
                    case <-timeout:                             92
                        break loop                              93
                    }                                           94
                }                                               95
                counts[i] = count                               96
                if err := n.CloseNode(); err != nil {           97
                    t.Errorf("Error closing node at %s: %v",    98
                        n.C.LocalAddr.String(), err)            99
                }                                               100
            }(i, node)                                          101
        }                                                       102
                                                                103
        for i := 0; i < msgNum; i++ {                           104
            sender.RoundRobinSend([]byte(strconv.Itoa(i)))      105
            time.Sleep(time.Duration(inter) * time.Second)      106
        }                                                       107
                                                                108
        if err := cmd.Wait(); err != nil {                      109
            t.Errorf("AMD uprof process finished with error: %v",  110
            err)                                                111
        }                                                       112
                                                                113
        wg.Wait()                                               114
                                                                115
        total := 0                                              116
        for _, cnt := range counts {                            117
            total += cnt                                        118
        }                                                       119
        if total != msgNum {                                    120
            t.Errorf("Total messages received %d != expected %d",  121
                total, msgNum)                                  122
        }                                                       123
    })                                                          124
```

```
      }                                                              125
    }                                                                126
}                                                                    127
```

Listing 1.28. Group load balance sending tests

References

1. Agha, G.: Actors: a model of concurrent computation in distributed systems. MIT Press (1986)
2. AMD: AMD uprof user guide (2023). https://www.amd.com/content/dam/amd/en/documents/developer/version-4-1-documents/uprof/uprof-ug-rev-4.1.pdf. Accessed 20 July 2024
3. Birman, K.: The promise, and limitations, of gossip protocols. SIGOPS Oper. Syst. Rev. **41**(5), 8–13 (2007). https://doi.org/10.1145/1317379.1317382
4. Broker vs. brokerless – ZeroMQ. http://wiki.zeromq.org/whitepapers:brokerless. Accessed 20 July 2024
5. Castro, M., Druschel, P., Kermarrec, A.M., Rowstron, A.: Scribe: a large-scale and decentralized application-level multicast infrastructure. IEEE J. Sel. Areas Commun. **20**(8), 1489–1499 (2002). https://doi.org/10.1109/JSAC.2002.803069
6. Castro, M., Druschel, P., Kermarrec, A.M., Rowstron, A.: One ring to rule them all: service discovery and binding in structured peer-to-peer overlay networks. In: Proceedings of the 10th Workshop on ACM SIGOPS European Workshop, pp. 140–145. EW 10, ACM, NY, USA (2002). https://doi.org/10.1145/1133373.1133399
7. Cesarini, F., Vinoski, S.: Designing for Scalability with Erlang/OTP. O'Reilly Media, Inc. (2016)
8. Chechina, N., Trinder, P., Ghaffari, A., Green, R., Lundin, K., Virding, R.: The design of scalable distributed erlang. In: Proceedings of the 24th Symposium on Implementation and Application of Functional Languages (IFL 2012), p. 461 (2012)
9. Cluster specification Akka documentation. https://doc.akka.io/docs/akka/current/typed/cluster-concepts.html. Accessed 20 July 2024
10. D'Amore, G.: NNG reference manual. Staysail Systems, Inc. (2018)
11. Das, A., Gupta, I., Motivala, A.: Swim: scalable weakly-consistent infection-style process group membership protocol. In: Proceedings International Conference on Dependable Systems and Networks, pp. 303–312. IEEE (2002)
12. Day, J., Zimmermann, H.: The OSI reference model. Proc. IEEE **71**(12), 1334–1340 (1983). https://doi.org/10.1109/PROC.1983.12775
13. Dossot, D.: RabbitMQ Essentials. Packt Publishing Ltd. (2014)
14. gob package. https://pkg.go.dev/encoding/gob. Accessed 20 July 2024
15. Hewitt, C., Bishop, P., Steiger, R.: A universal modular actor formalism for artificial intelligence. In: Proceedings of the 3rd International Joint Conference on Artificial Intelligence, pp. 235–245. IJCAI'73, Morgan Kaufmann Publishers Inc., San Francisco, CA, USA (1973)
16. Hunt, P., Konar, M., Junqueira, F.P., Reed, B.: {ZooKeeper}: wait-free coordination for internet-scale systems. In: 2010 USENIX Annual Technical Conference (USENIX ATC 2010) (2010)
17. Iyengar, J., Thomson, M.: QUIC: A UDP-Based Multiplexed and Secure Transport. RFC 9000 (2021). https://doi.org/10.17487/RFC9000

18. Jia, Y., Bodanese, E., Phillips, C., Bigham, J., Tao, R.: Improved reliability of large scale publish/subscribe based moms using model checking. In: 2014 IEEE Network Operations and Management Symposium (NOMS), pp. 1–8 (2014). https://doi.org/10.1109/NOMS.2014.6838311
19. Kreps, J., Narkhede, N., Rao, J., et al.: Kafka: a distributed messaging system for log processing. In: Proceedings of the NetDB, vol. 11, pp. 1–7. Athens, Greece (2011)
20. Langley, A., et al.: The quic transport protocol: Design and internet-scale deployment. In: Proceedings of the Conference of the ACM Special Interest Group on Data Communication, pp. 183–196. SIGCOMM 2017, Association for Computing Machinery, New York, NY, USA (2017). https://doi.org/10.1145/3098822.3098842
21. Mangos. https://pkg.go.dev/go.nanomsg.org/mangos/v3. Accessed 20 July 2024
22. Matsakis, N.D., Klock, F.S.: The rust language. In: Proceedings of the 2014 ACM SIGAda Annual Conference on High Integrity Language Technology, pp. 103–104 (2014)
23. Memberlist package. https://pkg.go.dev/github.com/hashicorp/memberlist. Accessed 20 July 2024
24. Nanomsg. https://nanomsg.org/index.html. Accessed 20 July 2024
25. Networking and RabbitMQ. https://www.rabbitmq.com/docs/networking. Accessed 20 July 2024
26. Postel, J.: User Datagram Protocol. Request for Comments (1980). https://doi.org/10.17487/RFC0768. https://www.rfc-editor.org/rfc/rfc768
27. Reilly, D., Reilly, M.: Java network programming and distributed computing. Addison-Wesley Professional (2002)
28. Steed, A., Oliveira, M.F.: Chapter 6 - sockets and middleware. In: Steed, A., Oliveira, M.F. (eds.) Networked Graphics, pp. 195–216. Morgan Kaufmann, Boston (2010). https://doi.org/10.1016/B978-0-12-374423-4.00006-9
29. van Steen, M., Tanenbaum, A.: Distributed Systems, 3rd edn (2017). https://www.distributed-systems.net
30. NNG - nanomsg-ng. https://nng.nanomsg.org/. Accessed 20 July 2024
31. NSQ docs 1.2.1 - a realtime distributed messaging platform. https://nsq.io/. Accessed 20 July 2024
32. NSQ docs 1.2.1 - design. https://nsq.io/overview/design.html. Accessed 20 July 2024
33. NSQ docs 1.2.1 - nsqd. https://nsq.io/components/nsqd.html. Accessed 20 July 2024
34. NSQ docs 1.2.1 - nsqlookupd. https://nsq.io/components/nsqlookupd.html. Accessed 20 July 2024
35. Vernon, V.: Reactive messaging patterns with the Actor model: applications and integration in Scala and Akka. Addison-Wesley Professional (2015)
36. Vinoski, S.: Advanced message queuing protocol. IEEE Internet Comput. **10**(6), 87–89 (2006). https://doi.org/10.1109/MIC.2006.116
37. Wenig, P., Papenbrock, T.: Actix-telepathy. In: Proceedings of the 10th ACM SIGPLAN International Workshop on Reactive and Event-Based Languages and Systems, pp. 14–24 (2023)
38. Williams, R., Roestenburg, R., Bakker, R.: Akka in action. Manning Publications (2016)

Author Index

C. Grelck et al. (eds.), *Promoting Sustainability as a Fundamental Driver in Software Development Training and Education*, LNCS 15670, p. 329, 2026.
https://doi.org/10.1007/978-3-032-22278-7